MIMO
From Theory to Implementation

MIMO
From Theory to Implementation

Alain Sibille
Télécom Paris Tech
46 rue Barrault, 75634 Paris Cedex 13, France

Claude Oestges
Université catholique de Louvain
ICTEAM Electrical Engineering, 3 place du Levant,
1348 Louvain-la-Neuve, Belgium

Alberto Zanella
National Research Council (CNR)
IEIIT, v.le Risorgimento 2, 40136, Bologna, Italy

AMSTERDAM • BOSTON • HEIDELBERG • LONDON
NEW YORK • OXFORD • PARIS • SAN DIEGO
SAN FRANCISCO • SINGAPORE • SYDNEY • TOKYO
Academic Press is an imprint of Elsevier

Academic Press is an imprint of Elsevier
30 Corporate Drive, Suite 400, Burlington, MA 01803, USA
525 B Street, Suite 1900, San Diego, CA 92101-4495, USA
The Boulevard, Langford Lane, Kidlington, Oxford, OX5 1GB, UK

Library of Congress Cataloging-in-Publication Data
MIMO: from theory to implementation / edited by Alain Sibille, Claude Oestges, Alberto Zanella.
p. cm.
Includes index.
ISBN 978-1-4933-0122-5 (paper back)
1. MIMO systems. I. Sibille, Alain. II. Oestges, Claude. III. Zanella, Alberto.
TK5103.2.M564 2010
621.384–dc22
2010037601

British Library Cataloguing-in-Publication Data
A catalogue record for this book is available from the British Library.

ISBN: 978-1-4933-0122-5

For information on all Academic Press publications visit our website at *books.elsevier.com*

Printed and bound in the USA
11 12 13 10 9 8 7 6 5 4 3 2 1

Contents

Foreword

From the early concepts 18 years ago, MIMO wireless is now witnessing widespread applications in modern radio access systems. After a brief phase of inevitable skepticism, rapid strides in MIMO research laid the theoretical foundations for this area and several experimental prototypes demonstrated the performance value in real environments. MIMO technology was then embedded into wireless standards fairly quickly, followed by some years of commercial development to finally reach widespread mass market adoption today. MIMO has greatly improved coverage and throughput of wireless networks, adding real value to all of us. MIMO along with a well matched OFDMA technology has become the ruling paradigm for current and future wireless systems.

MIMO offers many research and implementation challenges. In research, we first need a good understanding of the MIMO propagation channel that also incorporates the actual antennas that will be used. Channel knowledge helps design better antennas, size the antennas number and predict performance gains. Next, we need transmit encoding strategies that simultaneously maximize diversity and multiplexing rate performance, while keeping receiver complexity manageable. Receive decoding research attempts to find reliable methods to tame computational complexity with acceptable performance loss. As regards MIMO implementations, the first area of progress lies in algorithms for channel estimation and MIMO decoding, which can maximize performance – power tradeoffs. Power efficient architectures and protocol design can also buy big savings. WiFi and WiMAX systems that incorporate MIMO are already in the market, while HSPA+ and LTE will enter soon. The MIMO implementation practice is still early in its life cycle and many new ideas, yet to be discovered, will dramatically improve performance while reducing power.

While the early MIMO work focused on single user applications (base station to single subscriber or vice versa), MIMO concepts are now expanding to multi-user (base station to multiple users) and network applications (multi-base station to a single user). These applications offer new challenges spanning channel models to transmit coding and receiver decoding. Network MIMO has not yet entered the standardization phase. An even newer frontier for MIMO relates to its use in relays and ad hoc networks. Again, this segment has not matured to commercial deployment and many more efforts need be done. Yet another promising frontier is large MIMO systems, where the number of antennas is large enough (> 8) to support huge performance gains while also allowing fast iterative receiver decoding.

The authors have done a commendable job of pulling together outstanding researchers in the field to summarize key topics in MIMO theory and implementation, making it a valuable reference for engineers in both academia and industry.

I hope you will find the material as insightful and useful as I have.

Arogyaswami Paulraj
Professor (Emeritus), Stanford University (USA)

Preface

This book is one of the outcomes of the European Action of COoperation in Science and Technology COST 2100, entitled "Pervasive Mobile and Ambient Wireless Communications" (December 2006 – December 2010). The Action (see *www.cost2100.org* for details) involves researchers from more than 100 academic institutions, industrial groups, operators and small companies, located in about 30 countries (including US, Canada, China and Japan, besides almost all Europe). It deals with issues related to the radio channel, transmission techniques and network aspects of wireless terrestrial communications, spanning for example from body area networks to vehicular ad hoc networks and from broadcast to cellular systems. MIMO has been one of the hot topics of research in wireless communications since one decade and it has been one of the hot topics in COST 2100 as well, as the result of the high expectations for this novel radio communication paradigm.

The book nicely represents the spirit of the COST framework and, in particular, of the Action 2100. It is based on the views of prominent European researchers, who tightly exchanged on many theoretical and experimental aspects of MIMO systems within COST 2100 and its predecessor COST 273. In the spirit of a COST Action, the members presented and critically discussed their own research results and they shared the gained knowledge with the whole COST 2100 community. For those reasons, the book represents a concerted view of many more participants than the few writers of the chapters that follow.

The Authors and contributors who were actively involved in the book preparation were mostly lecturers of a COST2100 Training School on MIMO, held in Paris in March 2009, which attracted more than 150 attendees (PhD students, senior and junior researchers from both academia and industry). The School programme was defined by Alain Sibille, Claude Oestges and Alberto Zanella, who are also the Editors for this book. The event was highly appreciated by the attendees, according to their own sayings and to their responses to a blind post-school questionnaire, which motivated the Editors to invest the large effort spent in the following months to put the book together.

The School was composed of 10 lectures, some proposing a theoretical background (given by representatives of academia) and others providing a technical insight on the challenges that arise in the implementation phase of MIMO techniques (given by industry representatives, with specific attention dedicated to 802.11n, WiMax and LTE). This industry-academia mix is also reflected in the organization of the book and it represents maybe its strongest characteristic, giving the same relevance to theory and implementation aspects. The book then finishes with a Chapter highlighting MIMO roadmaps from both an academic and operator's perspective.

COST 2100 is by far not limited to MIMO; however, MIMO techniques very much reflect the COST 2100 cooperation paradigm, which emphasises the need for proper combination of skills in the area of channel modeling, signal processing and networking. Clearly, the challenges and promises offered by MIMO cannot be achieved unless of a design of the radio system efficiently addressing all these areas, together with their interactions. The book attempts to take this awareness into account, through its contents and an adequate structuring to cover these aspects as much as possible.

The Author of this Preface has the privilege to coordinate the huge amount of energy and scientific skills aggregated within COST 2100, a task which is facilitated by the positive attitude and the open-minded approach to science shared by all COST2100 participants, and the Authors and Editors of this book in particular.

Roberto Verdone

About the Editors, Authors and Contributors

Alain Sibille, Télécom Paris Tech, France
A. Sibille has been with France Telecom R&D for 13 years, then moved as Professor to ENSTA-ParisTech in 1992 and Télécom Paris tech in 2010. Since 1995, his research has focused on antennas and channels for single and multiple antennas and MIMO systems, including Ultra Wide band systems. He has been extensively involved in COST Actions 273 and 2100, particularly as Chairman of WG2 "Radio Channel" in COST 2100. He has also organized the 13th European Wireless Conference in Paris, 2007.

Claude Oestges, Université catholique de Louvain, Belgium
After holding various young researcher positions worldwide, C. Oestges is now Associate Professor at UCL and Research Associate of the Fonds de la Recherche Scientifique FNRS. His research interests cover wireless and satellite communications, with a specific focus on propagation. He has taken part in the European COST Actions 255, 273 and 2100 and is WP leader in the European network of excellence NEWCOM++. Claude Oestges received several awards and is the author/co-author of one book and more than 100 research papers.

Alberto Zanella, IEIIT-National Research Council (CNR), Italy
A. Zanella is senior researcher with IEIIT-CNR. His research interests include MIMO systems, adaptive antennas and mobile radio systems. He has published more than 60 international publications and conference communications and is co-recipient of some best paper awards at major international conferences. He serves as Editor for Wireless Systems, IEEE Transactions on Communications.

Main Authors

Ernst Bonek, Technical University of Wien, Austria

Alister Burr, University of York, United Kingdom

Luis M. Correia, Technical University of Lisbon, Portugal

Merouane Debbah, Supelec, France

Thomas Derham, Orange Labs Tokyo, Japan

Claude Desset, Imec, Belgium

Buon Kiong Lau, Lund University, Sweden

Rodolphe Legouable, Orange Labs, France

Martin Schubert, Fraunhofer German-Sino Lab for Mobile Communications MCI, Germany

Luc Vandendorpe, Université catholique de Louvain, Belgium

Roberto Verdone, University of Bologna, Italy

Guillaume Vivier, Sequans Communications, France
Zhipeng (Alexandre) Zhao, Comsis S.A.S, France

Contributors

Eduardo Lopez Estraviz, Imec, Belgium
Min Li, Imec, Belgium
Robert Fasthuber, Imec, Belgium
Bertrand Muquet, Sequans Communications, France
Serdar Sezginer, Sequans Communications, France
Amélie Duchesne, Sequans Communications, France
Ambroise Popper, Sequans Communications, France
C. Herzet, Université catholique de Louvain, Belgium
Onur Oguz, Université catholique de Louvain, Belgium
H. Sneessens, Université catholique de Louvain, Belgium
X. Wautelet, Université catholique de Louvain, Belgium

Introduction

Alberto Zanella, Claude Oestges, Alain Sibille

1 A BRIEF LOOK BACK TO THE STORY OF MIMO

The use of multiple antennas at both the transmit and receive ends has become one of the most important paradigms for the deployment of existing and emerging wireless communications systems. The importance of Multiple-Input Multiple-Output (MIMO) systems is witnessed by their presence in many recent standards, such as IEEE802.11n, Worldwide Interoperability for Microwave Access (WiMAX) and Long Term Evolution (LTE). Interestingly, like many other brilliant solutions developed in these years in the field of digital communications, the MIMO concept is not a new idea but dated back to the 1970s and was studied as a model for multipair telephone cables. Probably, the first contribution on this topic was the paper of Kaye and George [KG70], where an optimum linear receiver for Pulse-Amplitude Modulation (PAM) signals in a frequency selected MIMO system was proposed and investigated. A few years later, the information capacity limits of MIMO channels were investigated by Brandenburg and Wyner [BW74]. More specifically, an analytical expression for the information capacity of frequency selective MIMO channels with discrete-time signals with finite memory and perfect Channel State Information at the Transmitter (CSIT) was obtained in [BW74]. It is worth noting that eq. (9b) of [BW74] represents one of the first formalizations of the MIMO capacity as a function of the nonzero eigenvalues of $\mathbf{H}(f)\mathbf{H}(f)^\dagger$, where $\mathbf{H}(f)$ is the (frequency selective) channel matrix. The impact of linear equalization was studied by Amitay and Salz in the context of 2×2 doubly polarized terrestrial radio systems [AS84]. The transmitter optimization problem was also considered and an exact solution for the nonexcess bandwidth case was obtained. The analysis of the optimal receiver filters was then generalized by Salz to the case of multi-input, multi-output signals with the same number of inputs and outputs [Sal85], and by Salz and Wyner to the case of arbitrary inputs and outputs [SW85].

The first analysis of a MIMO cellular scenario composed by a base station equipped with multiple antennas and single-antenna mobile terminals, transmitting in the same frequency band, is due to [Win87]. In that paper, Minimum Mean Squared Error (MMSE) linear processing (also known as *optimum combining*) was applied to remove the co-channel interference caused by the contemporaneous transmission of the mobile terminals. That paper was also one of the first contributions to formulate the following equation:

$$C = \sum_i \log_2 (1 + \rho\ \lambda_i)\ \ [bit/s/Hz] \tag{1}$$

where C gives the MIMO channel capacity in the case of and perfect Channel State Information at the Receiver (CSIR). λ_i are the nonzero eigenvalues of $\mathbf{HH}^\dagger$ and ρ is the ratio between the energy spent to transmit one symbol per antenna and the thermal noise power spectral density.[1]

These few results reveal that, although the benefits of MIMO compared to the conventional Single-Input Single-Output (SISO) systems had been well-known for decades, the inherent complexity, in terms of Radio Frequency (RF) design and signal processing, discouraged their practical implementation. In these years, MIMO systems were considered to be very appealing, but in that period, unrealistic solutions to increase the spectral efficiency of wireless systems. To this regard, one of the first examples of practical application of MIMO is the patent of Paulraj and Kailath for application to broadcast digital TV [PK94].

In 1996, Foschini introduced the concept of Layered Space-Time (LST) architecture, which will be later referred as Bell Laboratories Layered Space-Time (BLAST). In his celebrated paper [Fos96], Foschini proposed an architecture for a point-to-point MIMO communication system where the data stream generated by the source was divided in several branches and encoded without sharing any information with each other. The association between bit-stream and transmit antennas was also periodically cycled to counteract the effects of bad paths acting on some channels. This idea was later denoted as *Diagonal Bell Laboratories Layered Space-Time Architecture* (D-BLAST). Unfortunately D-BLAST had the disadvantage of requiring efficient short layered coding schemes. To cope with these limitations, a new version of the BLAST architecture, called *Vertical Bell Laboratories Layered Space-Time Architecture* (V-BLAST), was proposed in [FGVW99]. In V-BLAST, every antenna transmitted its own independently encoded data and there was no need for short coding schemes. V-BLAST architecture was also simpler then that of D-BLAST, but at the price of a lower capacity compared to the older scheme. One of the most important results of those papers was that, under some fading conditions, up to $\min\{N_\text{T}, N_\text{R}\}$ (where N_T and N_R are the number of transmit and receive antennas, respectively) independent data streams can be simultaneously transmitted over the eigen-modes of the matrix channel $\mathbf{H}$ [FG98]. These results showed that very large spectral efficiencies could be achieved with MIMO systems of limited complexity.

Motivated by the results of Foschini, a large number of papers appeared in the open literature addressing the different aspects of the MIMO architectures. In their famous contribution, Tarokh, Seshadri and Calderbank designed a new class of channel codes to be used with multiple transmit antennas. This class of codes was named Space-Time Code (STC). In the same year, the use of a Discrete Multitone (DMT) technique to counteract the effects of frequency selective MIMO channels was proposed by Raleigh and Cioffi [RC98]. This contribution is considered the first proposal of an Orthogonal Frequency Division Multiplexing (OFDM)-like technique in the

[1] As indicated by [Win87], (1) can be obtained from [AS84, eq. (70)] in the case of flat fading channel. The expression [Win87, eq. (21)] is generalized to the case of arbitrary number of antennas.

context of MIMO channels. Telatar's paper [Tel99], which appeared in 1999 but was based on a technical memorandum written in 1995 [Tel95], provided a general framework for the investigation of the capacity of ergodic and nonergodic MIMO channels in a Rayleigh fading environment. This paper was one of the first contributions to focus on the link between MIMO performance in Rayleigh/Rice fading channels and eigenvalues' distribution of Wishart matrices. This relation between MIMO and the eigenvalues of random matrices will be extensively used in the following years to investigate several aspects of MIMO systems. Capacity bounds of MIMO systems with no CSI under the assumption of a slowly varying channel were given by Hochwald and Marzetta in [MH99]. Unfortunately, the analysis for an arbitrary but finite number of antennas does not lead to closed-form results in many cases of interest. This result has stimulated the investigation of scenarios in which the number of antennas approaches infinity. As an example of such developments, one of the first papers to apply the random matrix theory for MIMO analysis was [Tel99].

Since the double-directional channels characteristics are essential to MIMO performance, the assessment and modeling of MIMO channels were considered very early after the milestone papers cited above. The simple and elegant "Kronecker" channel model in particular, which is at the basis of many contemporary channel model standards, was proposed by Kermoal, et al. in 2002 [JLK+02]. More accurate models have been later proposed in order to better mitigate the Kronecker model deficiencies (see Chapter 2).

Further progress in MIMO concepts came up with the paper of Viswanath, Tse, and Anantharam [VTA01], which was one of the first contributions addressing MIMO multiple-access channels. MIMO multiantenna broadcast channels were also investigated in [CS03]. Exact expressions for the characteristic function of the capacity in the case of semi-correlated Rayleigh flat fading were given by Chiani, Win and Zanella in [CWZ03] and Smith, Roy and Shafi in [SRS03]. The paper of Zheng and Tse [ZT03] demonstrated the existence of a trade-off between the degree of diversity achievable by a point-to-point MIMO system and its multiplexing, that is the capability of transmitting independent information streams through the spatial channels. The diversity/multiplexing trade off was subsequently investigated in a multiuser scenario in [TVZ04].

After year 2000, other important topics in wireless communications have been investigated, but the interest on MIMO topics has been always high (see Figure 1). Some examples are: The impact of a limited feedback on MIMO performance, cellular MIMO, distributed STC, MIMO relaying channels and distributed MIMO schemes.

2 THEORETICAL LIMITS VS. REAL-WORLD DEPLOYMENTS

With MIMO technology now entering the deployment stage, it is worth considering the comparison between the promised benefits and various field trials, while simultaneously looking at the remaining open issues in MIMO research.

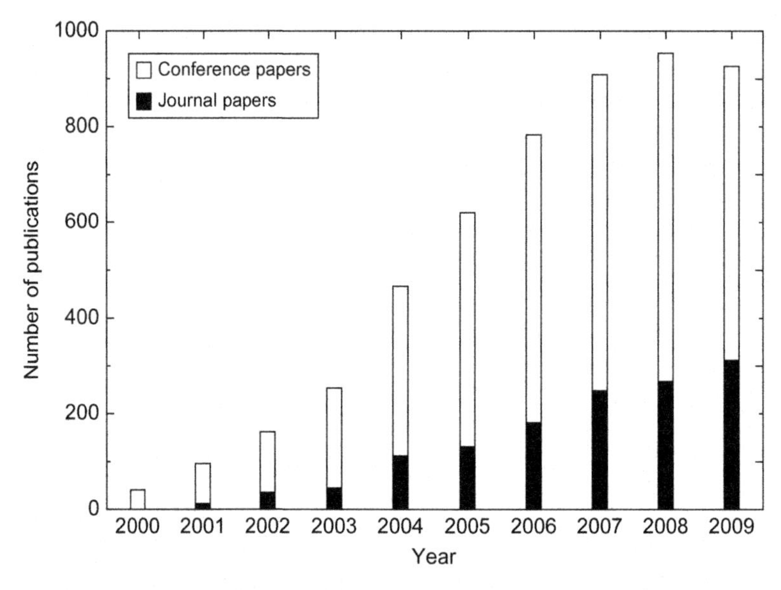

FIGURE 1

Publications per year on MIMO. (Keywords for the search: MIMO and wireless. Data taken from www.ieeexplore.org).

Impact of Propagation Channel on System Design

Many space-time techniques designed in the beginning of MIMO research often assumed perfectly uncorrelated channels (Rayleigh i.i.d.). Real-world experimental campaigns have since then shown that this was not the case. But how far is the capacity from this ideal assumption? As an example, demonstrations carried out by Alcatel-Lucent in Manhattan [CLW+03] have illustrated that 2×2 and 4×4 MIMO scenarios achieved spectral efficiencies within 90% of the i.i.d. Rayleigh capacity. Naturally, these values highly depend on the propagation scenario and the antenna configuration. More recent extensive field trials have been carried out by manufacturers for standards such as WiMAX, HSPA and LTE, clearly showing the benefits of MIMO transmissions, especially at cell edges where the throughput can be very low and the extra capacity can be very useful [FWRH10].

However, not all techniques are equally good in more adverse propagation conditions, and channel correlation could result in large performance losses. A better understanding on how the vector channel affects the performance of space-time coding was therefore required to design robust signal processing, such that ways to bridge the gap between theoretical and real-world performance had (and still have in some cases) to be developed to reach the full gains of MIMO. Still, any design has to remain sufficiently simple to enable a low-complexity implementation.

The use of MIMO is also currently investigated in other transmissions than simply cellular communications, sometimes in rather original media (body area networks, vehicular communications, etc). This motivates the development of solutions adapted to a large number of application scenarios, where MIMO could deliver substantial gains, provided that the wireless channel permits such gains to be achieved. In particular, the joint use of MIMO and UWB has recently attracted a renewed interest [Sib05] [Mol10] in order to increase the range of specific systems (e.g., body area networks).

Antenna Issues

The number of antenna elements and the array configuration are key parameters. While many research papers have considered (very) large arrays, practical realizations have limited this number to more modest values. Furthermore, optimal configurations which would suitably trade-off performance vs. complexity/cost/consumption/size have yet to be defined. While dual-polarized arrays have attracted some interest with respect to their compactness and the limited spatial correlation which result from the channel polarization behavior, they also have limitations in terms of performance [OCGD08]. In a large-scale demonstration of dual-polarized MIMO for fixed-wireless access systems that was reported in [ESBC04], experimental average spectral efficiencies were measured within 80–120% of the capacity of single polarization independent Rayleigh fading channels.

About Channel Knowledge

Full or partial channel knowledge at the transmitter is the requirement for many precoding schemes. However, acquiring this knowledge is not a straightforward task, especially in rapidly time-varying channels. In particular, the implementation of several real-time decoding algorithms (without CSIT) was compared to joint transmit and receive signal processing in [HFGS05]. On one hand, it was shown that nonlinear decoding is able to improve the throughput significantly, but also that error propagation is a major drawback. On the other hand, CSIT-based precoding performs best, provided that the channel is slowly varying.

Multilink MIMO

By far, most channel models and MIMO processing techniques have been developed as yet for the basic single-cell/single-user scheme. It turns out that these techniques can be extended to multiuser MIMO (MU-MIMO) schemes. One of the current impediments for satisfactorily achieving such schemes is that multiuser MIMO techniques still rely on simplified channel models and increasing efforts in multiuser channel modeling need be done. For this purpose, a multiuser MIMO field trial has been carried out in [KKGK09], using the Eurecom MIMO OpenAir Sounder (EMOS) testbed in a semi-urban hilly terrain in Southern France. The measurements showed that real-world channels are not uncorrelated Rayleigh distributed, and stressed the fundamental trade-off between channel information feedback and multiuser capacity:

A large user separation combined with codebooks adapted to the second order statistics of the channel provides a sum rate close to the theoretical limit, whereas a small user separation due to bad scheduling or a poorly adapted codebook severely impair the MU-MIMO gains.

Additionally, multicell MIMO has now become an important research area. Even more than in the multiuser case, models of multilink channels (i.e., from multiple users to multiple base stations) are barely emerging. As an example, the WINNER model [KMH+07] still recommends to represent the multicell channels as independent, owing to a number of contradictory results. Nevertheless, correlation between multiple links plays a crucial role in determining the global multicell capacity. The performance of multicell MIMO schemes is also heavily depending on the existing interference caused by multiple users and a proper MIMO interference model has yet to be developed. The real-world feasibility of Multilink MIMO transmissions has been validated in the framework of the EASY-C project [JTW+09, IMW+09]: Field trials show that coordinated multicell transmissions could achieve substantial capacity gains, under the condition that transmitters are tightly synchronized. This is an issue where physical and higher layers are strongly inter-related and impact the whole network architecture, implying that the truly successful implementation of multicell MIMO should rely on advances at all levels of the protocols, including the interaction between these levels and the design of the networks. This will be even truer as new solutions such as relays, femtocells and other network elements come into play.

As can be seen from the previous discussion, MIMO has entered a mature age, and has delivered a large part of the theoretical promises, even though a number of implementation problems had to be dealt with before a full-scale commercialization in a number of wireless standards. All the issues highlighted above are further dealt with in the various chapters of this book, and are introduced in the next section.

3 STRUCTURE OF THIS BOOK

Section 1 has highlighted a few major milestones in the story of MIMO for more than three decades. The immense research effort carried out worldwide since about 2000 has resulted in a number of conceptual discoveries, technical achievements and experimental demonstrations of MIMO systems, some of which being quoted in Section 2. The main goal targeted by this book is to provide readers with a synthetic, up to date view of the essential features of MIMO over those three aspects. For that reason, the structure of the book is two-fold, the first part covering fundamentals and the second implementation.

Not surprisingly since Shannon published his famous theorems, the deep foundations of the MIMO paradigm are to be sought into information theory. In this respect, Chapter 1 provides analytical expressions for the important outage capacity and it relates the capacity to ways to approach it through signal processing.

One of the most astounding facts at the basis of MIMO performance gain is the benefit brought by "imperfect" channels, i.e., channels with rich scattering properties which can beneficially be exploited towards both diversity and multiplexing gains. The description and modelling of these complex channel characteristics and the trade-off between accuracy and simplicity, which is one main concern in channel model standardization, are at the heart of Chapter 2.

Chapter 3 opens the door to the broad field of space-time coding, which is by far, when understood in a broad sense, the most popular and major way to achieve capacities that are not too far from the theoretical ones. The chapter defines the important notions of diversity and multiplexing gains and sets space-time coding as a general model for MIMO transmission schemes. The main practically important schemes are described and their performances compared in the context of the diversity-multiplexing trade-off.

Chapter 4 is another step into one of the very promising research topic for ensuring a wide performance gain from MIMO in cellular networks with many users. Extending the MIMO principle by considering several users communicating with a given base station opens the route to more optimality than considering the base station to each user independently. This general idea is investigated and various ways to achieve multi-user MIMO communications are described and evaluated in terms of, for example, capacity region. Interferences, which play a major role in multiuser communications, are taken into account through the convenient concept of interference function and are effectively taken into account in the performance optimization.

In Chapter 5, one first step towards implementation is initiated through a tutorial description of receiver designs that provide high MIMO performance. The involved architectures, which are based on the turbo principle, address coding, equalization and synchronization. Detailed performance results based on "realistic" simulations are presented.

Chapter 6 goes one step further in terms of MIMO implementation, by addressing the hardware design of MIMO techniques in a transceiver. The chapter focuses on one architectural and algorithmic approach dedicated to spatial division multiplexing in LTE and providing an excellent trade-off between performance and complexity. These "scalable list detectors" are indeed known to be slightly suboptimal with respect to maximum likelihood detection, but much less demanding in terms of hardware complexity.

Chapter 7 is dedicated to a specific, but important, implementation of MIMO into standards, namely IEEE 802.11n. The IEEE 802.11 standard and its various declinations is the basis on which WiFi alliance certifies commercial products that can be interoperated whatever the manufacturer. The final version of 802.11n was published in September 2009, although "Pre-N" products had been available since 2006. Chapter 8 describes both the transmitter and the receiver parts of the standard and provides simulation based performance results obtained from standardized channel models.

WiMAX is another set of several related IEEE 802.16-based standards, for which MIMO is considered to be an essential technique allowing to reach sufficient data rates and link quality in the relevant use cases. This standard is described as regards MIMO in Chapter 8, especially in relation with the uplink case and with the transceiver architectures.

One of the major upcoming cellular standards is LTE and its future version LTE-Advanced. For these standards again, MIMO will be essential to reach adequate throughput and QoS. Chapter 9 particularly tackles open-loop and closed-loop spatial multiplexing and highlights the enhancement brought to LTE by LTE-Advanced through techniques such as Coordinated Multipoint Processing or Multiuser MIMO.

In wireless communications, antennas are compulsory elements which are not so easily accounted for when their real use conditions are considered. This is even more the case for MIMO systems, since multiple antennas terminals are not easy to design for many reasons and are subject to a variety of imperfections with respect to ideal antenna systems, which are commonly assumed by MIMO theoreticians. Chapter 10 analyzes some of these major imperfections and provides clues for the proper design of compact multiple antennas terminals.

Finally, Chapter 11 attempts to delineate a roadmap for the incorporation of MIMO techniques into operational cellular networks, from both an operator's and a researcher's point of view. The main focus is logically IMT-Advanced and both LTE/LTE-Advanced and the upcoming versions of WiMAX are discussed. A "bird's eye view" of the future of MIMO is also proposed, where in all cases it appears that MIMO will be an essential part of wireless communications systems in the future.

PART

MIMO Fundamentals

I

CHAPTER

A Short Introduction to MIMO Information Theory

1

Merouane Debbah

1.1 THE SHANNON-WIENER LEGACY: FROM 1948 TO 2008

In 1948, two important landmark papers were published. The first one [Sha48] was published by Shannon, who provided and proved the expression of the achievable communication rate of a channel with noise. Quite remarkably, Shannon introduced a model (which is still of important use today) based on a statistical nature of the communication medium (see Figure 1.1). Hence, for a given probabilistic model of the medium, he provided the means to compute the exact maximum transmission rate of the information in the system. At the same time, Wiener [Wie48] derived the same capacity formula (without an explicit proof) and introduced the necessary notion of feedback in the communication scheme. Quite surprisingly, without the need of an explicit model of the "black box" (see Figure 1.2), one could theoretically control the output (which is determined by a specific target) based on the feedback mechanism, which provides a measurement of the error induced. The framework and the introduction of feedback is of great interest today in wireless communications, where one has only partial knowledge of the wireless medium. Control theory turns

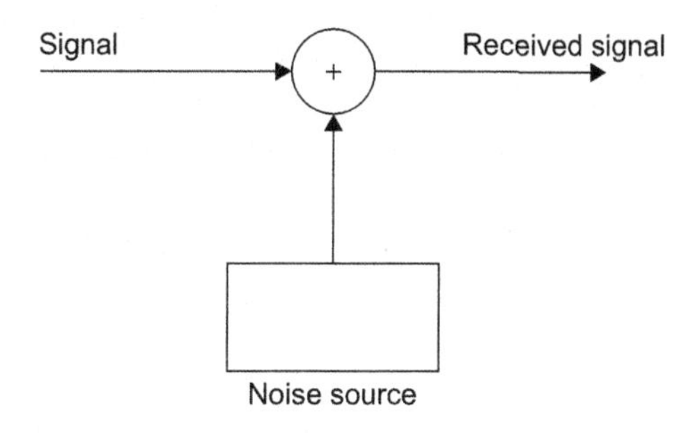

FIGURE 1.1

Shannon's approach.

MIMO. DOI: 10.1016/B978-0-12-382194-2.00001-0

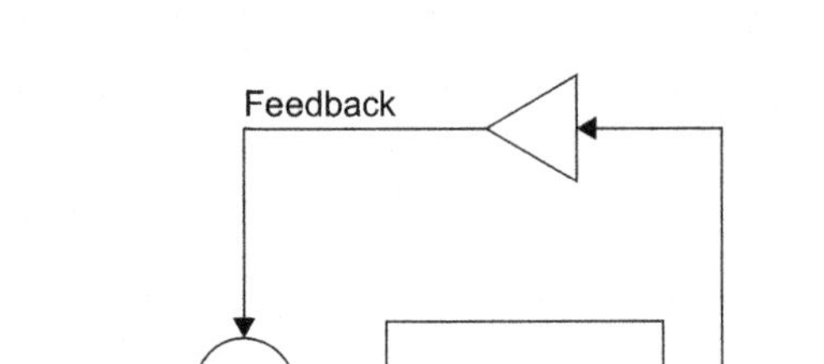

FIGURE 1.2

Wiener's approach.

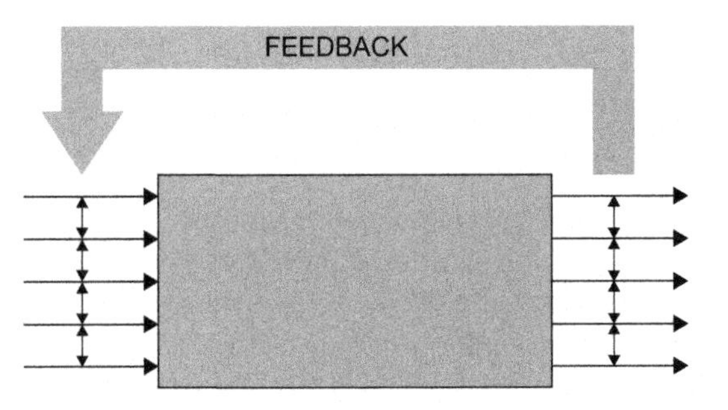

FIGURE 1.3

MIMO mobile flexible networks.

out to be a very neat way of designing the feedback (how many bits of feedback needed, analog or digital, and so on).

For the single-input single-output framework, these two papers were instrumental. Sixty years later, the MIMO Mobile Flexible Network framework in the realm of the cybernetic work of Wiener is much more general (see Figure 1.3) in the sense that the "black box" has multiple inputs and multiple outputs. The inputs are not necessarily connected or can be partially connected (input 1 can be connected to input 4, for example). The same holds for the outputs (the single user multiantenna case [Tel99] corresponds to the case where all the inputs and outputs are connected). Moreover, there is a lot of flexibility in the feedback mechanism (typically, for example, output 3 can be only connected to input 1). Finally, and this is a major difference with previous works, the designer must learn and control the "black box":

- within a fraction of time, and
- with finite energy.

Because of the user mobility, these constraints are at the moment extremely hard to cope with as the number of inputs and outputs (the dimensionality of the system) equal the time scale changes (in terms of number of time symbols) of the box.

1.2 PRELIMINARIES

Let us first introduce our notations. Consider two identical receivers at relative distance d. Let $c_1(t,\tau)$ and $c_2(t,\tau)$ be the fading channels characterizing the propagation between the transmitters and receivers 1 and 2, respectively. The correlation between the fading channels $c_1(t,\tau)$ and $c_2(t,\tau)$ depends on the geometry of the scattering elements and on the radiation patterns of transmit and receive antennas. For simplicity, we shall assume an isotropic radiation pattern for the receiving antennas and the planar scattering geometry of Figure 1.4, where both antennas receive scattered plane waves with uniform probability from all angles $\theta \in [-\pi,\pi)$. We assume that d is much smaller than the distance between the antennas and the scattering elements. Then, all scattered signals are received with the same delay τ_0 (the channel is frequency nonselective, since there is only one propagation path). Moreover, the fading Doppler spectrum is of Jakes' type, with time autocorrelation function $J_0(2\pi B_d \Delta t)$, with $B_d = f_c v/c$ being the maximum Doppler shift, where f_c and v are the carrier frequency and the relative speed between the scatterers and the receiving antennas.

Let $\lambda = c/f_c$ denote the carrier wavelength, and let $\Delta t = d/v$ be the time necessary to move from antenna 2 to antenna 1, at speed v. Then, the corresponding *space*

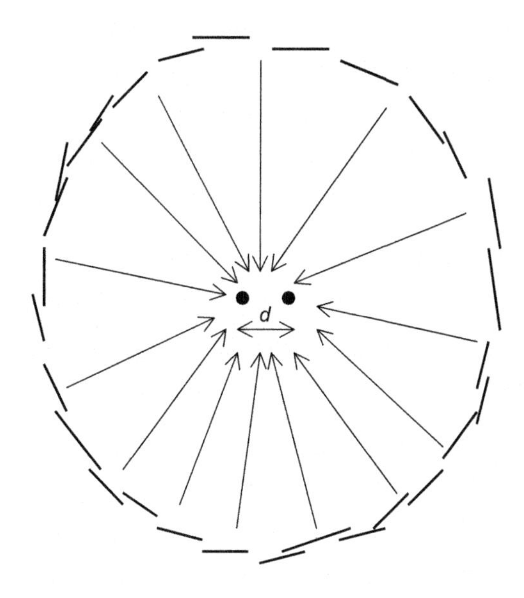

FIGURE 1.4

Isotropic planar scattering geometry.

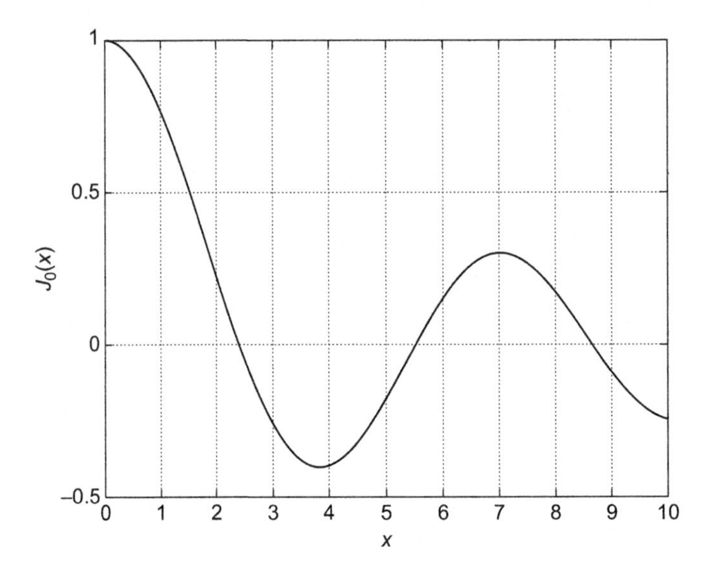

FIGURE 1.5

Bessel $J_0(x)$ function.

autocorrelation function of fading experienced by the two antennas is:

$$\begin{aligned}\mathbb{E}[c_1(t)c_2(t)^*] &= \mathbb{E}[c_1(t)c_1(t-\Delta t)^*] \\ &= J_0(2\pi B_d \Delta t) \\ &= J_0(2\pi d/\lambda) \end{aligned} \tag{1.1}$$

where we have used the fact that $c_2(t) = c_1(t - \Delta t)$. From Figure 1.5, representing the function $J_0(x)$, we see that for d/λ of the order of some unit, the fading at the two antennas is virtually uncorrelated.

The above consideration motivates *diversity* reception based on multiple antennas: Multiple observations of the transmit signal can be obtained from multiple receiving antennas. Each observation is characterized by different statistically independent fading. Hence, the probability that all channels fade at the same time is small and the chances that at least some of the observations have a good instantaneous Signal-to-Noise Ratio (SNR) is increased by increasing the number of receiving antennas. In the following, we make this statement more precise by considering an idealized simplified model, with receiving antennas having mutually independent frequency nonselective fading.

We assume that the transmission takes place between a mobile transmitter and a mobile receiver. The transmitter is equipped with n_t antennas and the receiver is equipped with n_r antennas. Moreover, we assume that the input transmitted signal passes through a time variant linear filter channel. Finally, we assume that the interfering noise is an additive white Gaussian random variable.

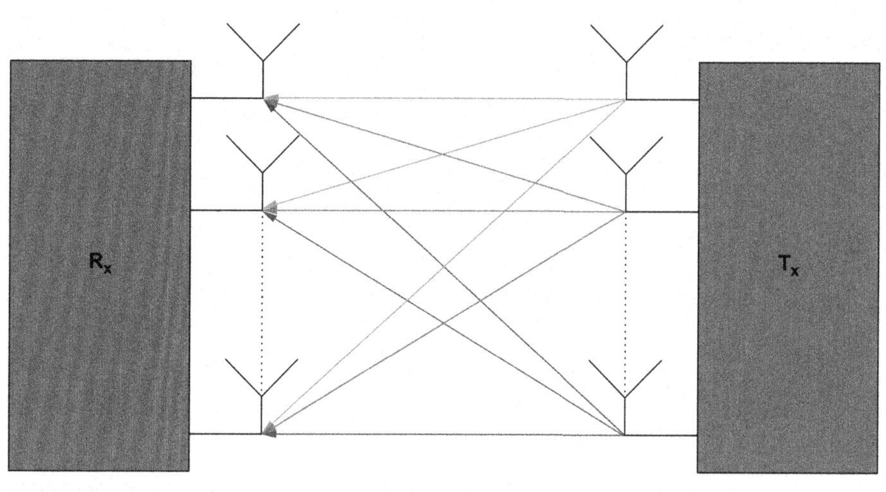

FIGURE 1.6

MIMO channel representation.

The transmitted signal and received signal are related as (see Figure 1.6):

$$\mathbf{y}(t) = \sqrt{\frac{\rho}{n_t}} \int \mathbf{C}_{n_r \times n_t}(\tau, t)\mathbf{x}(t-\tau)d\tau + \boldsymbol{\nu}(t) \tag{1.2}$$

with

$$\mathbf{C}_{n_r \times n_t}(\tau, t) = \int \mathbf{H}_{n_r \times n_t}(f, t)e^{j2\pi f\tau} df, \tag{1.3}$$

where ρ is the received SNR which represents the total transmit power per symbol versus total spectral density of the noise, t, f, and τ denote respectively time, frequency and delay, $\mathbf{y}(t)$ is the $n_r \times 1$ received vector, $\mathbf{x}(t)$ is the $n_t \times 1$ transmit vector, $\boldsymbol{\mu}(t)$ is an $n_r \times 1$ additive standardized white Gaussian noise vector.

In the rest of this chapter, we will only be interested in the frequency domain channel (knowing that the impulse response matrix can be accessed through an inverse Fourier transform according to relation 1.3). For simplicity, we drop the time and frequency indices.

1.3 INFORMATION THEORETIC ASPECTS

1.3.1 Information Theoretic Metrics

Before starting any discussion on MIMO capacity, let us first review the pioneering work of Telatar [Tel95] (later published as [Tel99]) that triggered research in multiantenna systems[1]. In this paper, Telatar develops the channel capacity of a general MIMO channel. Assuming perfect knowledge of the channel matrix $\mathbf{H}$ at the receiver (in this case, there is no mobility), the mutual information I^M between input and output is given by[2]:

$$\begin{aligned} I^{\mathrm{M}}(\mathbf{x};(\mathbf{y},\mathbf{H})) &= I^{\mathrm{M}}(\mathbf{x};\mathbf{H}) + I^M(\mathbf{x};\mathbf{y} \mid \mathbf{H}) \\ &= I^{\mathrm{M}}(\mathbf{x};\mathbf{y} \mid \mathbf{H}) \\ &= \mathrm{Entropy}(\mathbf{y} \mid \mathbf{H}) - \mathrm{Entropy}(\mathbf{y} \mid \mathbf{x},\mathbf{H}) \\ &= \mathrm{Entropy}(\mathbf{y} \mid \mathbf{H}) - \mathrm{Entropy}(\boldsymbol{\nu} \mid \mathbf{H}) \end{aligned}$$

When the entries have a covariance ($\mathbf{Q} = \mathbb{E}(\mathbf{x}\mathbf{x}^H)$), since $\mathbf{y} = \sqrt{\frac{\rho}{n_t}}\mathbf{H}\mathbf{x} + \boldsymbol{\nu}$, we have:

$$\mathbb{E}(\mathbf{y}\mathbf{y}^H) = \mathbf{I}_{n_r} + \frac{\rho}{n_t}\mathbf{H}\mathbf{Q}\mathbf{H}^H$$

$$\mathbb{E}(\boldsymbol{\nu}\boldsymbol{\nu}^H) = \mathbf{I}_{n_r}$$

If the $x_i, i = 1, \ldots, n_t$ are Gaussian[3], the mutual information which will be denoted in this case by $C(\mathbf{Q}) \triangleq I^{\mathrm{M}}$ is:

$$\begin{aligned} C(\mathbf{Q}) &= \mathbb{E}\left[\mathrm{Entropy}(\mathbf{y} \mid \mathbf{H} = H) - \mathrm{Entropy}(\nu \mid \mathbf{H} = H)\right] \\ &= \mathbb{E}\left[\log_2 \det\left(2\pi e\left(\mathbf{I}_{n_r} + \frac{\rho}{n_t}\mathbf{H}\mathbf{Q}\mathbf{H}^H\right)\right) - \log_2 \det(2\pi e \mathbf{I}_{n_r})\right] \\ &= \mathbb{E}\left[\log_2 \det\left(\mathbf{I}_{n_r} + \frac{\rho}{n_t}\mathbf{H}\mathbf{Q}\mathbf{H}^H\right)\right] \end{aligned}$$

As a consequence[4], the ergodic capacity of an $n_r \times n_t$ MIMO channel with Gaussian entries and covariance matrix $\mathbf{Q}$ ($\mathbf{Q} = \mathbb{E}(\mathbf{x}\mathbf{x}^H)$)[5] is:

$$\overline{C} = \max_{\mathbf{Q}} \mathbb{E}_{\mathbf{H}}(C(\mathbf{Q})) \tag{1.4}$$

where the maximization is over a set of positive semidefinite hermitian matrices $\mathbf{Q}$ satisfying the normalized average power constraint trace $(\mathbf{Q}) \leq n_t$, and the

[1] For contribution [Tel99], Telatar received the 2001 Information Theory Society Paper Award. After the Shannon Award, the IT Society Paper Award is the highest recognition award by the IT society.
[2] Note that the channel is entirely described with input $\mathbf{x}$ and output $(\mathbf{y},\mathbf{H}) = (\mathbf{H}\mathbf{x} + \mathbf{n}, \mathbf{H})$.
[3] The differential entropy of a complex Gaussian vector $\mathbf{x}$ with covariance $\mathbf{Q}$ is given by $\log_2 \det(\pi e \mathbf{Q})$.
[4] We only derived the mutual information with Gaussian entries, but did not prove that it achieves capacity. This stems from the fact that, for a given covariance $\mathbf{Q}$, the entropy of $\mathbf{x}$ is always inferior to $\log_2 \det(\pi e \mathbf{Q})$ with equality, if and only if $\mathbf{x}$ is complex Gaussian.
[5] In the general case where the noise is Gaussian with a covariance matrix $\mathbf{Z}$, the capacity is given by: $C(\mathbf{Q},\mathbf{Z}) = \log_2 \frac{\det(\mathbf{Z} + \frac{\rho}{n_t}\mathbf{H}\mathbf{Q}\mathbf{H}^H)}{\det(\mathbf{Z})}$.

expectation is with respect to the random channel matrix. Therefore, the ergodic capacity is achieved for a particular choice of the matrix $\mathbf{Q}$.

In the original paper [Tel95], Telatar exploits the isotropic property of Gaussian i.i.d. $\mathbf{H}$ channels to show that in this case the ergodic capacity is achieved with $\mathbf{Q} = \mathbf{I}_{\mathbf{n_t}}$. However, this result has been proved only for Gaussian i.i.d. channel matrices and was not extended to other types of matrices. In correlated fading, $C(\mathbf{I}_{\mathbf{n_t}})$ is called the average mutual information with covariance $\mathbf{Q} = \mathbf{I}_{\mathbf{n_t}}$. It has never been proved that capacity was close to this mutual information except for certain particular cases (see [Tel99, GJJV03]). $C(\mathbf{I})$ underestimates the maximum achievable rate.[6] Indeed, even though the channel realization is not known, the knowledge of the channel model (Is it independent and identically distributed (i.i.d.) Rayleigh fading?; Is it i.i.d. Rice Fading?; Is it correlated Rayleigh fading with a certain covariance matrix?) can be taken into account in order to optimize the coding scheme at the transmitter. There is no reason why one should transmit independent substreams on each antenna. It is as if one stated that the space-time codes designed through the rank and determinant criterion [GSS+03] were optimal for *all kinds of MIMO channels*.

Note that for a wireless content provider the most important criterion is the quality of service to be delivered to customers. This quality of service can be quantified through measures such as the outage capacity: If $q = 1\%$ is the outage probability of having an outage capacity of R, then this means that the provider is able to ensure a rate of R for 99% of the random channel realizations. Since the channels are rarely ergodic, the derivations of ergodic capacities are of limited use to content providers. We give the definitions of the ergodic capacity and the outage capacity:

- If the time varying channels are ergodic, $C(\mathbf{Q})$ can be averaged over many channel realizations and the corresponding capacity is defined as: $\overline{C} = \max_{\mathbf{Q}} \mathbb{E}(C(\mathbf{Q}))$.
- If the channels are static, there is only one channel realization, and an outage probability for each positive rate of transmission can be defined. The outage capacity is:[7]

$$C_q = \max_{\mathbf{Q}} \sup\{R \geq 0 : \Pr[C(\mathbf{Q}) < R] \leq q\}.$$

1.3.2 Analytical Expressions

A common assumption when deriving the outage probability or the outage mutual information (and not the outage capacity) is the Gaussian input covariance matrix

[6] However, note that although not optimum, the mutual information with covariance $\mathbf{Q} = \mathbf{I}_{\mathbf{n_t}}$ can be useful in the analysis of systems where the codebook cannot be changed according to the wireless environment and therefore remains the same during the whole transmission.

[7] Note that the covariance matrix $\mathbf{Q}$ which optimizes the ergodic capacity does not necessarily achieve the outage capacity.

$\mathbf{Q} = \mathbf{I}_{\mathbf{n_t}}$. This assumption can be realistic if the channel distribution is unknown at the transmitter side. In [Chi02], the author gives the analytical expression of the moment generating function of the mutual information assuming uniform power allocation over the transmit antennas. Let $n_{\min} = \min\{n_r, n_t\}$ and $r = \max\{n_r, n_t\} - n_{\min}$. The moment generating function of $C(\mathbf{I}_{\mathbf{n_t}})$ was proven to be equal to:

$$\Phi(z) \triangleq \mathbb{E}[e^{Cs}] = b^{-1}\det(\mathbf{G}(z)), \tag{1.5}$$

where $b = \prod_{i=1}^{n_{\min}} \Gamma(r+i)$[8], and $\mathbf{G}(z)$ is a $n_{\min} \times n_{\min}$ Hankel matrix with (i,k) entry:

$$g_{i,k}(z) = \int_0^{+\infty} \left(1 + \frac{\rho}{n_t} x\right) x^{i+k+r} e^{-x} \mathrm{d}x, \tag{1.6}$$

for $i,k \in \{1,\ldots,n_{\min}\}$. The authors of [Chi02] derived the exact analytical expression of the outage probability by using the fact that the cumulative distribution function (cdf) of $C(\mathbf{I}_{\mathbf{n_t}})$ is equal to the inverse Laplace transform of $z^{-1}\Phi(z)$:

$$P_{\text{out}}(\mathbf{I}_{\mathbf{n_t}}, R) = \frac{1}{2\pi j} \int_{\xi - j\infty}^{\xi + j\infty} z^{-1}\Phi(z) e^{zR} \mathrm{d}z, \tag{1.7}$$

where ξ is a fixed positive number.

This result was then generalized by [CWZ03] to the case of semicorrelated Rayleigh fading with correlation at the receiver and $n_t \geq n_r$ (or, equivalently, correlation among transmit antennas and $n_t < n_r$). The case of correlation at the receiver and $n_t < n_r$ (or correlation at the transmitter and $n_t \geq n_r$) was investigated in [SRS03]. However, these exact expressions are intractable and difficult to analyze. When deriving the outage capacity, if the channel distribution is known at the transmitter side, then it is possible to optimize the covariance matrix of the transmitting signal. However, this is not an obvious task even in the case of i.i.d. Gaussian channel models where for a fixed transmission rate $R > 0$, the optimal covariance matrix that minimizes the outage probability, $P_{\text{out}}(\mathbf{Q}, R) = \Pr[C(\mathbf{Q}) < R]$ is unknown in general. Telatar conjectured in [Tel99] that the transmit strategy that minimizes the outage probability is to allocate uniform power only over a subset of $\ell \in \{1,\ldots,n_t\}$ antennas, such that the optimal covariance matrix is a diagonal matrix with ℓ nonzero entries: $\mathbf{Q} = \frac{n_t}{\ell}\mathrm{diag}(1,\ldots,1,0,\ldots,0)$. The optimal number of active antennas depends on the channel parameters and on the target rate R and was also conjectured by Telatar: For a fixed ρ, the higher the target rate R, the smaller the optimal number of active antennas. Note again that the covariance matrix that achieves the ergodic capacity $\mathbf{I}_{n_t}$ does not necessarily minimize the outage probability. This conjecture was solved in [KS07] in the particular case where the transmitter is equipped with two antennas, $n_t = 2$, and the receiver has a single antenna, $n_r = 1$. In [JB06], the conjecture was proven in a more general case where the transmitter has an arbitrary number of

[8] $\Gamma(z)$ denotes the Gamma function $\Gamma(z) \triangleq \int_0^{+\infty} x^{z-1} e^{-x} \mathrm{d}x$.

antennas $n_t \geq 2$. The outage probability for the Multiple-Input Single-Output (MISO) channel can be written as:

$$\begin{aligned} P_{\text{out}}(\mathbf{Q},R) &= \Pr\left[\log_2\left(1+\frac{\rho}{n_t}\mathbf{h}\mathbf{Q}\mathbf{h}^H\right) \leq R\right] \\ &= \Pr\left[\log_2\left(1+\frac{\rho}{n_t}\sum_{i=1}^{n_t} p_i|h_i|^2\right) \leq R\right] \\ &= \Pr\left[\sum_{i=1}^{n_t} p_i|h_i|^2 \leq n_t\gamma\right] \end{aligned} \tag{1.8}$$

where $\{p_i\}_{i\in\{1,\ldots,n_t\}}$ represent the eigenvalues of the covariance matrix and $\gamma = \frac{2^R-1}{\rho}$. In this case, the following theorem holds:

Theorem 1 [JB06] Under the Gaussian i.i.d. MISO channel model, the optimal outage probability is: $\sup_{\mathbf{Q}}\{1-P_{out}(\mathbf{Q},R)\} = g^{(\ell)}(\gamma), \text{if} \gamma \in (\gamma_\ell, \gamma_{\ell-1}]$ where $g^{(\ell)}(x) = 1-\Pr\left[\frac{1}{\ell}\sum_{k=1}^{n_t}|h_k|^2 \leq x\right]$ and the fixed positive constants $\{\gamma_\ell\}_{\ell=1}^{n_t-1}$ are taken such that $1 \leq \gamma_{n_t-1} \leq \ldots \leq \gamma_1 \leq 2$ with the convention $\gamma_0 = \infty$, $\gamma_{n_t} = 0$. Thus, the optimal covariance matrix is of the form $\mathbf{Q} = \frac{n_t}{\ell}\text{diag}(1,\ldots,1,0,\ldots,0)$ if $\gamma \in (\gamma_\ell, \gamma_{\ell-1}]$.

The positive constants $\{\gamma_\ell\}_{\ell=1}^{n_t-1}$ can be computed numerically by considering the properties of the probability density function (pdf) of weighted sums of Chi-square random variables. To give an insight on this result, for a fixed target rate R, these constants correspond to SNR thresholds. If the SNR level is sufficiently high, the optimal strategy is to spread the available power uniformly over all the transmit antennas. If the SNR level is sufficiently low, then using only one transmit antenna (beamforming power allocation policy) minimizes the outage probability. In other words, the optimal number of active antennas is increasing with the SNR level. The results obtained in the extreme SNR regimes have been extended in [JB06] to the general MIMO case.

Having a general explicit expression based on different channel models taking into account correlation [MÖ2, CTKV02, HG03, Gra02] or not [Tel95] is a difficult task. Moreover, in the finite case, very few works have been devoted to the outage probability and outage capacity [KHX02, HMT02, MSS02] or deriving the capacity distribution [SM02, Sca02]. Another approach is based on asymptotic random matrix theory, which provides a good approximation of the performance measures even for a small number of antennas. Indeed, recently, using tools from random matrix theory, it was shown in [DM05] that in many cases the mutual information of MIMO models has an asymptotically Gaussian behavior. Random matrices were first proposed by Wigner in quantum mechanics to explain the measured energy levels of nuclei in terms of the eigenvalues of random matrices. When Telatar [Tel95] (in the context of multiantenna channel capacity analysis) and then nearly simultaneously Tse & Hanly [TH99] and Verdu & Shamai [VS99] (for the analysis of uplink unfaded Code Division Multiple Access (CDMA) equipped with certain receivers) introduced random matrices, the random matrix theory entered the field of telecommunications[9]. From that time, random matrix theory has been successively extended to other cases such as uplink CDMA fading channels [SV01], OFDM [DHLdC03], downlink CDMA [CHL02], multiuser detection [MV01], etc. One of the useful features of random matrix theory is the ability to predict, under certain conditions, the behavior of the empirical eigenvalue distribution of products or sums of matrices. The results are striking in terms of closeness to simulations with reasonable matrix sizes and enable to derive linear spectral statistics for these matrices with only few meaningful parameters.

Main Results

In the case of the i.i.d. Gaussian channel, the following theorem holds:

Theorem 2 Assuming the Gaussian i.i.d. channel model, as $n_t \to \infty$ with $n_r = \beta n_t$, $\ln\det\left(\mathbf{I}_{n_r} + \frac{\rho}{n_t}\mathbf{H}\mathbf{H}^H\right) - n_t\mu_{\text{iid}}(\beta,\rho)$ converges in distribution to a Gaussian $\mathcal{N}(0, N_{\text{iid}}(\beta,\rho))$ random variable where[10]:

$$\mu_{\text{iid}}(\beta,\rho) = \beta\ln(1+\rho-\rho\alpha_{\text{iid}}(\beta,\rho)) + \ln(1+\rho\beta-\rho\alpha_{\text{iid}}(\beta,\rho)) - \alpha_{\text{iid}}(\beta,\rho)$$

and

$$N_{\text{iid}}(\beta,\rho) = -\ln\left[1 - \frac{\alpha^2{}_{\text{iid}}(\beta,\rho)}{\beta}\right]$$

with

$$\alpha_{\text{iid}}(\beta,\rho) = \frac{1}{2}\left[1+\beta+\frac{1}{\rho} - \sqrt{\left(1+\beta+\frac{1}{\rho}\right)^2 - 4\beta}\right]$$

Proof. We give here only a sketch of the proof for deriving the value of μ_{iid}, which corresponds to the asymptotic mutual information per transmitting antenna:

$$\begin{aligned}
\mu_{\text{iid}} &= \lim_{n_t\to\infty}\frac{1}{n_t}\ln\left[\det\left(\mathbf{I}_{n_r} + \frac{\rho}{n_t}\mathbf{H}\mathbf{H}^H\right)\right] \\
&= \lim_{n_t\to\infty}\frac{1}{n_t}\ln\left[\det\left(\mathbf{I}_{n_t} + \frac{\rho}{n_t}\mathbf{H}^H\mathbf{H}\right)\right] \\
&= \lim_{n_t\to\infty}\frac{1}{n_t}\sum_{i=1}^{n_t}\ln(1+\rho\lambda_i) \\
&= \lim_{n_t\to\infty}\int\ln(1+\rho\lambda)dF_{n_t}(\lambda).
\end{aligned}$$

[9]It should be noted that in the field of array processing, Silverstein already used in 1992 random matrix theory [SC92] for signal detection and estimation.

[10]ln is the natural logarithm such as $\ln(e) = 1$. When this notation is used, the mutual information is given in nats/s. When the notation $\log_2(x) = \frac{\ln(x)}{\ln(2)}$ is used, the results are given in bits/s.

The second equality comes from the determinant identity $\det(I+AB) = \det(I+BA)$. The parameters λ_i, $i \in \{1,\ldots,n_t\}$, are the eigenvalues of the matrix $\frac{1}{n_t}\mathbf{H}^H\mathbf{H}$ and $F_{n_t}(\lambda)$ is the empirical eigenvalue distribution function of $\frac{1}{n_t}\mathbf{H}^H\mathbf{H}$ defined by: $dF_{n_t}(\lambda) = \frac{1}{n_t}\sum_{i=1}^{n_t}\delta(\lambda-\lambda_i)$. It is now well established that the empirical eigenvalue distribution $F_{n_t}(\lambda)$ converges weakly to a nonrandom distribution defined by:

$$f_{\text{iid}}(x) = (1-\beta)\delta(x) + \frac{1}{2\pi x}\sqrt{\left((1+\sqrt{\beta})^2 - x\right)\left(x-(1-\sqrt{\beta})^2\right)}$$

defined in the interval $\left[(1-\sqrt{\beta})^2,(1+\sqrt{\beta})^2\right]$.

The asymptotic mean value is therefore equal to:

$$\begin{aligned}\mu_{\text{iid}}(\beta,\rho) &= \int_{(1-\sqrt{\beta})^2}^{(1+\sqrt{\beta})^2} \ln(1+\rho\beta)f(\lambda)d\lambda \\ &= \beta\ln(1+\rho-\rho\alpha(\beta,\rho)) + \ln(1+\rho\beta-\rho\alpha(\beta,\rho)) - \alpha(\beta,\rho)\end{aligned}$$

with

$$\alpha(\beta,\rho) = \frac{1}{2}\left[1+\beta+\frac{1}{\rho} - \sqrt{\left(1+\beta+\frac{1}{\rho}\right)^2 - 4\beta}\right].$$

Remark 1 For realistic models, it can be shown that the mean mutual information scales at high SNR as:

$$\mathbb{E}(C(\mathbf{I_{n_t}})) = \min(n_t,n_r,s)\log_2(\rho)$$

where s expresses the number of scatterers in the environment. The factor $\min(n_t,n_r,s)$ is also known as the multiplexing gain.

Interestingly, the environment is the limiting factor (through the scatterers and their correlation). As a consequence, roughly speaking, there is no use in using more antennas than the number of degrees of freedom the environment is able to provide.

Deriving the Outage Mutual Information

Let q denote the outage probability and I_q^{M} the corresponding outage mutual information with covariance $\mathbf{Q} = \mathbf{I_{n_t}}$, then:

$$\begin{aligned}q &= \text{P}\left(I^{\text{M}} \leq I_q^{\text{M}}\right) \\ &= \int_{-\infty}^{I_q^{\text{M}}} dI^{\text{M}} p(I^{\text{M}}) \\ &\approx \frac{1}{\sqrt{2\pi N_{\text{iid}}}} \int_{-\infty}^{I_q^{\text{M}}} dI^{\text{M}} e^{-\frac{\left(I^{\text{M}}-n_t\mu_{\text{iid}}\right)^2}{2N_{\text{iid}}}}\end{aligned}$$

$$\approx 1 - Q\left(\frac{I_q^{\mathrm{M}} - n_t\mu_{\mathrm{iid}}}{\sqrt{N_{\mathrm{iid}}}}\right)$$

$$I_q^{\mathrm{M}} \approx n_t\mu_{\mathrm{iid}} + \sigma Q^{-1}(1-q).$$

We define[11]:

$$Q(x) = \frac{1}{\sqrt{2\pi}} \int_x^\infty \mathrm{d}t e^{\frac{-t^2}{2}}.$$

Therefore, in the large system limit, only the knowledge of the mean and variance of the mutual information distribution is needed for deriving the outage mutual information.

1.4 SIGNAL PROCESSING ASPECTS

The MMSE receiver has several attributes that makes it appealing for use. The MMSE receiver is known to generate a soft decision output, which maximizes the received Signal-to-Interference plus Noise Ratio (SINR), see [MH94], whereas in an Additive White Gaussian Noise (AWGN) channel with no interference, the match filter maximizes the output SNR. This advantage combined with the low complexity implementation of the receiver (due in part to its linearity) has triggered the search for other MMSE-based receivers such as the MMSE Decision Feedback Equalization (DFE). The MMSE DFE [CDEF95a, CDEF95b] is at the heart of very famous schemes such as BLAST [GFVW99]. BLAST was invented by Foschini at Bell Labs in 1998. It is highly related to spatial multiplexing, introduced in a 1994 Stanford University patent by A. Paulraj [PK94]: The idea is based on successive interference cancellation where each layer is decoded, re-encoded and subtracted from the transmitted signal. This approach has triggered several implementation research schemes as it was shown by Varanasi and Guess [VG97][12] to be optimal. Therefore, the MMSE receiver is at the heart of many schemes and studying the SINR distribution enables to design and understand the performance of many MMSE-based receivers (e.g., MMSE DFE [CDEF95a, CDEF95b, ADC95], MMSE parallel interference cancellation [CW01]).

1.4.1 The SIMO Case

Assume the Single-Input Multiple-Output (SIMO) channel where the transmitter is equipped with a single antenna, $n_t = 1$. The discrete-time signal received at the i-th antenna is given by:

$$y_i = \sqrt{\mathcal{E}} h_i x + v_i$$

[11] Usual programming tools use the complementary error function defined as $\mathrm{erfc}(x) = \frac{2}{\sqrt{\pi}} \int_x^\infty e^{-t^2} dt$. In this case, $Q(x) = \frac{1}{2}\mathrm{erfc}\left(\frac{x}{\sqrt{2}}\right)$.

[12] The optimality follows in fact directly from a simple determinant identity.

where $v_i \sim \mathcal{N}_C(0, N_0)$ and x is the input symbol. Noise contributions at different antennas are mutually independent. At this point, the multiple observations obtained from the n_r receiving antennas are totally analogous to the multiple observations obtained from the fingers of a Rake receiver (this is seen immediately by replacing the antenna index i by the path index p). The observation vector $\mathbf{y} = (y_1, \ldots, y_{n_r})^T$ can be written as:

$$\mathbf{y} = \sqrt{\mathcal{E}}\mathbf{h}x + \boldsymbol{\nu} \tag{1.9}$$

where we define $\mathbf{h} = (h_1, \ldots, h_{n_r})^T$ and $\boldsymbol{\nu} = (\nu_1, \ldots, \nu_{n_r})^T$. The optimal receiver consists of a Maximal Ratio Combining (MRC) followed by either a symbol-by-symbol detector (under uncoded modulation) or a soft decoder.

After normalization, the output of the MRC is given by:

$$y = \sqrt{\mathcal{E}}\,\tilde{h}x + \nu$$

where $\tilde{h} = \sqrt{\sum_{i=1}^{n_r} |h_i|^2}$ and $\nu \sim \mathcal{N}_C(0, N_0)$.

The performance of uncoded modulation with MRC can be calculated as for the Rake receiver, either by using the technique based on Hermitian Quadratic Form of complex Gaussian Random Variables (HQF-GRV), or by using the technique based on the integral expression of the Gaussian Q-function.

The quality of wireless links with very slowly-varying fading (quasi-static fading) is often characterized by the probability that the SNR at the detector input falls below a certain threshold s_0, that depends on the coding and modulation scheme used. The SNR threshold determines the required Quality of Service (QoS) of the link. If SNR $< s_0$, the required QoS is not met and we say that the link is in an *outage* state. The outage probability $P_{\text{out}} = \Pr(\text{SNR} < s_0)$ is directly obtained by the SNR cumulative distribution function c.d.f., evaluated at s_0.

The Cumulative Distribution Function of the SNR at Output of the MRC

Suppose independent normalized Ricean fading for all antennas; that is, $h_i \sim \mathcal{N}_C(\sqrt{K/(1+K)}, 1/(1+K))$. Then, the SNR:

$$\text{SNR}_{\text{mrc}} = \frac{\mathcal{E}}{N_0} \sum_{i=1}^{n_r} |h_i|^2$$

is distributed according to the noncentral Chi-squared probability density function (pdf) with $2r$ degrees of freedom:

$$f_{\text{mrc}}(z) = \frac{1+K}{\rho} \left(\frac{z(1+K)}{\rho n_r K} \right)^{(n_r-1)/2} \times \exp\left(-(1+K)\left(\frac{n_r K}{1+K} + \frac{z}{\rho} \right) \right) I_{n_r-1}\left(2\sqrt{\frac{z n_r K(1+K)}{\rho}} \right)$$

where we let $\rho = \mathcal{E}/N_0$, and where $I_{n_r-1}(x)$ is the modified Bessel function of the first kind of order $n_r - 1$.

For $K = 0$ (Rayleigh fading), the SNR pdf becomes a central Chi-squared with $2n_r$ degrees of freedom:

$$f_{\text{mrc}}(z) = \frac{1}{\rho}\frac{1}{(n_r-1)!}\left(\frac{z}{\rho}\right)^{n_r-1} e^{-z/\rho}$$

For $K \to \infty$, the channel becomes AWGN, and the SNR converges in distribution to the constant value $n_r\rho$.

The SNR cdf, for independent identically distributed Ricean fading, is given by:

$$F_{\text{mrc}}(z) = 1 - Q_{n_r}\left(\sqrt{2n_rK}, \sqrt{\frac{2z(1+K)}{\rho}}\right)$$

where $Q_{n_r}(a,b)$ is the generalized Marcum Q-function [SA98], defined by:

$$Q_{n_r}(a,b) = \int_b^{\infty} x(x/a)^{n_r-1}e^{-(x^2+a^2)/2}I_{n_r-1}(ax)dx$$

The cdf of the output SNR for i.i.d. Rayleigh fading is given by:

$$F_{\text{mrc}}(z) = 1 - e^{-z/\rho}\sum_{\ell=0}^{n_r-1}\frac{1}{\ell!}\left(\frac{z}{\rho}\right)^{\ell}.$$

Selection Combining

A suboptimal but simpler combining scheme consists of selecting the output of the antenna with largest instantaneous SNR. This method is called *selection combining.* With selection combining, the detector input is given by:

$$y = h_{\max}x + \nu$$

where $h_{\max}$ is defined as the fading coefficient with maximum squared magnitude, overall $i = 1,\ldots,n_r$, and where $\nu \sim \mathcal{N}_{\mathbb{C}}(0,N_0)$. The resulting output SNR is given by (we drop again the time index):

$$\text{SNR}_{\text{sel}} = \frac{\mathcal{E}}{N_0}|h_{\max}|^2.$$

This technique does not "combine" the multiple observations; however, it can be seen as a linear combiner with suboptimal combining coefficients equal either to 0 or to 1.

If the fading coefficients of all antennas are i.i.d., we can write the SNR cdf as:

$$\begin{aligned} F_{\text{sel}}(z) &= \Pr(\text{SNR} \le z) \\ &= \Pr\left(\rho \max_{i=1,\ldots,n_r} |h_i|^2 \le z\right) \\ &= \Pr\left(|h_1|^2 \le z/\rho,\ldots,|h_{n_r}|^2 \le z/\rho\right) \end{aligned}$$

$$= \prod_{i=1}^{n_r} \Pr\left(|h_i|^2 \leq z/\rho\right)$$

$$= F_g(z/\rho)^{n_r} \tag{1.10}$$

where g denotes the power gain of any of the n_r channels (they are identically distributed) and $F_g(z)$ denotes its cdf.

The SNR pdf is obtained by deriving its cdf with respect to z:

$$f_{\text{sel}}(z) = \frac{n_r}{\rho}\left(F_g(z/\rho)\right)^{n_r-1} f_g(z/\rho)$$

For example, in the case of normalized Rayleigh fading we have:

$$f_{\text{sel}}(z) = \frac{n_r}{\rho}\sum_{\ell=0}^{n_r-1}\binom{n_r-1}{\ell}(-1)^{\ell}e^{-z(\ell+1)/\rho}$$

for $z \geq 0$, and $f_{\text{sel}}(z) = 0$ for $z < 0$.

Figure 1.7 shows the SNR pdf for MRC and selection combining in the case of Rayleigh fading with $n_r = 1, 2, 4, 8$ independent antennas.

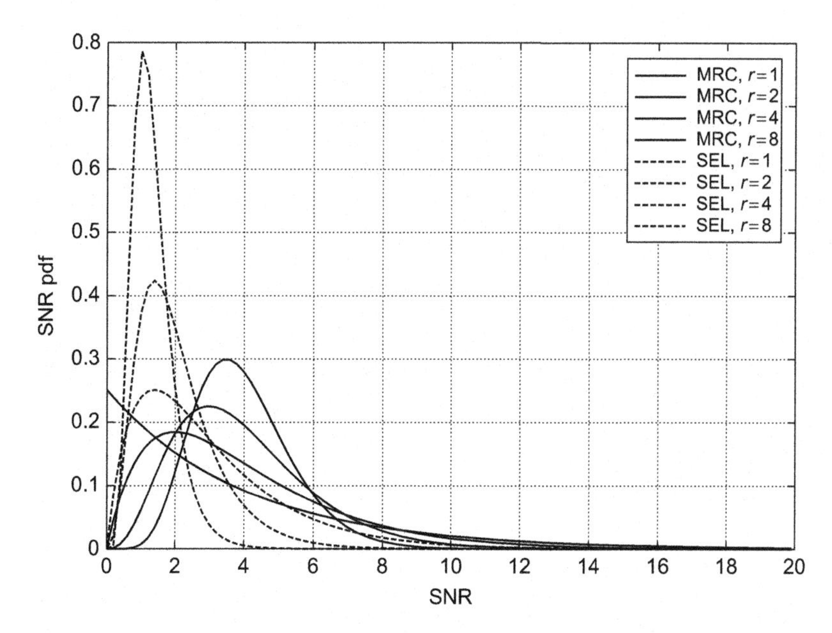

FIGURE 1.7

Output SNR pdf for MRC and selection combining for independent Rayleigh fading.

1.4.2 The MIMO Case

MMSE Formulation

As far as the MMSE SINR is concerned, the output of the MMSE detector $\hat{\mathbf{x}}_{\text{mmse}} = [\hat{x}_{\text{mmse},1},\ldots,\hat{x}_{\text{mmse},n_t}]^T$ is given by:

$$\hat{\mathbf{x}}_{\text{mmse}} = \mathbb{E}\left(\mathbf{x}\mathbf{y}^H\right)\left[\mathbb{E}\left(\mathbf{y}\mathbf{y}^H\right)\right]^{-1}\mathbf{y} \tag{1.11}$$

$$= \sqrt{\frac{\rho}{n_t}}\mathbf{H}^H\left(\frac{\rho}{n_t}\mathbf{H}\mathbf{H}^H + \mathbf{I}_{n_r}\right)^{-1}\mathbf{y} \tag{1.12}$$

$$= \sqrt{\frac{\rho}{n_t}}\mathbf{H}^H(\mathbf{A})^{-1}\mathbf{y} \tag{1.13}$$

with $\mathbf{A} = \frac{\rho}{n_t}\mathbf{H}\mathbf{H}^H + \mathbf{I}_{n_r}$. Each component $\hat{x}_k$ of $\hat{\mathbf{x}}$ is corrupted by the effect of both the thermal noise and the multi-user interference due to the contributions of the other symbols $\{x_\ell\}_{\ell \neq k}$.

Let us now derive the expression of the SINR at one of the n_t outputs of the MMSE detector. Let $\mathbf{h}_k$ be the column of $\mathbf{H}$ associated to element x_k, and $\mathbf{U}$ the $n_r \times (n_t - 1)$ matrix which remains after extracting $\mathbf{h}_k$ from $\mathbf{H}$.

The component $\hat{x}_{\text{mmse},k}$ after MMSE equalization has the following form:

$$\hat{x}_{\text{mmse},k} = \eta_{\mathbf{h}_k} x_k + \sigma_k$$

where

$$\eta_{\mathbf{h}_k} = \frac{\rho}{n_t}\mathbf{h}_k^H(\mathbf{A})^{-1}\mathbf{h}_k \tag{1.14}$$

and

$$\sigma_k = \frac{\rho}{n_t}\mathbf{h}_k^H(\mathbf{A})^{-1}\mathbf{H}[x_1,\ldots,x_{k-1},0,x_{k+1},\ldots,x_{n_t}]^T + \sqrt{\frac{\rho}{n_t}}\mathbf{h}_k^H(\mathbf{A})^{-1}\nu.$$

The variance of σ_k is given by: $V = \mathbb{E}(|\sigma_k|^2|\,\mathbf{H})$. Knowing that $\mathbf{U}\mathbf{U}^H = \mathbf{H}\mathbf{H}^H - \mathbf{h}_k\mathbf{h}_k^H$, we get:

$$\begin{aligned}
V &= \frac{\rho^2}{n_t^2}\mathbf{h}_k^H(\mathbf{A})^{-1}\mathbf{U}\mathbf{U}^H(\mathbf{A})^{-1}\mathbf{h}_k + \frac{\rho}{n_t}\mathbf{h}_k^H(\mathbf{A})^{-1}(\mathbf{A})^{-1}\mathbf{h}_k \\
&= \frac{\rho}{n_t}\left(\mathbf{h}_k^H(\mathbf{A})^{-1}\left[\frac{\rho}{n_t}\mathbf{H}\mathbf{H}^H - \frac{\rho}{n_t}\mathbf{h}_k\mathbf{h}_k^H + \mathbf{I}_{n_r}\right](\mathbf{A})^{-1}\mathbf{h}_k\right) \\
&= \frac{\rho}{n_t}\left(\eta_{\mathbf{h}_k} - \eta_{\mathbf{h}_k}^2\right) \\
&= \frac{\rho}{n_t}\eta_{\mathbf{h}_k}(1 - \eta_{\mathbf{h}_k}).
\end{aligned}$$

The signal to interference plus noise ratio $\mathrm{SINR}^{\mathrm{k}}$ at the output k of the MMSE detector can thus be expressed as:

$$\begin{aligned}\mathrm{SINR}^{\mathrm{k}} &= \frac{\mathbb{E}[\mid \eta_{\mathbf{h}_k} x_k \mid^2 \mid \mathbf{H}]}{\mathbb{E}[\mid \sigma_k \mid^2 \mid \mathbf{H}]} \\ &= \frac{(\eta_{\mathbf{h}_k})^2}{\eta_{\mathbf{h}_k}(1-\eta_{\mathbf{h}_k})} \\ &= \frac{\eta_{\mathbf{h}_k}}{1-\eta_{\mathbf{h}_k}}\end{aligned}$$

Writing $\mathbf{H}\mathbf{H}^H = \mathbf{U}\mathbf{U}^H + \mathbf{h}_k\mathbf{h}_k^H$ and invoking the matrix inversion lemma[13], after some simple algebra we get another useful expression for this SINR (see e.g. [TH99]):

$$\mathrm{SINR}^{\mathrm{k}} = \mathbf{h}_k^H \left(\mathbf{U}\mathbf{U}^H + \frac{n_t}{\rho}\mathbf{I}_{n_r}\right)^{-1} \mathbf{h}_k. \tag{1.15}$$

For a fixed n_r and n_t, it is extremely difficult to get insight on the performance of the MMSE receiver from the expressions (1.15) and (1.14). As a consequence, in order to obtain interpretable expressions, we will focus on an asymptotic analysis of the SINR. Moreover, it has been shown in [PV97] and [ZCT01] that the additive noise σ_k can be considered as Gaussian when n_t and n_r are large enough. In this case, σ_k is an asymptotically zero mean Gaussian noise of variance V and one can easily derive performance measures such as Bit Error Rate (BER) or spectral efficiency with MMSE equalization. More precisely, the following results can be obtained.

Main Results

These results were derived by Tse et al. [TZ00]:

Theorem 3 Assuming the Gaussian i.i.d. channel model, as $n_t \to \infty$ with $n_r = \beta n_t$,

$$\sqrt{n_t}\left(\mathrm{SINR}^{\mathrm{k}} - \mu_{\mathrm{iid}}(\beta,\rho)\right)$$

converges in distribution to a $\mathcal{N}(0, N_{\mathrm{iid}}(\beta,\rho))$ random variable where:

$$\mu_{\mathrm{iid}}(\beta,\rho) = \frac{(\beta-1)}{2}\frac{\rho}{\beta} - \frac{1}{2} + \sqrt{\frac{(1-\beta)^2}{4}\frac{\rho^2}{\beta^2} + \frac{(1+\beta)}{2}\frac{\rho}{\beta} + \frac{1}{4}}$$

and

$$N_{\mathrm{iid}}(\beta,\rho) = \frac{2\mu_{\mathrm{iid}}(1+\mu_{\mathrm{iid}})^2}{1+\frac{\beta(1+\mu_{\mathrm{iid}})^2}{\rho}} - 2\frac{\mu_{\mathrm{iid}}^2}{\beta}$$

[13] The matrix inversion lemma states that for any invertible matrix $\mathbf{F}$ and $\mathbf{E}$: $(\mathbf{D}^{-1} + \mathbf{F}\mathbf{E}^{-1}\mathbf{F}^H)^{-1} = \mathbf{D} - \mathbf{D}\mathbf{F}(\mathbf{E} + \mathbf{F}^H\mathbf{D}\mathbf{F})^{-1}\mathbf{F}^H\mathbf{D}^H$.

This theorem extends previous results from Tse et al. [TH99][14] where it is shown that in a large system limit, the SINR of a user at the output of the MMSE receiver converges to a deterministic limit. The proof uses ideas of Silverstein [Sil90]. In fact, for many linear receivers (MMSE, matched filter, decorrelator), asymptotic normality of the receiver output (in general the output decision statistic) can be proven [GVR02, ZCT01].

Outage BER

Hence, let q denote the outage probability and p_q the corresponding outage probability of error with Quadrature Phase-Shift Keying (QPSK) uncoded constellations. In this case, we have:

$$
\begin{aligned}
1-q &= P\left(Q(\sqrt{\text{SINR}}) \leq p_q\right) \\
&= P\left(\text{SINR} \leq \left(Q^{-1}(p_q)\right)^2\right) \\
&\approx \int_{-\infty}^{(Q^{-1}(p_q))^2} \sqrt{\frac{n_r}{2\pi N_{\text{iid}}}} e^{-\frac{n_r}{2N_{\text{iid}}}(\text{SINR}-\mu_{\text{iid}})^2} d\text{SINR} \\
&\approx 1-Q\left(\frac{\sqrt{n_r}\left(Q^{-1}(p_q)\right)^2-\mu_{\text{iid}})}{\sqrt{N_{\text{iid}}}}\right)
\end{aligned}
$$

Therefore,

$$
p_q \approx Q\left(\sqrt{\frac{\sigma}{\sqrt{n_r}} Q^{-1}(q)+\mu}\right)
$$

In the case of the MMSE receiver, the Gaussian behavior of the SINR is appealing as error-control codes which are optimal for the AWGN channel will also be optimal for the MIMO channel.

Spatial Multiplexing: The V-BLAST Scheme

The idea of *spatial multiplexing* is very simple: n_t independent data streams are sent in parallel, at the same time, from n_t transmit antennas. If the number of receiving antennas n_r is sufficiently large, the data streams can be separated at the receiver and independently detected successfully. Eventually, the n_t-input n_r-output MIMO channel behaves as a set of n_t parallel channels, thus increasing the effective transmission rate by a factor n_t.

The first proposed and best known scheme for spatial multiplexing is the so-called V-BLAST system [GFVW99]. Since the scheme applies to uncoded transmission, we drop the time index once again and let $\mathbf{x} = (x_1, \ldots, x_{n_t})^T$ be the vector of modulation

[14] For this contribution, Tse and Hanly received the IEEE Communications and Information Theory Society Joint Paper Award in 2001.

symbols (taking values into a signal set $\mathcal{A}$) to be transmitted at time t from the n_t transmit antennas and $\mathbf{H} \in \mathbb{C}^{n_r \times n_t}$ be the channel matrix such that $[\mathbf{H}]_{i,j} = h_{i,j}$ is the complex fading coefficient between transmit antenna j and receiving antenna i. Then, the received signal vector is given by:

$$\mathbf{y} = \sqrt{\mathcal{E}/n_t}\mathbf{H}\mathbf{x} + \mathbf{v}. \tag{1.16}$$

Notice that the energy term is divided by n_t since, in order to keep the total transmit energy per symbol interval equal to $\mathcal{E}$, with n_t symbols sent in parallel over the n_t antennas, each symbol must have average energy $\mathcal{E}/n_t$.

The Maximum Likelihood (ML) detection of the symbol vector $\mathbf{x}$ when the receiver has perfect knowledge of $\mathbf{H}$ is immediately given by:

$$\widehat{\mathbf{x}} = \arg\min_{\mathbf{x} \in \mathcal{A}^t} \left| \mathbf{y} - \sqrt{\mathcal{E}/n_t}\mathbf{H}\mathbf{x} \right|^2. \tag{1.17}$$

The above detection rule might be too complex for large n_t and/or large constellation size $|\mathcal{A}|$. In fact, this requires the evaluation of $|\mathcal{A}|^{n_t}$ squared Euclidean distances. By comparing (1.16) with the vectored form of the fading Intersymbol Interference (ISI) channel, we notice that both problems are formally analogous. Therefore, decision-feedback as well as linear equalization can be used as a suboptimal but simpler detection method.

By replicating the derivation of the MMSE-DFE and of the Zero Forcing ZF-DFE we obtain the following receiver schemes, commonly known as *nulling and canceling*, since for each symbol in $\mathbf{x}$ the already detected symbols are subtracted while the undetected interfering symbols are filtered (nulled) out by the receiver.

MMSE-DFE V-BLAST Receiver

Assume that symbol detection is performed in the "bottom-up" order, i.e., $n_t, n_t - 1, \ldots, 1$. Then, if symbols $n_t, n_t - 1, \ldots, i+1$ have been successfully detected, they can be subtracted exactly form the received signal $\mathbf{y}$, so that the signal after subtraction is given by:

$$\mathbf{y} - \sqrt{\mathcal{E}/n_t} \sum_{\ell=i+1}^{n_t} \mathbf{h}_\ell x_\ell = \underbrace{\sqrt{\mathcal{E}/n_t}\mathbf{h}_i x_i}_{\text{useful signal}} + \underbrace{\sqrt{\mathcal{E}/n_t} \sum_{\ell=1}^{i-1} \mathbf{h}_\ell x_\ell}_{\text{interference}} + \underbrace{\mathbf{v}}_{\text{noise}} \tag{1.18}$$

where we let $\mathbf{h}_\ell$ denote the ℓ-th column of $\mathbf{H}$ and where we put in evidence the useful signal for the detection of x_i. The (unbiased) MMSE filter for the detection of x_i with observation given by (1.18) is given by:

$$\mathbf{f}_i = \frac{\sqrt{\mathcal{E}/n_t}}{\mu_i} \boldsymbol{\Sigma}_i^{-1} \mathbf{h}_i \tag{1.19}$$

where

$$\boldsymbol{\Sigma}_i = N_0 \mathbf{I}_{n_r} + \frac{\mathcal{E}}{n_t} \sum_{\ell=1}^{i-1} \mathbf{h}_\ell \mathbf{h}_\ell^H$$

and where $\mu_i = \frac{\mathcal{E}}{n_t}\mathbf{h}_i^H \mathbf{\Sigma}_i^{-1}\mathbf{h}_i$ is the instantaneous SINR conditioned with respect to the channel matrix at the input of the detector for symbol i.

Hence, the V-BLAST MMSE detection scheme works as follows: For $i = n_t, n_t - 1, \ldots, 1$, compute the MMSE estimate of symbol x_i as:

$$z_i = \mathbf{f}_i^H \left(\mathbf{y} - \sqrt{\mathcal{E}/n_t} \sum_{\ell=i+1}^{n_t} \mathbf{h}_\ell \widehat{x}_\ell \right)$$

where $\widehat{x}_\ell$ denote the symbol decisions already available. Then, treat z_i as if it was the output of a virtual Gaussian noise channel with SNR equal to μ_i, and detect symbol x_i as:

$$\widehat{x}_i = \arg\min_{a \in \mathcal{A}} |z_i - a|^2$$

As far as performance analysis is concerned, unfortunately, even for the simple case where $\mathbf{H}$ has i.i.d. elements $\sim \mathcal{N}_{\mathbb{C}}(0,1)$, the statistics of the SINR μ_i is not known analytically, so that it is impossible to compute in closed form the **SEP!** (**SEP!**) $\Pr[\widehat{x}_i \neq x_i]$ (even assuming no error propagation in the decision feedback).

ZF-DFE V-BLAST Receiver

Assume that $n_r \geq n_t$. Then, for any continuous fading distribution the channel matrix $\mathbf{H}$ has rank n_t with probability 1. In this case, we apply the QR decomposition $\mathbf{H} = \mathbf{QR}$ where $\mathbf{Q}$ is $n_r \times n_t$ such that the columns of $\mathbf{Q}$ are orthonormal, and $\mathbf{R}$ is a $n_t \times n_t$ upper triangular matrix.

By multiplying $\mathbf{y}$ by $\mathbf{Q}^H$, we obtain:

$$\mathbf{y}^{(w)} = \sqrt{\mathcal{E}/n_t}\mathbf{R}\mathbf{x} + \mathbf{v} \tag{1.20}$$

where $n = \mathbf{Q}^H \boldsymbol{\mu}$ is still white with entries $\mathcal{N}_{\mathbb{C}}(0, N_0)$. In fact,

$$E[\mathbf{v}\mathbf{v}^H] = E[\mathbf{Q}^H \boldsymbol{\nu}\boldsymbol{\nu}^H \mathbf{Q}] = N_0 \mathbf{Q}^H \mathbf{Q} = N_0 \mathbf{I}_{n_r}$$

It is possible to show that $\mathbf{y}^{(w)}$ is the analogous of the output of the Whitened Matched Filter (WMF) for ISI channels, generalized to the case where the channel matrix $\mathbf{H}$ is not necessarily Toeplitz.

It is also possible to see easily that the matrix $\mathbf{Q}$ implements the zero-forcing forward filter for the ZF-DFE. Indeed, the i-th component of $\mathbf{y}^{(w)}$ contains only the contribution of symbols $i, \ldots, n_t$. We have:

$$y_i^{(w)} = \sqrt{\mathcal{E}/n_t}\, \mathbf{R}_{i,i}\, x_i + \sqrt{\mathcal{E}/n_t} \sum_{\ell=i+1}^{n_t} \mathbf{R}_{i,\ell}\, x_\ell + n_i.$$

By detecting the symbols in the order $n_t, n_t - 1, \ldots, 1$, we can subtract from $y_i^{(w)}$ the decisions about the interfering symbols. The V-BLAST ZF detection works as follows: For $i = n_t, n_t - 1, \ldots, 1$, compute the estimate of x_i as:

$$z_i = y_i^{(w)} - \sqrt{\mathcal{E}/n_t} \sum_{\ell=i+1}^{n_t} r_{i,\ell}\, \widehat{x}_\ell.$$

Then, treat z_i as if it were the output of a virtual Gaussian noise channel, and detect symbol x_i as:

$$\widehat{x}_i = \arg\min_{a \in \mathcal{A}} \left| z_i - \sqrt{\mathcal{E}/n_t}\, r_{i,i}\, a \right|^2.$$

As far as analysis in independent Rayleigh fading is concerned, we have the following result:

Lemma. Let $\mathbf{h}_1, \ldots, \mathbf{h}_{i-1}, \mathbf{h}_i$ be n_r-dimensional vectors with i.i.d. entries $\sim \mathcal{N}_{\mathbb{C}}(0,1)$ (we assume $n_r \geq i$). Let $\mathbf{P}$ be the orthogonal projector onto the orthogonal complement of the subspace generated by $\mathbf{h}_1, \ldots, \mathbf{h}_{i-1}$. Then, the random variable $|\mathbf{P}\mathbf{h}_i|^2$ is central Chi-squared with $2(n_r - i + 1)$ degrees of freedom, i.e., its pdf is given by:

$$p_{\mu_i}(z) = \frac{z^{n_r - i} e^{-z}}{(n_r - i)!}, \quad z \geq 0. \tag{1.21}$$

Proof. Notice that $\mathbf{P}$ is independent of $\mathbf{h}_i$. Then, by definition of orthogonal projector, $\mathbf{P}$ can be written as $\mathbf{U}\mathbf{U}^H$, where the columns of $\mathbf{U}$ are an orthonormal basis for the orthogonal complement of the subspace generated by $\mathbf{h}_1, \ldots, \mathbf{h}_{i-1}$. With probability 1 this subspace has dimension $n_r - i + 1$, therefore, $\mathbf{g}_i = \mathbf{U}^H\, \mathbf{h}_i$ is a $n_r - i + 1$ dimensional Gaussian vector with i.i.d. entries $\sim \mathcal{N}_{\mathbb{C}}(0,1)$. Finally, it is well known that $|\mathbf{g}_i|^2$ is Chi-squared distributed with $2(n_r - i + 1)$ degrees of freedom.

Then, notice that the i-th detector input z_i, when there are no errors in the decision feedback, is conditionally Gaussian with instantaneous SNR given by:

$$\mathrm{SNR}_i = \frac{\mathcal{E}}{n_t N_0} |r_{i,i}|^2$$

where $r_{i,i} = \mathbf{P}\mathbf{h}_i$; therefore, SNR_i is Chi-squared distributed with $2(n_r - i + 1)$ degrees of freedom. The pairwise error probability assuming no decision feedback errors can be bounded from above by:

$$P(x_i \to x_i') \leq \frac{1}{2}\left(1 + \frac{\rho}{n_t}\frac{d_i^2}{4}\right)^{-(n_r - i + 1)} \tag{1.22}$$

where, as usual, we let $\mathcal{E}/N_0 = \rho$, and $d^2 = |x_i - x_i'|^2$. We see that the i-th symbol enjoys a diversity order of $n_r - i + 1$. The overall error probability is then dominated by the first detection, which has diversity order $n_r - n_t + 1$.

1.5 WIENER VS. SHANNON: AN EVER CLOSER UNION

The MMSE receiver plays a central role in telecommunications. Recently, it was shown that the MMSE estimator, which is rooted in signal processing, is also fundamental in information theory. Nice discussions on the Shannon [Sha48, Sha49] versus Wiener [Wie48, Wie49] legacy are given by Forney [GDF03] and Guo [GSV04].

It may seem strange that it took more than fifty years to discover quite fundamental relationships between the input-output mutual information and the minimum mean square error of an estimate. Astonishingly, it is shown in [GSV04] that the derivative of the mutual information (nats) in the SISO case with respect to the SNR is equal to half the MMSE. In the MIMO case, similar results can be proven [SJL05].

Indeed, assuming that **x** and **n** are uncorrelated with one another, we have:

$$\mathbf{R}_y = \mathbb{E}(\mathbf{y}\mathbf{y}^H) = \frac{\rho}{n_t}\mathbf{H}\mathbf{Q}\mathbf{H}^H + \mathbf{I}_{\mathbf{n_r}} \tag{1.23}$$

$$\mathbf{R}_{xy} = \mathbb{E}(\mathbf{x}\mathbf{y}^H) = \sqrt{\frac{\rho}{n_t}}\mathbf{Q}\mathbf{H}^H. \tag{1.24}$$

From the previous paragraph, the MMSE estimate of **x** and its covariance matrix are given by:

$$\hat{\mathbf{x}} = \mathbf{R}_{xy}\mathbf{R}_y{}^{-1}\mathbf{y} \tag{1.25}$$

and the covariance matrix of the MMSE receiver is:

$$\mathbf{R}_{\text{MMSE}} = \mathbb{E}\left[(\mathbf{x}-\hat{\mathbf{x}})(\mathbf{x}-\hat{\mathbf{x}})^H\right] \tag{1.26}$$

$$= \mathbf{R}_{xy}\mathbf{R}_y{}^{-1}\mathbf{R}_{xy}^H - 2\mathbf{R}_{xy}\mathbf{R}_y^{-1}\mathbf{R}_{xy}^H + \mathbf{Q} \tag{1.27}$$

$$= \mathbf{Q} - \mathbf{R}_{xy}\mathbf{R}_y{}^{-1}\mathbf{R}_{xy}^H \tag{1.28}$$

It follows from the inversion lemma that:

$$\mathbf{R}_{\text{MMSE}}^{-1} = \mathbf{Q}^{-1} + \mathbf{Q}^{-1}\mathbf{R}_{xy}\left(\mathbf{R}_y - \mathbf{R}_{xy}^H\mathbf{Q}^{-1}\mathbf{R}_{xy}\right)^{-1}\mathbf{R}_{xy}^H\mathbf{Q}^{-1} \tag{1.29}$$

$$= \mathbf{Q}^{-1} + \frac{\rho}{n_t}\mathbf{H}^H\left(\mathbf{R}_y - \mathbf{H}\mathbf{Q}\mathbf{H}^H\right)^{-1}\mathbf{H} \tag{1.30}$$

$$= \mathbf{Q}^{-1} + \mathbf{H}^H\mathbf{H} \tag{1.31}$$

Finally, the capacity is given by:

$$C = \log_2\det\left(\mathbf{I}_{\mathbf{n_r}} + \frac{\rho}{n_t}\mathbf{H}\mathbf{Q}\mathbf{H}^H\right) \tag{1.32}$$

$$= \log_2\det\left(\mathbf{I}_{\mathbf{n_r}} + \frac{\rho}{n_t}\mathbf{H}^H\mathbf{H}\mathbf{Q}\right) \tag{1.33}$$

$$= \log_2\det(\mathbf{Q}) + \log_2\det\left(\mathbf{Q}^{-1} + \mathbf{H}^H\mathbf{H}\right) \tag{1.34}$$

Hence, the channel capacity can be rewritten:

$$C = \log_2 \frac{\det(\mathbf{Q})}{\det(\mathbf{R}_{\text{MMSE}})} \tag{1.35}$$

The expression relates, in a simple manner, the channel capacity to the covariance matrix of the MMSE estimate of $\mathbf{x}$. In this form, the channel capacity formula has an intuitive appeal. In fact, the MMSE estimate $\hat{\mathbf{x}}$ lies (with high probability) in a "small cell" centered around the codeword $\mathbf{x}$. The volume of the cell is proportional to $\det(\mathbf{R}_{\text{MMSE}})$. The volume of the codebook space (in which $\mathbf{x}$ lies with high probability) is proportional to $\det(\mathbf{R}_{\mathbf{x}})$. The ratio $\delta = \frac{\det(\mathbf{R}_{\mathbf{x}})}{\det(\mathbf{R}_{\text{MMSE}})}$ gives the number of cells that can be packed into the codebook space without significant overlapping. The "center" of each such cell, the codeword, can be reliably detected using $\hat{\mathbf{x}}$. As a consequence, one can communicate reliably using a codebook of size δ, which contains $\log_2(\delta)$ information bits. This provides an intuitive motivation to the capacity formula, in the same vein as [S49].

CHAPTER

MIMO Propagation and Channel Modeling

2

Ernst Bonek

2.1 INTRODUCTION

The channel is that part of a communication system that cannot be engineered. *Propagation* is at the heart of any wireless system; it sets the ultimate limits for other fields of communications engineering. However, we engineers should not see the propagation channel only as a limitation. We can devise methods to exploit the opportunities that this channel offers. In particular, we can put to work the *spatial domain* for our goal to improve wireless information transfer. If "Smart Antenna" was the catchword of the past in the area where signal processing meets electromagnetics, "MIMO" has taken its place today. Or is it just hype? While it is true that MIMO systems can, in theory, provide the seemingly paradoxical arbitrary multiplication of Shannon's capacity, and while nobody doubts that MIMO will be *the* enabling technology for high-speed wireless data, many questions remain.

The flood of MIMO papers in the past ten years, presumably intended to shed light on the topic, has led to quite a considerable confusion about what can actually be achieved. This paper will try to clarify the role of the *propagation* channel.

MIMO may offer three different benefits, namely beamforming gain, spatial diversity, and spatial multiplexing (Figure 2.1).

By *beamforming*, the transmit and receive antenna patterns can be focused into a specific angular direction by the appropriate choice of complex baseband antenna weights. The more *correlated* the *antenna signals*, the better for beamforming. Under Line of Sight (LOS) channel conditions, the Receiver Rx and Transmitter Tx gains may add up, leading to an upper limit of $m \cdot n$ for the beamforming gain of a MIMO system (n and m being the number of antenna elements of Rx and of Tx, respectively).

Multiple replicas of the radio signal from different directions in space give rise to spatial *diversity*, which can be used to increases the transmission reliability of the fading radio link. For a spatially white MIMO channel, that is, completely *uncorrelated antenna signals*, the diversity order is limited to $m \cdot n$. Spatial correlation will reduce the diversity order and is therefore an important channel characteristic.

MIMO. DOI: 10.1016/B978-0-12-382194-2.00002-2

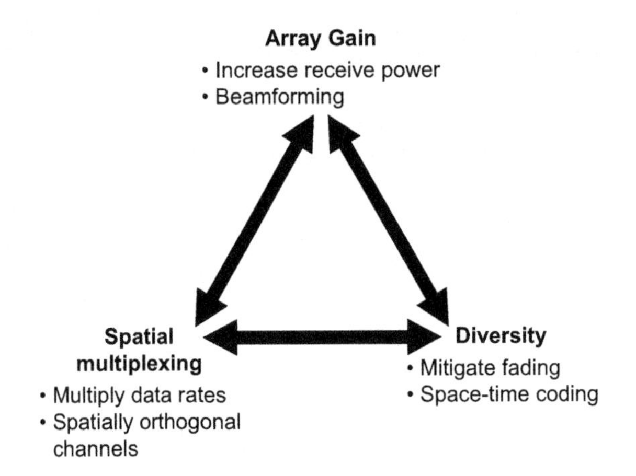

FIGURE 2.1

Benefits that MIMO offers in principle.

Most MIMO systems are designed for moving terminals, so we will have to discuss the topic "correlation" both in space and in time. MIMO channels can support parallel data streams by transmitting and receiving on orthogonal spatial channels (*spatial multiplexing*). The number of usefully multiplexed streams depends on the rank of the instantaneous channel matrix $\mathbf{H}$, which, in turn, depends on the spatial properties of the radio environment. The spatial multiplexing gain may reach $\min(m,n)$ in a sufficiently rich scattering environment. So we have to consider what sufficiently rich scattering means and what its relative importance compared to SNR is.

Knowledge about the state of the instantaneous MIMO channel is essential for the system engineers. Whether this Channel State Information (CSI) is available at the receiver, the transmitter or both, further whether this information relates to the actual, instantaneous channel or is only information about a statistical average of the channel, will result in entirely different receive and transmit strategies.

Beamforming, diversity, and multiplexing are rivaling techniques. To highlight the role of the propagation channel, the threefold trade-off between beamforming, diversity, and multiplexing can be broken down into several dichotomous trade-offs [Wei03], see Figure 2.2.

First, the optimal trade-off between beamforming on one hand and diversity/-multiplexing on the other hand is mainly dictated by the channel properties, similar to the trade-off between beamforming and diversity in the MISO case. A directive channel favors beamforming, a *diverse* (nondirective) channel allows for diversity and/or multiplexing. Second, there is the trade-off between diversity and multiplexing. Partly, this trade-off is also dictated by the channel. If one side of the MIMO link is purely directive no multiplexing is possible, but diversity can be exploited at the uncorrelated link end. However, if both link-ends of the MIMO channel are (partly)

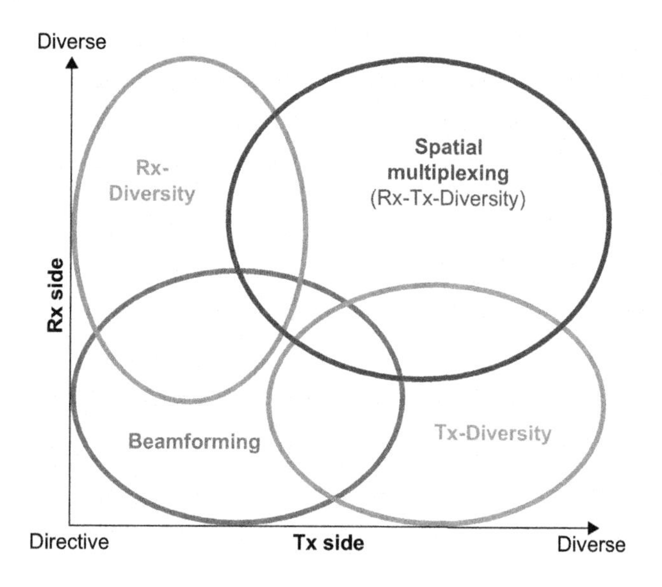

FIGURE 2.2

How directivity or diversity at the receive and transmit ends of a MIMO link determine to which extent the channel supports beamforming, diversity, or multiplexing.

decorrelated both features, diversity and multiplexing, are possible. In this case the optimal trade-off between diversity and multiplexing [ZT03] is purely determined by system requirements, i.e., desired data rate and reliability of transmission. High data rates can be achieved by employing multiplexing to full extent, high reliability is attained by diversity.

The partial overlap of the ellipses in Figure 2.2 indicates that there is a gradual transition between the pure realizations of a certain MIMO benefit. In summary, it is the *propagation environment* that determines what can be gained by MIMO techniques; however, this does not mean that it will be gained in actual MIMO operation. The idea that "proper" signal processing can convert any channel into $\min(m,n)$ parallel channels, still voiced by parts of the MIMO community, is far too optimistic and must be dismissed.

2.2 MODEL CLASSIFICATION

How shall we describe or model the MIMO channel? First, this depends what we want to model for. For MIMO system deployment and network planning, we need *site-specific* models, for MIMO algorithm development, and system design and testing we need *site-independent* models. Then, what kind of model serves our purpose best? Given that the number of MIMO models is huge, and still increasing, this is a difficult question that is not always decided rationally. System engineers pressure for models that are "simple," sometimes ridiculously so[1], while radio engineers strive to capture the details of the radio channel. Before adopting a MIMO model, one should have answers ready on these questions:

- At which level should the model function? Propagation, channel, link, system?
- Which aspect of MIMO shall be modeled? Multiplexing, beamforming, or diversity gain?
- Has the model been validated, and if yes, how?

Figure 2.3 tries to provide a compass to navigate through the jungle of MIMO models.

Electromagnetic wave propagation provides the basis for *propagation or physical models*. The final result of physical modelling is the *characterization of the environment* on the basis of propagation. Reference scenarios are agreed-on environments that make comparison of models and their performance much easier. Specifying a system *bandwidth* and *antenna arrays* at both link ends by setting the number of antenna elements, their geometrical configuration, and their polarizations turns the

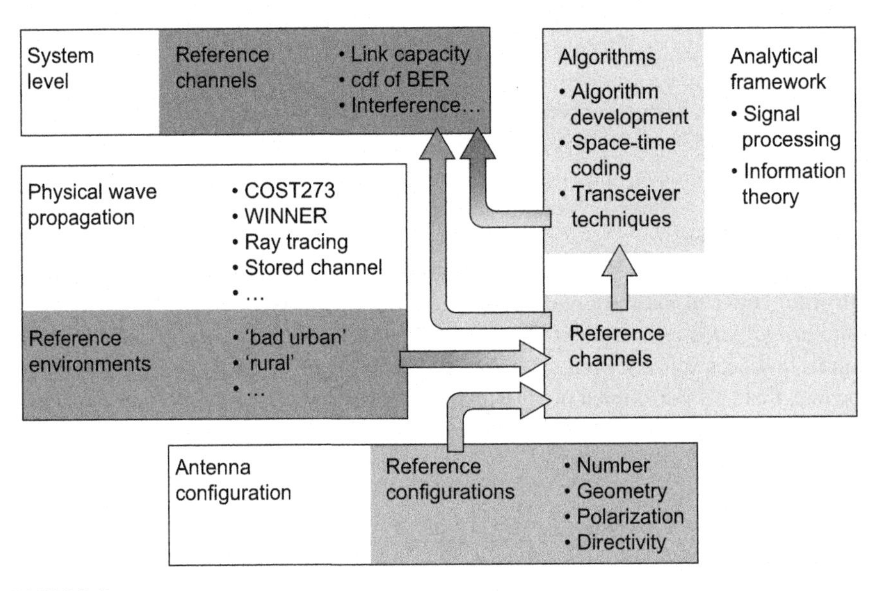

FIGURE 2.3

MIMO models—an overview.

[1]There have been attempts to persuade standardization committees to adopt MIMO channel models without a spatial component.

propagation model into a MIMO *channel model*. Accounting for mobility of a terminal complements a channel model on link level. Such a model then provides MIMO *channel matrices* as an *analytical framework* for designing transmit and receive techniques for a MIMO link, e.g., space-time codes. How MIMO channel models on the link level may be combined to model a MIMO implementation on network level is an important question, but will not be discussed here.

MIMO models appear in a confusing variety of names. Here, I adopt a nomenclature that has evolved in the European Cooperation in Science and Technology (COST) Actions 231, 259, and 273.

Among the *physical* models, we found it convenient to distinguish between:

- deterministic models (for example, ray tracing, stored measurements);
- geometry-based stochastic models (for example, the GSCM of COST 259/273 [Cor01], [Cor06]);
- stochastic models (for example, without geometrical input, Hao Xu, et al. [Xea04], Zwick [TCW02]).

Deterministic models boast good agreement with physically existing results (site-specific!) and are easily reproducible. However, they may not be really representative of the environment modeled, require large data bases characterizing the propagation environment, which are expensive to produce, and their parameters cannot be changed easily. By the way, the computational complexity of ray tracing is not the prohibitive argument against their use—it is the environmental data bases they require.

Stochastic models describe the radio channel by (multidimensional) pdfs of the desired parameters. Good models of this type derive their pdfs from extensive measurements in clearly distinct environments and distinguish these environments. Throwing together channel measurements from different environments into one ensemble will inevitably result in a Gaussian model (central-limit theorem!) of a virtual environment that is never encountered in practice. Simulation runs with them are fast because of legacy tapped-delay-line realizations of SISO models, a fact that makes such stochastic models popular, but they are difficult to parameterize over large areas. Parameterization can be sped up by feeding in a single site-specific MIMO measurement and randomizing the parameters of the multipath components [AMM+02].

Geometry-based stochastic channel model (*GSCM*) try to combine the best of the two worlds by starting with geometrical input about the environment and then superimposing statistical information. They excel in simulating both the spatial and the temporal evolution of the channel when terminals or scatterers move.

Analytical models specify MIMO *channel matrices* directly. I see two kinds. One starts from the correlation properties of the channel right away, extracted from MIMO measurement campaigns (Information Society Technologies (IST)-METRA or "Kronecker" model [JLK+02] [DGMJ00], Weichselberger [Wea06] [Wei03]) or theoretical considerations alone, like the "i.i.d. model" with random Gaussian

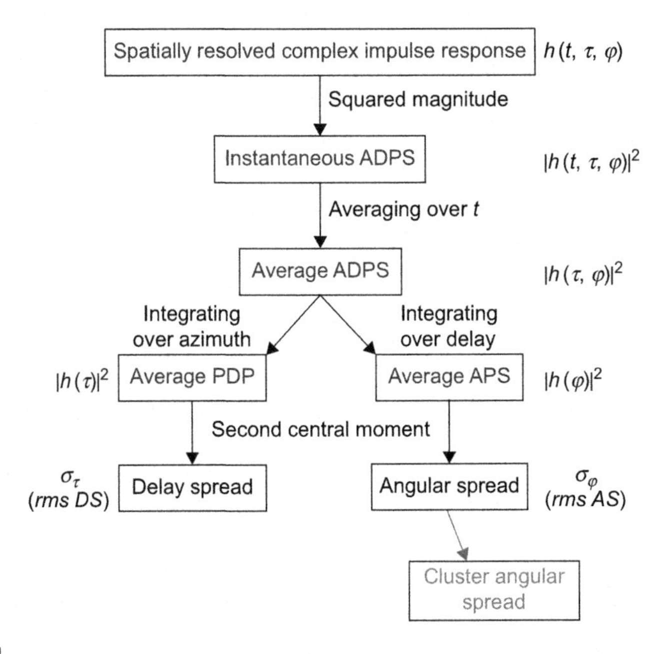

FIGURE 2.4

Aggregate parameters of MIMO radio channel models derived from the spatially resolved impulse response. Angular Delay Power Spectrum (ADPS), Angular Power Spectrum (APS), and Cluster Angular Spread (CAS).

entries in the **H** matrix, very popular with theoreticians. Other analytical models are motivated by propagation principles, for example, the "finite scatterers model" [Bur03], or the "virtual channel representation" [A.M02].

Models adopted by standardization bodies such as 3GPP [3gp03] and IEEE 802.11n [Eea04] combine elements of both physical and analytical modeling (not necessarily in an optimal way), and prescribe temporal and spatial characteristic of the MIMO radio channel rigorously by *aggregate* parameters (see Figure 2.4).

2.3 PARAMETERS OF THE MIMO RADIO CHANNEL

The channel is characterized by a time- and delay-dependent MIMO channel matrix, $\mathbf{H}(t, \tau)$ that contains the impulse responses from each transmit to each receive antenna (for both polarizations), calculated from dual-polarized double-directional wave propagation. Alternatively, the equivalent time- and frequency-dependent MIMO transfer function matrix can be used instead of the channel impulse response matrix.

If, for simplicity sake, we assume a narrowband system and adopt complex baseband representation, the matrix **H** is composed of the complex-valued entries $h_{i,j}$, which characterize the impulse response from the jth transmit to the ith receive antenna. Of particular interest is the Wishart form, $\mathbf{HH}^H$. It determines the *mutual* *information*, I, [BT06]:

$$I = \log_2 \det \left(\mathbf{I}_n + \frac{\text{SNR}}{m} \mathbf{HQH}^H \right) \quad (2.1)$$

of that particular realization of the MIMO channel. Here, $\mathbf{I}_n$ is the $n \times n$ identity matrix, $\mathbf{Q}$ the signal covariance matrix and SNR the prevalent signal-to-noise ratio. Applying the expectation operator over an ensemble of realizations and optimizing $\mathbf{Q}$ according to the channel statistics ($\mathbf{Q}$ is usually taken to be the identity matrix) yields $\mathbf{HH}^H$ in (2.1) as the very popular ergodic capacity formula of the MIMO channel. I hasten to say that capacity is neither the only interesting quantity nor a particularly meaningful one to assess a model's goodness. If a model does not render "capacity" correctly, the model is poor and should be discarded. However, if capacity is modeled to within $\pm 10\%$ of the true value, this does not say very much. Practically all models that are published achieve this figure easily.

Which parameters do we have to consider when we want to model a MIMO radio channel in detail for system design? In *delay* domain, we are interested in the *complex-valued impulse* response and derived parameters. Its length and temporal clusters will guide our choice of the system's symbol length. As the secret power of MIMO is based upon spatial samples of the spatially varying electromagnetic field, taken or imposed by the array antennas, the *angularly resolved impulse response* is what distinguishes MIMO from conventional channel modelling. Direction of Arrival (DoA) and Direction of Departure (DoD) of Multipath Components (MPC) and their clustering in direction are paramount. From this information we can derive number and configuration of Tx and Rx antenna arrays to exploit best what the channel offers. For time-varying channels, we additionally have to model the average duration of fades and a Doppler profile to choose appropriate frame and block lengths and the interleaving depth. Common aggregate Power Delay Profile (PDP) parameters are then usually derived in the way of Figure 2.4.

The parameters of the time domain and the delay domain are well known from the traditional SISO radio channel. We will focus here on the spatial domain and note that polarization offers advantages particular to MIMO. Further I want to highlight the *cluster angular spread*[2] [AME02], because channel situations that will produce widely different MIMO responses may have the same (global) angular spread, derived from the entire environment. Figure 2.5 illustrates how the very same global rms angular spread of 20° results from two totally different propagation situations, characterized by their APS, which can make a world of difference in a MIMO system exploiting spatial multiplexing.

Also, a note on the double-directional viewpoint of the MIMO propagation channel is in order here. Intuition and numerous papers on high-resolution algorithms

[2]International Telecommunications Union–Radio (ITU-R) and 3rd Generation Partnership Project (3GPP) use the terms intra-cluster Angular Spread (AS) and *component* AS for this quantity. The global root mean square (rms) angular spread is termed total AS or *composite* AS.

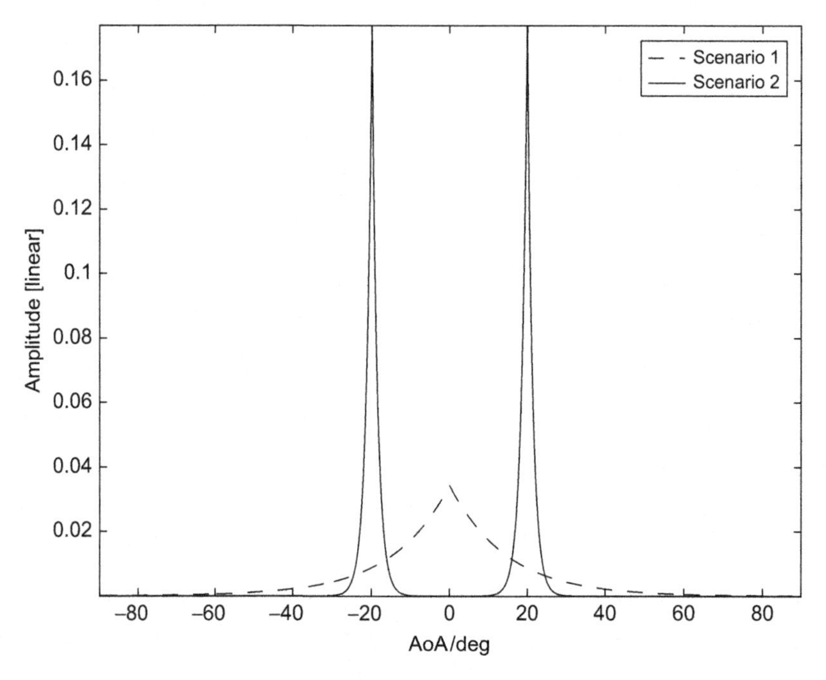

FIGURE 2.5

The two different APS shown produce the same global angular spread of 20°. Reprinted with permission of IEEE, [CYz+07].

for DoA determination have made clear the concept of electromagnetic waves arriving from specific directions on a receiving antenna. What, however, is the direction of *departure* of an omnidirectional antenna? The original "double-directional" paper [MAE01] clarifies the term. The propagation paths between transmitter and receiver can be traced, experimentally proven in [MDG+00], and a DoD is a direction that can be followed until it terminates in a valid DoA at the receiver.

2.4 CSI AND CHANNEL RANDOMNESS

The type and amount of knowledge about the channel state determine the MIMO strategy that will be implemented. When instantaneous channel knowledge is available, the eigenvalues of the **H** matrix matter. When full CSIT is not available, second order statistics are used instead, and the long-term properties of the channel become important. Transmit precoding then matches the signals to the eigenmodes of the spatial Tx correlation matrix. Other chapters in this book elaborate on the subject in detail. Here, we restrict the discussion on the question of whether CSI is known at the transmitter *instantaneously* or only on average, that is, by its statistics, in favorable cases expressed by a MIMO correlation matrix, $\mathbf{R_H}$ (see next section). For the modeling task this means that we have to represent or measure $\mathbf{R_H}$.

For clarification of terms:

- If we take a snapshot (a measurement) of the MIMO channel, this gives us the *instantaneous* MIMO channel.
- If we determine one snapshot of the MIMO channel, this snapshot is *deterministic*.
- If this MIMO channel does not change and stays the same, the MIMO channel is *static*. (However, this is rarely the case or, at best, an approximation.)
- In general, the MIMO channel is *random*. Some people call it the *fading* MIMO channel.
- One snapshot of the MIMO channel is a single realization of the random MIMO channel.

What makes the MIMO channel random? We observe six different sources of MIMO randomness (R. Bultitude in [Cor06]):

1. Type I: random changes that result when any of the antennas used in a MIMO link moves *continuously* in an otherwise physically static environment.
2. Type II: random changes that result when Interacting Objects (IOs) in the critical region[3] of MIMO link move, but all antennas remain static.
3. Type III: random changes that result when both Type I and Type II activity occurs.
4. Type IV: random differences among instantaneous physical link transfer functions between physically static antenna elements (this is spatial variation, not fading!).
5. Type V: random changes that occur when either the Tx array or the Rx array or both are moved *in steps* to different locations within a local area throughout which shadowing by obstructions remains constant.
6. Type VI: random changes that occur when either the Tx array or Rx array or both are moved in steps *beyond* the boundaries of a (local) area with stationary characteristics.

When taking MIMO measurements, it is important to keep these differences in mind. Otherwise, the results of the measurements might be interpreted erroneously. A common mistake is to throw results, gathered in Type I and Type VI random environments, together and treat them as one statistical ensemble.

As an example, Figure 2.6 illustrates Type VI randomness.

When the race car moves out of the stationary environment defined by the two large scatterers "house" and "forklift" (MPCs as full lines) in another local area, in which only the house is a relevant scatterer, it enters a different stationary environment (MPCs as dotted lines). The wrench is considered to be never in the critical region, from which MPC can be received by the car antennas.

[3] The critical region is loosely defined as that part of the environment from which significant multipath reaches the Rx.

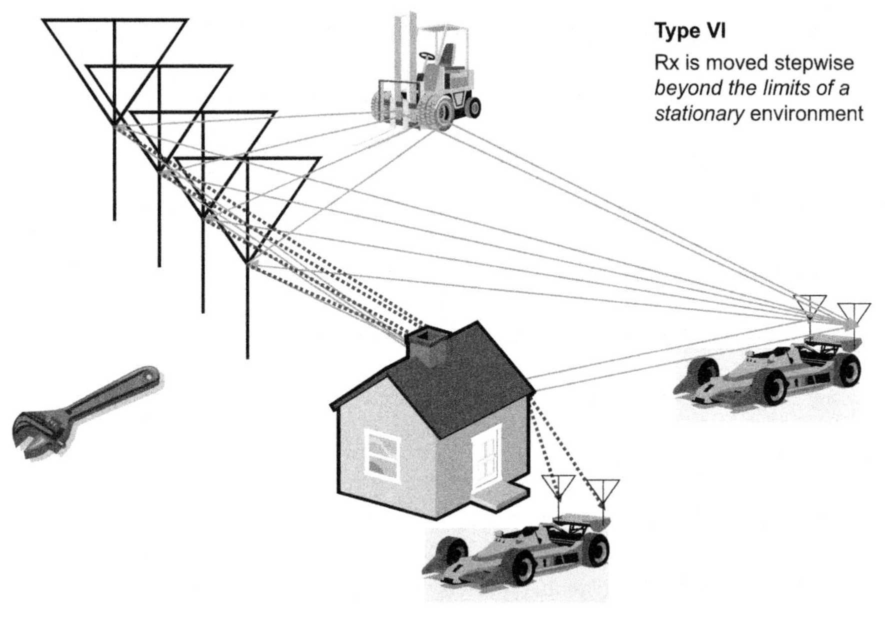

FIGURE 2.6

Type VI MIMO channel randomness. When the race car moves behind the house, the stationary environment characterized by the MPCs shown as full lines is left for a new environment (MPCs shown as dotted lines).

2.5 WHAT KIND OF CORRELATION IN MIMO?

Temporal correlation is important for modeling the channel for terminals in motion, and this subject is well known from SISO channels. *Spatial* correlation, on the other hand, is a feature that entered the scene by the application of array antennas in MISO or SIMO situations. There, the correlation of the antenna signals matters, but of course only at one end of the radio link.

In MISO and SIMO, only antenna signals are available that can be compared and only in one location. For SIMO, the correlation of the receive antenna signals, available at the Rx only, is the receive correlation.[4] For MISO, the correlation between the antenna signals is the transmit correlation (at the Tx).

In contrast, with MIMO it is always a *correlation between channels*, which are established between each individual antenna element at the Tx and each element at the Rx. When one measures the signals at the Rx, the phase relation of the Tx signals has to be known. This is one of the reasons why MIMO measurements are so tricky—Rx and Tx must be synchronized, either by cables (cumbersome!), by expensive rubidium clocks, or satellite signals.

[4]Oestges & Clerckx [OC07, p. 45], remind correctly of the fact that this is actually also a channel correlation, namely between the two Tx channels that share the same receive antenna.

If the channel can be characterized completely by second-order statistics, and only then, the complete way to describe the spatial MIMO correlation is by the full correlation matrix, $\mathbf{R_H}$:

$$\mathbf{R_H} = \mathbb{E}\left\{\text{vec}(\mathbf{H})\,\text{vec}(\mathbf{H})^H\right\} \tag{2.2}$$

where the vec() operation stacks all columns of $h_{i,j}$ into a single column vector. Elements of $\mathbf{R_H}$ describe correlation between any pair of $\mathbf{H}$ elements. However, the elements of $\mathbf{R_H}$ are difficult to interpret physically, except diagonal elements. Above all, the full correlation matrix is very large. We will come back to meaningful approximations of this $(n \cdot m) \times (n \cdot m)$ matrix shortly.

By the way, correlation has been identified as the show-stopper of MIMO very early [DGMJ00]. Any model that disregards spatial correlation, as for instance the "i.i.d. model," very popular among theoreticians, will give too optimistic capacity values. Any of its use should be properly justified.

To determine the spatial correlation in a specific environment, we have to measure the full MIMO system with specific Tx and Rx arrays in place. Analytical models start from the full channel correlation matrix, $\mathbf{R_H}$ (2.2) and become simpler, when making assumptions about propagation[5].

A popular approach, the so-called Kronecker model, assumes—though only implicitly—complete independence of the Tx and Rx propagation environments [DGMJ00], [JLK+02]. The MPC arriving at the receiver have "forgotten" how they have been sent off from the transmitter. Then, the MIMO correlation properties are characterized by separate correlation matrices at the receiver, $\mathbf{R}_{\text{Rx}}$, and the transmitter, $\mathbf{R}_{\text{Tx}}$:

$$\mathbf{R}_{\text{Rx}} = \mathbb{E}\left\{\mathbf{H}\mathbf{H}^H\right\} \tag{2.3}$$

$$\mathbf{R}_{\text{Tx}} = \mathbb{E}\left\{\mathbf{H}^T\mathbf{H}^*\right\} \tag{2.4}$$

giving the full correlation matrix $\mathbf{R_H}$ as their Kronecker product; hence, the name of this model. However, I want to stress that this approach neglects the correlation terms across the link ("cross correlation," "joint correlation"), which do matter, at least in some indoor scenarios. This joint correlation makes MIMO to more than just the sum of SIMO and MISO. Another surprising result, emphasizing that separate Rx and Tx correlation matrices are not able to completely describe MIMO channels, was found in [CA04]: so called "diagonal correlations" may boost the ergodic capacity beyond the previously accepted upper limit of—totally uncorrelated—i.i.d. random entries of $\mathbf{H}$.

The Weichselberger model, taken as an example for a particularly well validated analytical model (see Subsection 2.8.2), makes use of transmit and the receive correlations, but of the joint correlation as well.

[5] A hidden assumption of correlation-based models is that they are strictly valid only for Gaussian channels (i.e., that have Rayleigh fading statistics of the $h_{i,j}$).

2.6 MIMO MEASUREMENTS

2.6.1 Characterization of Multipath Components

The double-directional characterization of MPC in the radio channel [MAE01], supported by double-directional channel sounding with multielement near-omni-directional arrays (Figure 2.7), is closely related to MIMO radio channel modeling.

A review paper [Tea05] recently summarized the state-of-the-art of double-directional channel sounding. Besides DoDs, DoAs, and delay, even polarization and Doppler of MPC has been measured. To increase the accuracy of the channel transfer matrix's spatial and temporal structure, high-resolution parameter estimation techniques (Multiple Signal Classification (MUSIC), Estimation of Signal Parameters via Rotational Invariance Techniques (ESPRIT), (Space Alternating Generalized Expectation maximization (SAGE)) have been pushed to their limits [FJS02]. Figure 2.8 gives a nice example of what can be achieved with respect to angular, delay, and polarization resolution.

2.6.2 Diffuse Multipath

I hasten to say that, while high-resolution algorithms have shown amazing ability to detect individual MPC, not the total power is captured. In addition to discrete, identifiable MPC, measurements have shown that there exist typically a large number of MPC that cannot be modeled discretely. Andreas Richter, who investigated this phenomenon [Ric05] in detail, named the residual power "diffuse multipath component,"

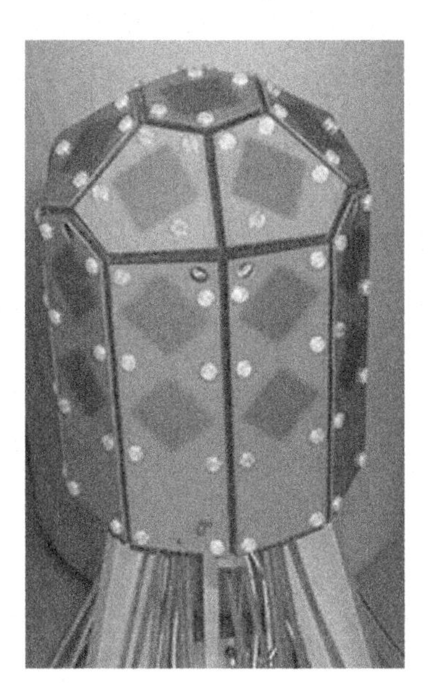

FIGURE 2.7

Multielement 5.3 GHz antenna with nearly omni-directional antenna pattern comprising dual-polarized patch elements. Courtesy Elektrobit Oy.

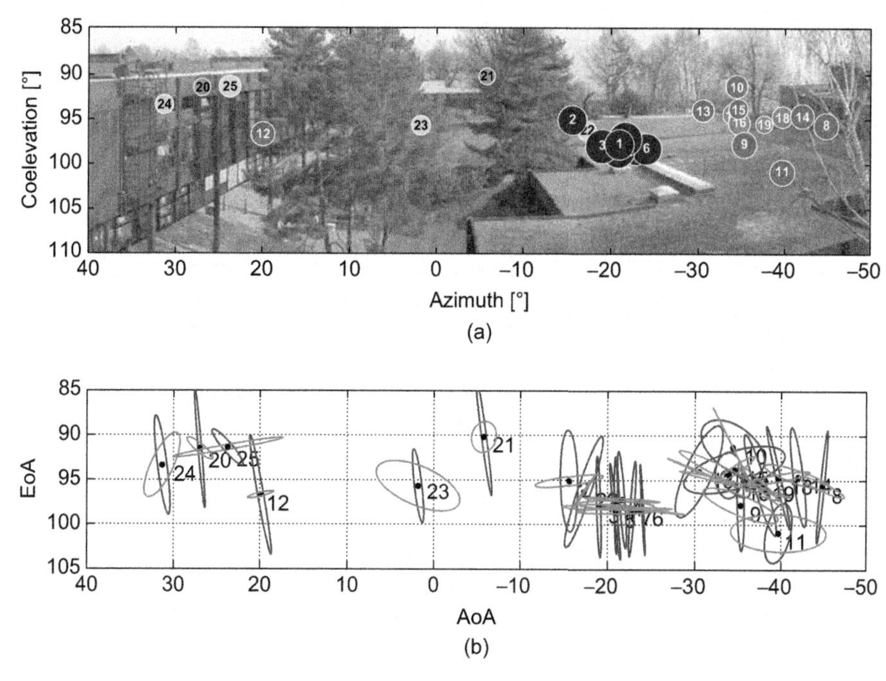

FIGURE 2.8

(a) Multipath as seen from the receiver, numbered by delay. Lighter dots also indicate longer delay. Signal strength is indicated by dot size (large: strong MPC). (b) Polarization of received MPCs. Full lines: originally vertically transmitted, shaded lines: originally horizontally transmitted. Courtesy B. Fleury, Aalborg.

which represent the contributions of diffuse scattering, of higher-order reflections, and of diffraction[6]. We model this experimental observation by superimposing an exponentially decaying, random spatially white i.i.d. channel matrix to the channel matrix resulting from clusters and discrete components. The relative power of this i.i.d. component has to be specified beforehand (between 10% in LOS and 70% in Non-Line-Of-Sight (NLOS)[Ric05]).

2.6.3 Clusters

Clustering of MPC helps reducing the number of MIMO model parameters, but has been frowned upon as long as clusters were identified by "visual inspection," that is, utilizing the huge processing power of the human brain. Such identification is, of course, subject to individual idiosyncrasy, although different human brains seemed to work amazingly similar. Recently, [CBH+06] has introduced a practical *automatic*

[6] In a furnished room, diffraction cannot be reliably modeled today, and I doubt whether rigorous diffraction modeling is meaningful in time-varying environments at all.

clustering framework to MIMO modeling. This complete solution to the problem of how to parameterize cluster-based stochastic MIMO channel models from measurement data, with minimum user intervention, will be given as the example shown in Fig. 2.16. The method is particularly useful to process the huge amount of data that measurement campaigns tend to produce.

2.6.4 Normalization

Normalization of the channel matrices to a predefined average power value is a common pitfall in analyzing measured data[BW06]. The received power of MIMO systems varies considerably with position and direction of Tx and Rx arrays [Hea02]. Figure 2.9 illustrates how average mutual information may vary by moving the Tx by just a few wavelengths. Normalization of each measured matrix *separately* will level out pathloss information and thus mask this variation. (This procedure corresponds to the common assumption of constant receive SNR.)

In contrast, pathloss information is kept by normalizing all matrices measured in one location by the same factor. But only normalizing all matrices of all locations of an entire measurement campaign with one and the same factor will preserve location-dependent information about capacity and the like. Figure 2.9(b) shows how low pathloss correlates with high mutual information.

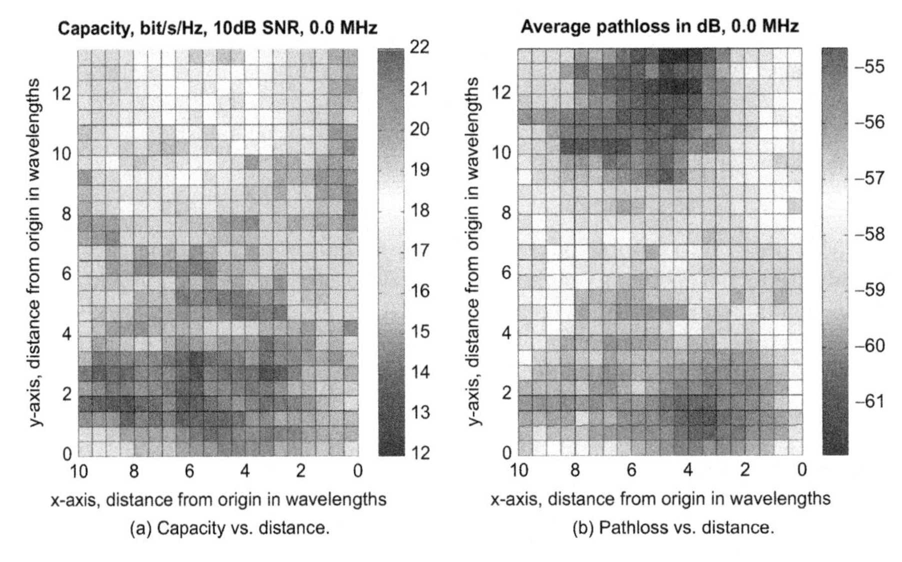

FIGURE 2.9

(a) Average mutual information as a function of Tx array position. (b) Average pathloss at a single frequency (0.0 MHz means lowest measurement frequency within a 120 MHz interval around 5.3 GHz) High average MI correlates with low pathloss. (8×8 MIMO, 5.3 GHz measurements in an office environment: Measurement details can be seen in [Özc04] and [CYz+07]).

It was shown that proper normalization of the measured MIMO matrices reveals the expected linkage between capacity and pathloss [HMRE03]. Also, it makes a significant difference whether one considers a constant-Tx-power or a constant-receive-SNR MIMO system: different normalization of the measured instantiations of the MIMO channel is appropriate for these two totally different MIMO schemes.

Proper normalization is also the key to meaningful comparison of dual-polarized MIMO schemes with single-polarized ones.

2.6.5 Which Channel Has Actually Been Measured?

While the difficult, time-consuming MIMO measurements are tricky and produce GByte of data, their proper interpretation is even trickier. It is the interpretation of measurements that is important for modeling.

Which type of randomness was prevalent when the measurements were taken? Have the measurements been performed in a *stationary* environment? Or have environments been measured that have different statistics and the data combined into a single ensemble? If the data taken in a number of different environments are all thrown into one ensemble, the central limit theorem will produce a Gaussian-like channel. This so-modeled Gaussian channel, however, will barely occur in any given situation. So, actually, one could have saved oneself the trouble of measuring in first place.

2.6.6 Rich Multipath or High Receive Power?

There is still a discussion ongoing whether a large number of eigenvalues ("rich multipath" in NLOS) or high receive SNR (usually in LOS) are more important[7]. Equation (2.1) shows that both factors contribute to mutual information. A high SNR associated with LOS may imply a low degree of scattering. A key to resolving controversial opinions could be fact that some LOS environments, but not all, also have rich multipath around the Rx [SW03].

2.6.7 Measurement Summary

In summary measurements reveal that:

- DoA depend on DoDs, leading to the double-directional viewpoint;
- DoAs and DoDs may be different for different delays;
- Strong discrete multipath components appear in clusters;
- But there may be also considerable diffuse power;
- Doppler profiles do not automatically obey Jakes' spectrum. Actually, they rarely do.

[7]A concrete measure to characterize "rich multipath" was proposed by [AdN05] in the form of the *richness function* (see the following).

As a consequence, the benefits of MIMO such as capacity exhibit:

- local variation,
- time dependence,
- and frequency selectivity.

2.7 WHAT MAKES A GOOD CHANNEL MODEL?

2.7.1 Requirements and Validation

First, here are some basic requirements for any type of modeling. Is the model accurate? Does it reflect measurements? Is the model specific, that is, does it model what was intended to be modeled? Is the model sensitive, that is, does variation of a certain parameter make a difference? And the most crucial of all questions for all the MIMO models that mushroomed in the last decade, has the model been *independently* validated? In the last point, I see large deficits.

On the experimental side of the modeling spectrum, researchers performed measurements and created a new model from their data, sometimes quite limited. Validation efforts were often confined to testing the new model with exactly the same data that originally served as the basis for the model. Other models were constructed by arranging two antenna arrays and by positioning scatterers, governed by intuition or by computational convenience. Then, by assuming un-attenuated scattering off these scatterers, the resulting MPCs are summed up at the receiver, and parameters such as spatial, temporal or frequency correlation are derived (see for example, [ZG08]. Validation is by simulation only. Only recently, multi-institution projects (e.g., COST 273, (WINNER)) have earnestly pursued a number of measurement campaigns to validate the MIMO channel models developed by these groups.

2.7.2 Metrics

Specifically for MIMO, we have to look for metrics that capture the benefits that this scheme offers. Not every metric will reflect equally well spatial multiplexing, diversity, or beamforming capabilities.

I discourage *ergodic* capacity as a model metric. It implicitly involves assumptions about clever coding, over an infinite number of channel samples, a field apart from radio propagation. Input-output Mutual Information (*MI*), Equation (2.1), in contrast, can always be defined and readily measured. In a random channel, MI plotted as a cumulative distribution function (cdf) of all measured or simulated channel realizations is a convenient and versatile metric. The outage MI (or *outage capacity*) is particularly useful to specify a desired degree of system performance at a specified level in bits/s/Hz. Figure 2.10 illustrates how outage capacity might differ between models in a different way than the median MI differs.

To describe multiplexing capability, eigenvalue profiles are useful. J. Bach Andersen [AdN05] defined a *richness function* by the eigenvalues λ_i of $\mathbf{HH}^H$ as

$$\mathbf{R}(k) = \sum \log_2 (\lambda_i) . \tag{2.5}$$

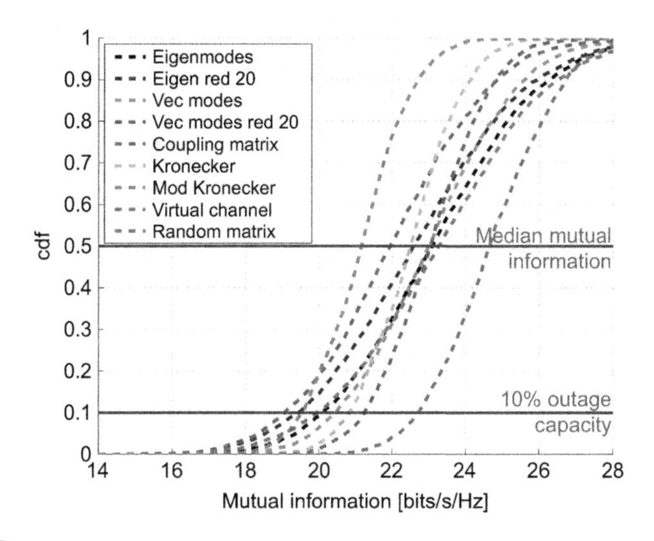

FIGURE 2.10

Definition of median mutual information and outage capacity.

To describe diversity, Ivrlac and Nossek [MJ03] found the *Diversity Metric*:

$$\Psi\left(\mathbf{R_H}\right) = \frac{\left(\sum_{i=1}^{K} \lambda_i\right)^2}{\sum_{i=1}^{K} \lambda_i^2} \tag{2.6}$$

useful, where the λ_i now are the K eigenvalues of $\mathbf{R_H}$ (see (2.2)).

Whether this full correlation matrix is well predicted by a model can be assessed by the Correlation Matrix Distance (CMD) [ME04]. Other common metrics to test the validity of propagation models are fading statistics and various correlation functions. The problem with the latter is that their numerical outcomes give rise to ambiguity. The difficulty to agree on what is a "high" or a "low" correlation makes such metrics subjective.

Actually, the spatial structure of the channel is what matters, because it gives intuitive insight to diversity, multiplexing, and beamforming opportunities. However, it is again difficult to quantify objectively whether a model reflects a measurement well, or to which degree. Figure 2.11 gives a 2D example of the spatial power distribution of a MIMO channel. Marginalizing to Rx and Tx sides separately, we obtain the Angular Power Spectrum (APS) or Power Angular Spectrum (PAS). These give the angular power distribution as seen from Rx or Tx.

2.7.3 Polarization

Polarization diversity is a means to reduce the size of MIMO-equipped terminals and base stations [KCVW02]. Orthogonally polarized antennas give rise to low correlation even if co-located.

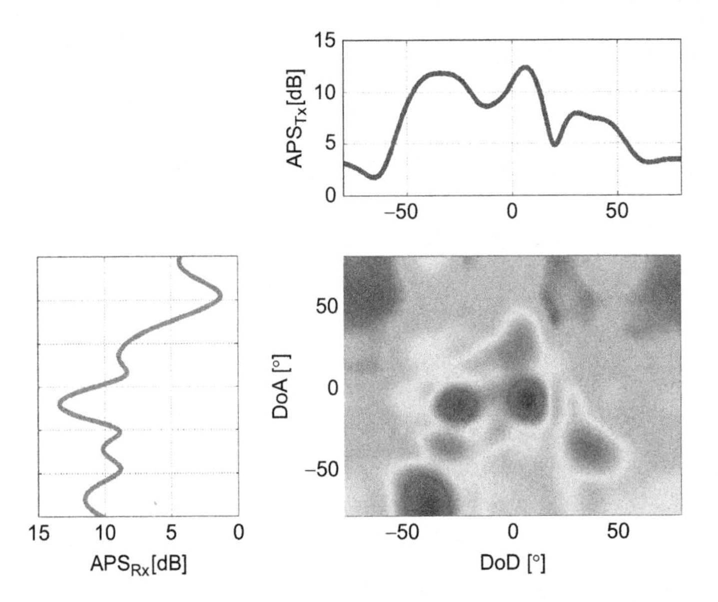

FIGURE 2.11

Example of the power distribution in a measured indoor MIMO channel, reflecting its spatial structure (lower right-hand picture). High power as seen from Tx (DoD) and Rx (DoA) is dark, low power light. The marginal distributions left and above are the APS.

When modeling polarization for MIMO channels, two different origins of cross-polarization have to be considered: The incomplete Cross-Polarization Discrimination (XPD) of the antennas at Tx and Rx, and the de-polarization that electromagnetic waves undergo when they propagate through the environment. Every single interaction with objects (reflection, scattering, refraction) changes the polarization state, in general to a three-parameter elliptical polarization, unless propagation direction, original polarization and interacting object are in extremely special arrangement.

As a consequence, cross-polarization ratio is an essential parameter of any good model. Polarization modeling for MIMO is a highly current research topic [FOHD].

2.8 EXAMPLES OF MIMO RADIO CHANNEL MODELS

Having established some basic considerations about MIMO radio channels, we will now turn to three examples of models that excel for various reasons. (An entire recent book deals with the art and science of MIMO channel models [CH10].)

2.8.1 WINNER II

The first examples is the WINNER II model [KMH$^+$07], called after the corresponding European Union (EU)-funded project. It is a *stochastic geometry-based* radio channel model, similar to the 3GPP Spatial Channel Model (SCM) [3gp03] and COST 273 [Cor06] models. Being based on the *double-directional* view it separates the antennas from the propagation environment. In this way, the best possible arrangement of antenna arrays can be found by simulation, given a certain environment.

Figure 2.12 shows the details of a MIMO link as modeled by WINNER II. Transmit array 1 consists of S antenna elements that transmit via N paths to U receive antenna elements comprising Array 2. Each "path" combines M "sub-paths" (or MPC or "rays," modeled as plane waves), shown in principle for Path 1, to create small-scale fading in conjunction with an implicit Doppler spectrum given by the geometry of the scenario. The parameters delay, DoD, DoA[8], gain, and cross-polarization define each path completely. Also the antenna patterns are accounted for by the corresponding antenna patterns in both polarizations separately. The term *cluster* is used alternatively for paths, but there is a slight inconsistency in this equalization. Actually the model places "clusters" in such a way as to generate prescribed and separate azimuth power spectra at Tx and at Rx. A path may be completed via more than one cluster, as shown in the bottom of Figure 2.12. The direct path between Tx and Rx, not numbered in Figure 2.12, has to be modeled, too, of course.

Structurally, the WINNER II model has several levels of randomness [KMH+07]. At first, Large Scale (LS) parameters such as pathloss, delay spread, angular spreads of departure and arrival, shadow fading variance, K-factor (if any) are drawn randomly from tabulated distribution functions. Next, the small scale parameters like actual delays, powers and directions arrival and departure of MPC are drawn randomly according to tabulated distribution functions and random LS parameters

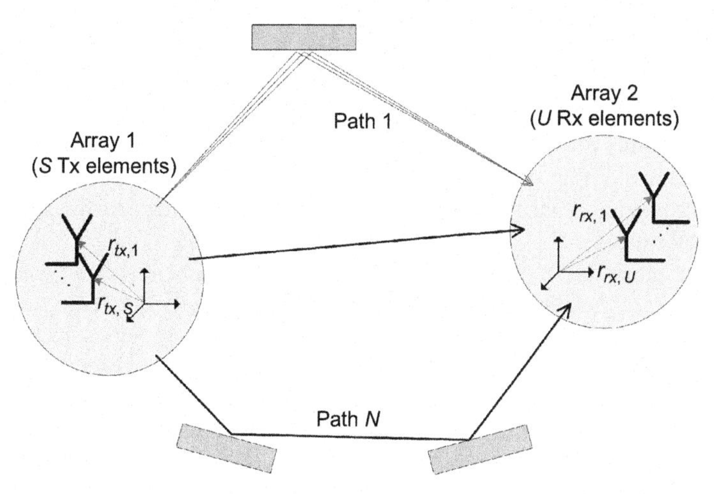

FIGURE 2.12

Structure of WINNER II model. Courtesy Elektrobit Oy.

[8]The original documents [KMH+07] speak of *angles* of arrival and departure (Angle of Arrival (AoA), Angle of Departure (AoD)), but I prefer the word "directions" according to [Fle00].

(second moments). At this stage the geometric setup is fixed and the only free variables are the random initial phases of the MPC. When also these initial phases are fixed, the model is fully determined.

A large number of measurements supplied the data for parameterization of no less than 18 different scenarios: Outdoor, indoor, outdoor2indoor; with and without LOS, and even high-speed scenarios. Thus, one should speak of the *family* of WINNER II models. Equation (2.7) gives the double-directional impulse response of the channel from Tx antenna element s to Rx element u for cluster n as:

$$\begin{aligned}
&\mathbf{H}_{\mathbf{u,s,n}}(t,\tau) \\
&= \sum_{m=1}^{M} \begin{bmatrix} F_{rx,u,V}(\varphi_{n,m}) \\ F_{rx,u,H}(\varphi_{n,m}) \end{bmatrix}^T \begin{bmatrix} \alpha_{n,m,VV} & \alpha_{n,m,VH} \\ \alpha_{n,m,HV} & \alpha_{n,m,HH} \end{bmatrix} \begin{bmatrix} F_{tx,s,V}(\phi_{n,m}) \\ F_{tx,s,H}(\phi_{n,m}) \end{bmatrix} \\
&\times \exp(j2\pi\lambda_0^{-1}(\bar{\varphi}_{n,m}\cdot\bar{r}_{rx,u}))\exp(j2\pi\lambda_0^{-1}(\bar{\phi}_{n,m}\cdot\bar{r}_{tx,s})) \\
&\times \exp(j2\pi\nu_{n,m}t)\delta(\tau-\tau_{n,m})
\end{aligned} \tag{2.7}$$

where M is the number of rays in the cluster, $F_{rx,u,V}$ and $F_{rx,u,H}$ are the antenna element u field patterns for vertical and horizontal polarizations, $\alpha_{n,m,VV}$ and $\alpha_{n,m,VH}$ are the complex gains of vertical-to-vertical and horizontal-to-vertical polarizations of ray n,m respectively[9]. Further λ_0 is the wavelength of the carrier frequency, $\bar{\phi}_{n,m}$ is DoD unit vector, $\bar{\varphi}_{n,m}$ is DoA unit vector, $\bar{r}_{rx,u}$ and $\bar{r}_{tx,s}$ are the location vectors of element s and u, respectively, and $\nu_{n,m}$ is the Doppler frequency component of ray n,m. The first term in (2.7) accounts duly for the 4-element matrix polarization characteristic of each ray. The second term describes the phase shifts due to antenna locations, the third term the phase shift due to movement of arrays.

Channel realizations of the link between Tx and Rx are generated by picking (randomly) different initial phases and by summing the contributions of all M rays and all N clusters. This superposition does not only result in temporal fading, but also creates the spatial correlation between antenna elements. Once a channel realization has been randomly drawn (such a realization is called a *drop*), it can be used to simulate the channel during movements of the terminal(s) over a short distance. How to simulate smooth temporal evolution of the channel by WINNER is still an interesting topic for future research.

Isn't the procedure to obtain the drop parameters time-consuming and computationally very demanding? Yes and no. The computational complexity that comes with any *spatial* radio channel model cannot be avoided. So, either the WINNER way (ray superposition) or the 3GPP way (introducing spatial correlation a priori) require of the order of 10^7 real operations to generate the parameters of one drop. When the number of antennas is higher, the correlation method has higher complexity. However, the essential step is the channel convolution, which is equal to Finite Impulse Response (FIR) filtering the transmitted signal with an L tap filter. And here the number of required real operations soars to 10^{10} (see Figure 2.13 [KJ07])! While

[9]Here, n and m are the numbering indices of rays, not to be confused with the number of transmit and receive antennas in the rest of this chapter.

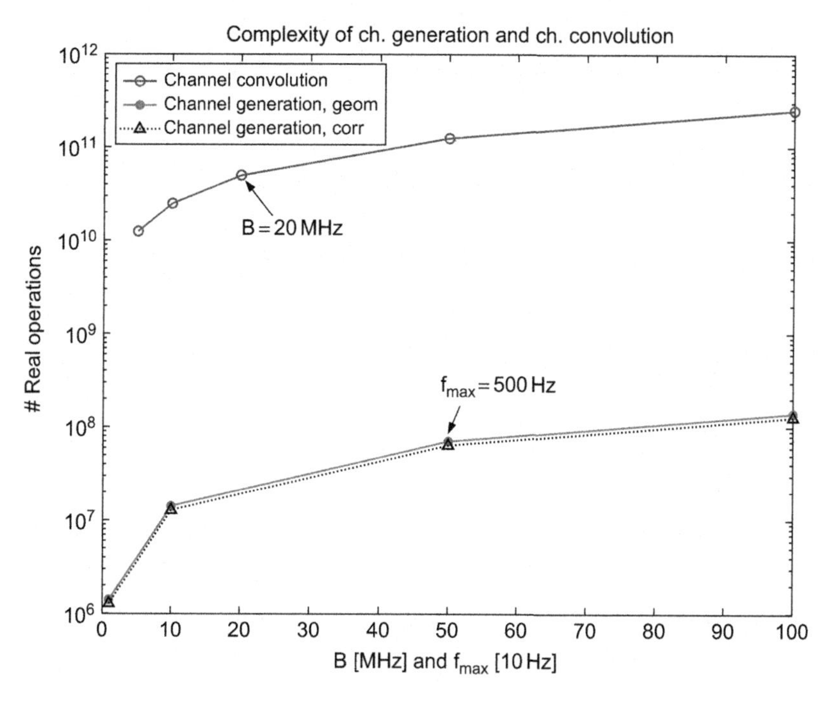

FIGURE 2.13

Computational complexity of generating a single drop of a 4 × 4 MIMO model, measured as the number of real operations. Lower curves represent the numbers required to generate the parameters of the model, the upper curve the number for channel convolution. Reprinted by permission IEEE, [KJ07].

the computational complexity of channel simulation (i.e., convolution) is a linear function of system bandwidth in wideband beyond Third Generation (3G) systems, the channel coefficient generation scales only with the maximum Doppler frequency [KJ07]. Therefore, computation of the drop parameters only marginally affects the overall computational complexity.

A note of the number of taps in WINNER: This number depends on the scenario, but each tap is modeled as one single cluster (which might be one of the few weaknesses of the model, when different DoAs arrive with the same delay).

The WINNER models are suitable for interference simulations on network level; then, the same procedure has to be carried out for all links involved. WINNER II focuses on global channel properties rather than on cluster properties and describes channel variability well due to the careful selection of pdfs of the parameters. This and the fact that the WINNER model is very general and covers many scenarios helped to achieve a breakthrough for this MIMO model: it has been selected by ITU-R for International Mobile Telecommunications (IMT)-Advanced candidate testing [M2108], and thus constitutes the *standard model* worldwide today. With interest and pride we note that three essential ingredients of the WINNER II models originated in the COST 231/259/273/2100 Actions: The *geometry-based stochastic* modeling

[JP98] and the *clustering* approaches, and the *double-directional* viewpoint that separates the antennas from the propagation environment.

2.8.2 An Analytical MIMO Channel Model (Weichselberger)

We now introduce an example from the class of *analytical* models. Such models are useful when the antenna arrays of the MIMO link are already known or fixed, and they provide realizations of the **H** matrix directly for simulation. The *Weichselberger model* [Wei03, Wea06], as it is sometimes called, starts from an eigenmode analysis of the full correlation matrix $\mathbf{R_H}$ (see Equation 2.2). Its strength is to give immediate and intuitive insight into which MIMO benefit can be reaped from a certain environment and which not. Furthermore, several independent measurement and validation campaigns have proven this model to render ergodic capacity and diversity better than other correlation-based analytical models [Don07], [WH07], [ETM06], [WMA+08].

In the eigenmode point-of-view we have to distinguish between eigenbases and eigenvalues. Eigenbases have some nice properties: They fade independently, they are orthogonal, their eigendecomposition is unique, the principal eigenmode maximizes power, and they provide the smallest number of modes possible, but their number does not exceed $n \cdot m$.

The spatial correlation of transmit weights (complex excitation of the Tx array elements) determines how much power is radiated into which directions (and polarizations). The spatial *eigenbases reflect the radio environment*, i.e. number, positions, and strengths of the scatterers, and are not affected by the transmit weights. The *eigenvalues*, on the other hand, do depend on the transmit weights. They show *how the scatterers* are *illuminated* by the radio waves propagating from the transmitter. Radiating in certain directions, for example, may illuminate only certain scatterers and leave others "dark." This will affect, of course, the spatial correlation at the Rx array[10].

The true eigenmodes of **H** are matrices, but antenna arrays can only be excited by vectors. Weichselberger solves this problem by approximating the eigenmodes by vector modes. Further, a structure is imposed in these vector modes, meaning they have to comply with the correlation matrices $\mathbf{R}_{\mathrm{Rx}}$ and $\mathbf{R}_{\mathrm{Tx}}$ at receive and transmit sides separately. The Weichselberger model performs an eigendecomposition of the correlation matrices $\mathbf{R}_{\mathrm{Rx}}$ and $\mathbf{R}_{\mathrm{Tx}}$. These matrices enter the model only via their eigenmodes $\mathbf{U}_{\mathrm{Rx}}$ and $\mathbf{U}_{\mathrm{Tx}}$:

$$\mathbf{H} = \mathbf{U}_{\mathbf{R}_{\mathrm{x}}} \left(\tilde{\mathbf{\Omega}} \odot \mathbf{G} \right) \mathbf{U}_{\mathbf{T}_{\mathrm{x}}}{}^{T}. \tag{2.8}$$

Equation (2.8) gives the formal definition of how to generate a realization of the Weichselberger model. Here, $\mathbf{\Omega}$ is the so-called the coupling matrix, **G** is a random i.i.d. matrix, and the dotted circle denotes element-wise multiplication. $\tilde{\mathbf{\Omega}}$ is the element-wise square root of $\mathbf{\Omega}$.

[10]A shortcoming of the Kronecker model is the assumption that the receive correlation (2.3) is always the same, irrespective of what the transmitter sends.

The structure of $\mathbf{\Omega}$ is a direct consequence of the spatial arrangement of scattering objects. It tells us *how many parallel data streams* can be multiplexed, which degree of *diversity* is present at side A and at side B, and how much *beamforming* gain can be achieved.

Figure 2.14 shows some archetypical examples of structures of $\mathbf{\Omega}$ of a 4×4 MIMO system and the corresponding physical radio environments. A full square designates a high value of this $\mathbf{\Omega}$ element, open squares correspond to zero (or small) entries. The columns show the Tx eigenmodes, the rows the Rx eigenmodes.

In order to aid intuition, think of the eigenmodes as discrete directions (center parts of Figure 2.14). Such an interpretation of eigenmodes is not generally correct, but it facilitates their visualization. The number of eigenmodes present in the channel considered equals the number of resolvable multipath components, which evidently is a lower bound to all multipath components present. The right-hand parts of Figure 2.14 tells which benefit of MIMO can be exploited in the exemplary channels shown. For instance, $\mathbf{\Omega}_2$ shows a single Tx eigenmode coupling into all four Rx eigenmodes, useful for beamforming at the Tx and diversity reception at Rx.

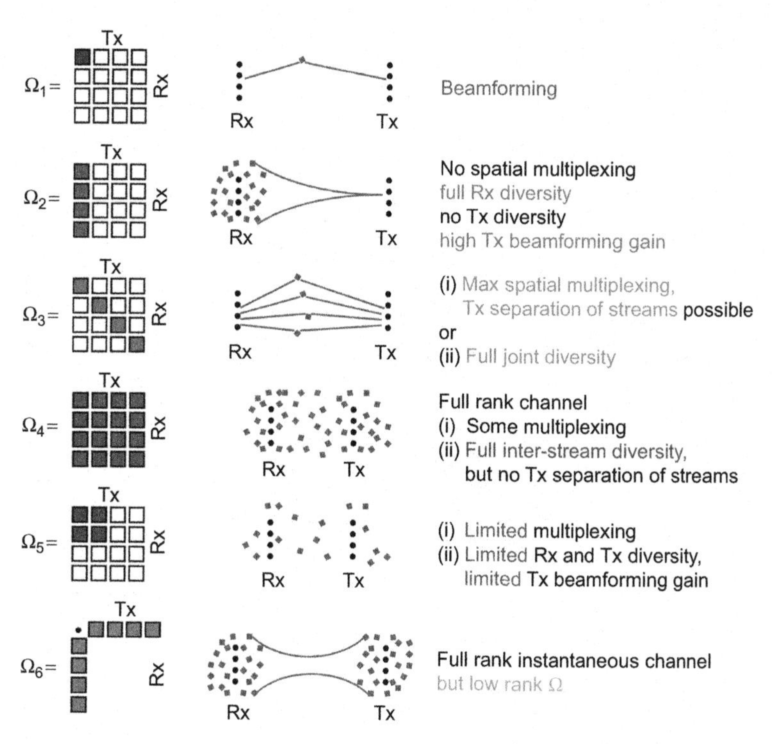

FIGURE 2.14

Sample structures of coupling matrix. A full square designates a high value of this $\mathbf{\Omega}$ element, open squares correspond to zero (or small) entries. The columns show the Tx eigenmodes, the rows the Rx eigenmodes. Courtesy W. Weichselberger.

Actual coupling matrices of a channel measured [CYz+07] with an 8×8 MIMO system in an office environment are shown in Figures 2.15(a)–2.15(c) (linear scale, arbitrary units).

It is evident that only three parallel channels are available for spatial multiplexing in Figure 2.15(b), although this is an 8×8 MIMO system; not all the eigenmodes are present. Ignoring results of this kind might end up in bitter disappointment about the capabilities of 8×8 MIMO, now under consideration for LTE Advanced.

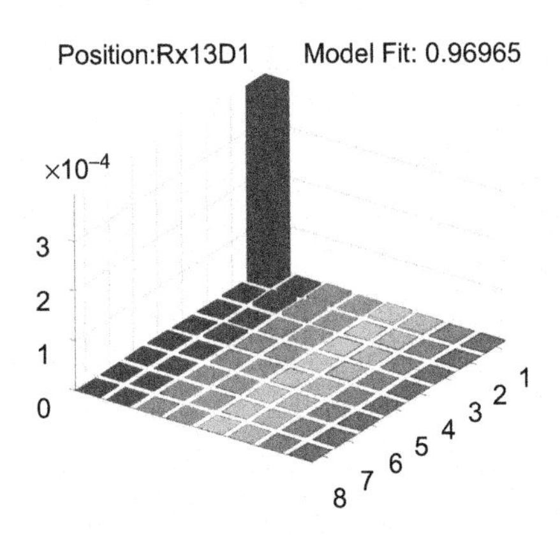

(a) Tx eigenmode 1 couples only to Rx eigenmode 1: good for beamforming at both link-ends.

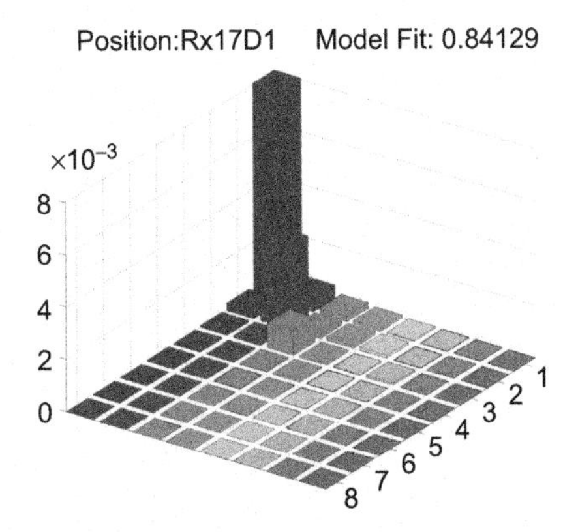

(b) Three Tx eigenmodes couple to three *Rx* eigenmodes each: good for spatial multiplexing of three independent streams.

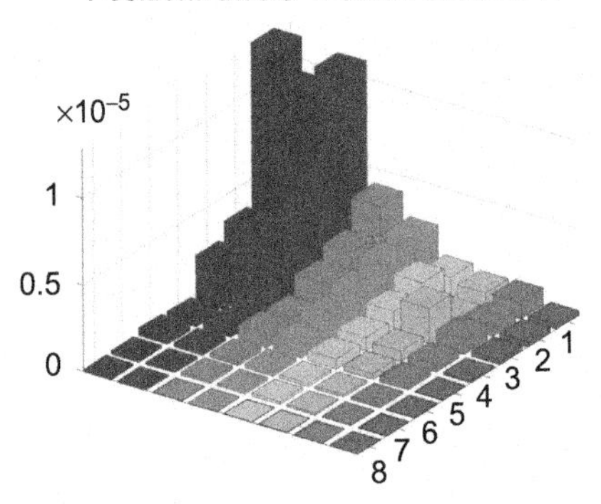

(c) Seven Tx eigenmodes couple strongly to six Rx eigenmodes. Proper space-time coding will exploit either multiplexing or diversity. This channel is not Kronecker separable.

FIGURE 2.15

Examples of actual coupling matrices $\mathbf{\Omega}$, measured at 5.3 GHz in an office environment with an 8×8 MIMO system (linear scale, arbitrary units). Courtesy W. Weichselberger.

Table 2.1 How to Obtain the Coupling Matrices from Measurement

Simple estimator:	$\hat{\mathbf{R}}_{R_x} = \frac{1}{T}\sum_{t=1}^{T} \mathbf{H}(t)\mathbf{H}^H(t)$ $\hat{\mathbf{R}}_{T_x} = \frac{1}{T}\sum_{t=1}^{T} \mathbf{H}^T(t)\mathbf{H}^*(t)$
Eigendecomposition:	$\hat{\mathbf{R}}_{R_x} = \hat{\mathbf{U}}_{R_x}\hat{\mathbf{\Lambda}}_{R_x}\hat{\mathbf{U}}_{R_x}^H$ $\hat{\mathbf{R}}_{T_x} = \hat{\mathbf{U}}_{T_x}\hat{\mathbf{\Lambda}}_{T_x}\hat{\mathbf{U}}_{T_x}^H$
Auxiliary matrix:	$\mathbf{K}(t) \triangleq \hat{\mathbf{U}}_{R_x}^H\mathbf{H}(t)\hat{\mathbf{U}}_{T_x}^*$
Coupling matrix	$\hat{\mathbf{\Omega}} = \frac{1}{T}\sum_{t=1}^{T} \mathbf{K}(t) \odot \mathbf{K}^*(t)$

The comment *model fit* shall remind that the Weichselberger model is an approximation of the full correlation matrix; model fit equalling unity would be a perfect match. By the way, since these measured positions were only a few wavelengths apart, Figures 2.15(a)–2.15(c) corroborate the notion that the MIMO radio channel is a very local phenomenon, and so is capacity!

If the theory behind the Weichselberger model sounded difficult, obtaining the coupling matrix from measurements is simple and straightforward. Given an ensemble of T samples of the channel is available, the corresponding coupling matrix is only four clicks in MATLAB away by executing the steps of Table 2.1.

Weichselberger's eigenmode analysis has been successfully extended to time-varying channels [WGH07, Wil10] and to wideband, frequency-dispersive channels [CH08]. The initial goal of the eigenbase analysis was to parameterize a correlation-based model directly from channel measurements or estimates obtained during system operation. The next example of MIMO radio channel models takes this approach even one step further.

2.8.3 The Random Cluster Model

The Random Cluster Model (RCM) creates clusters randomly, but the distribution of their parameters strictly follow probability density functions (pdfs) measured in specific environments [CZN+09]. Together with a procedure that allows to define clusters *automatically, without user intervention*, from multiantenna measurements, the RCM offers the possibility to create a channel model that fits measured data well, but still keeps its random nature. The RCM is not tied to a certain antenna geometry, the array response can be user defined and thus optimized. An example of parameter pdfs measured in two different indoor office environments is given in Figure 2.16.

Also the number of clusters existing in the specific environment (model order) is automatically determined by an algorithm that lets the user set a threshold for clustering. All clusters must show a specific minimum power with respect to the total power of the scenario, to be regarded as important. The number of clusters, N_c, is

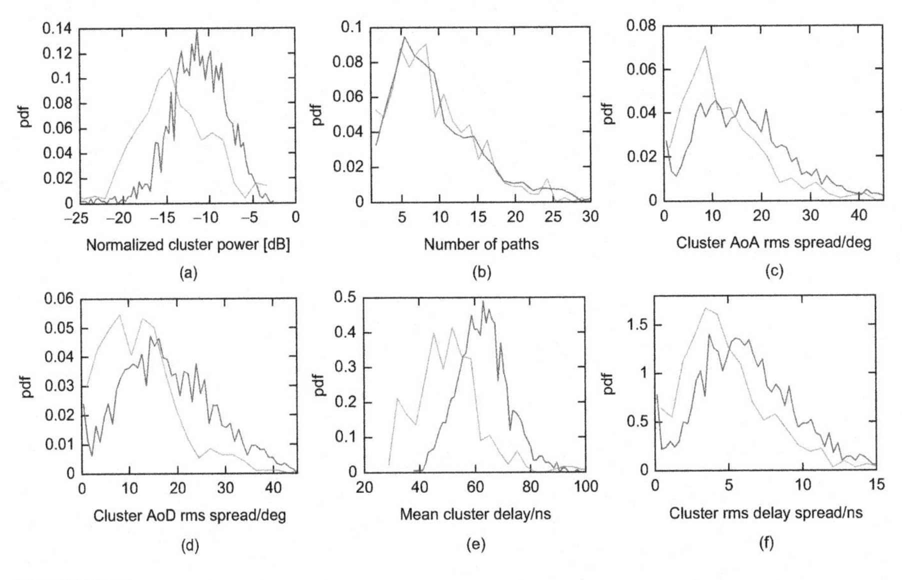

FIGURE 2.16

Example of pdfs of the cluster parameters for the parameterization of the RCM. Measurements were taken in LOS (light) and in obstructed LOS (dark) indoor environments at the University of Oulu, Finland, at 2.55 GHz. Reprinted by permission IEEE GLOBECOM 2006 [CBH+06].

thus mathematically defined by a minimum power threshold. The automatic cluster identification algorithm also includes a Kalman filter for tracking the cluster parameters (cluster position and delay) over time. Doing so, the cluster speed is derived, which directly corresponds to the cluster Doppler shift.

Having identified the parameters of all clusters occurring in the environment, one can estimate the joint probability density function of their parameters. This so-called environment pdf is at the heart of the RCM. From this environment pdf $p_{\theta_c}(\theta_c)$, new realizations of the same environment are randomly drawn, thus generating similar propagation channels. In summary, creating this multidimensional $p_{\theta_c}(\theta_c)$ takes the steps of Table 2.2.

After drawing new clusters and paths from the environment pdf, the channel's complex impulse response has to be calculated. This is done by simply applying an antenna filter and a system bandwidth filter implemented in frequency domain. We arrive at the entire channel's (complex) impulse response in baseband notation:

$$\mathbf{H}(t, \Delta f) = \sum_{c=1}^{N_c} \sum_{p=1}^{N_{p,c}} \gamma_{cp} \mathbf{a}_{\mathrm{Rx}}(\varphi_{\mathrm{Rx},cp}, \theta_{\mathrm{Rx},cp}) \mathbf{a}_{T_{\mathrm{x}}}^{T}(\varphi_{\mathrm{Tx},cp}, \theta_{\mathrm{Tx},cp}) e^{-j2\pi \Delta f \tau_{cp}}. \quad (2.9)$$

The RCM framework boasts a wealth of novel features, such as power-weighted automatic clustering, and a math-based criterion for the number of clusters. The

Table 2.2 The RCM at a Glance

MIMO channel measurement
$\Downarrow$
impulse responses
$\Downarrow$
path estimation
$\Downarrow$
cluster identification
$\Downarrow$
cluster parameters
$\Downarrow$
density estimation
$\Downarrow$
multidimensional environmental pdf $p_{\theta_c}(\theta_c)$ to parameterize the RCM

inherent tracking of clusters makes it well suited for simulating smoothly time-variant channels. Reference [CRB$^+$08] specifies a method how to introduce Dense Multipath Component (DMC) modeling supplementarily in the RCM. In comparison to models that randomly draw or describe MPC, it is directly parameterized from measurements, which significantly reduces the number of external parameters.

2.9 SOME CONCLUSIONS

Considering the last of the three exemplary MIMO channel models, do we have to expect (or fear) that MIMO channel modeling will be taken from engineers and handed over to computers? No, I think not. The core of modeling, which is the selection of scenarios, the decision of what should be measured, and the interpretation of what has been measured, is the creative and intelligent work of an engineer. Having made the right decisions, we can leave the number crunching then confidently to the dumb machines.

With the impending widespread deployment of MIMO systems, both singleuser and multiuser, interference will become a pressing issue. Thus, *interference mitigation and interference exploitation* will be the topics of LTE and LTE Advanced in the next years. Available MIMO radio channel models do have the capability to account for such schemes, but good measurements are lacking. In principle, such measurements have to be taken with several multiantenna channel sounders in place, but I am aware of only a single campaign so far (WILATI Project) [Kea07]. There are workarounds available that avoid the use of more than one channel sounder, but they have yet to prove their viability in practical modeling [CBVV$^+$08].

To model the upcoming applications of MIMO in *sensor networks* [VDCM08], *on-body networks* [Hea07], and *vehicle2vehicle* and *vehicle2infrastructure* systems [AMI07], [PKC$^+$], many more measurements campaigns have to be designed, performed, and evaluated.

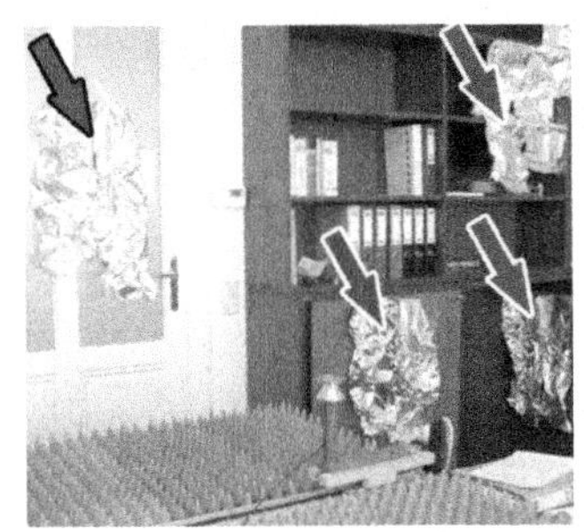

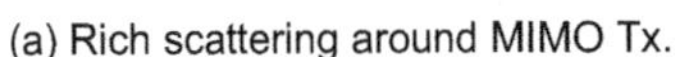

(a) Rich scattering around MIMO Tx. (b) Rich scattering around MIMO Rx.

FIGURE 2.17

Enhancing rich scattering around MIMO Tx (a), and MIMO Rx (b) by wrinkled aluminium foils. Courtesy C. Mecklenbräuker, Technische Universität Wien.

Should the informed reader have missed the catchwords "pinhole" or "keyhole" so far, the answer is readily given: Such channels exist only under circumstances that have to be created artificially and with great effort [PFA03].

The very first sentence of this chapter defined the channel as that part of a communication system that cannot be engineered. Have a look at Figure 2.17(a) and 2.17(b) showing Ralf Müller at Forschungszentrum Telekommunikation Wien (ftw)'s first office, which he equipped with wrinkled aluminium foil to create rich scattering during a MIMO measurement. Maybe we can engineer the MIMO propagation channel after all.

In the distant past, we engineered systems with the *worst case* model in view. This approach sometimes resulted in over-provisioning. More recently, we modeled the *average* propagation channel and stochastic variation of it. This is an advance over the past, but is it sufficient? I think not. Some people still devise algorithms for this average channel, which of course rarely occurs in reality. The real challenge is modelling the *dynamic* channel. Meeting this challenge we will get the reward of really exploiting the MIMO radio channel as it exists here and now.

Acknowledgment

I gratefully acknowledge the ingenuity and the hard work of my former PhD students who shaped my views of the MIMO radio channel. Particularly, I thank Niki Czink and Werner Weichselberger for their help with the manuscript. The sponsorship of their PhD projects by Elektrobit (N.C.) and by Nokia Research Center, Helsinki, (W.W.), is gratefully acknowledged. Mobilkom Austria AG, the Austrian mobile operator committed to the advancement of mobile communications, sponsored most of the other PhD students of whom results have been used in this paper. The participants of the COST groups 231/259/273 have been a continuous source of inspiration about mobile communications, for which I also thank them very much.

CHAPTER

Space Time Codes and MIMO Transmission

3

Alister Burr

3.1 INTRODUCTION

Having introduced the MIMO wireless channel, and the remarkable capacities that are in principle available to exploit it, in this chapter we introduce some of the practical techniques that can be used to transmit data upon it, hence realizing this capacity.

In particular, we discuss the concept of *space-time coding*. This is in its own right one of the most important transmission techniques, but we will also use it as a general mathematical tool to analyze the performance of any transmission technique for the MIMO channel. As such we will apply it to other transmission techniques which are not usually thought of as space-time codes, notably spatial multiplexing.

However, first, in Section 3.2 we discuss the performance benefits of the MIMO channel, usually characterized in terms of *diversity* and *multiplexing*. We will define these, both in terms of the straightforward intuitions on which the terms are based, and in the more abstract but quantitative terms that have become conventional. We will see that they are in a sense a property of the underlying MIMO channel, but that the MIMO transmission technique must be carefully designed to optimally exploit them.

To demonstrate this in Section 3.3 we will explore the theory of space-time coding, mentioned above, and use it to explain the criteria that can be used to ensure diversity and multiplexing benefits are fully exploited. We will also discuss the so-called *diversity-multiplexing trade-off*, which defines the extent to which the two benefits can be obtained simultaneously.

Drawing on this theoretical background, we will describe in Sections 3.4 to 3.6 the three main MIMO transmission techniques: space-time codes, *spatial multiplexing*, and *precoding*. The distinction between these can also be expressed in terms of diversity and multiplexing: space-time codes have been designed to maximize diversity, and spatial multiplexing (as the name might suggest) to maximize diversity benefit. However, we will see that there is no sharp dividing line, and that the first two approaches can be used to exploit both benefits to a large extent, and in fact spatial multiplexing can also be treated in terms of space-time coding. The third approach,

MIMO. DOI: 10.1016/B978-0-12-382194-2.00003-4

precoding, is widely used in emerging fourth generation wireless standards, especially to allow adaptation to the MIMO channel. However, this approach, too, is not exclusive to adaptive MIMO schemes, and we will see that both space-time codes and spatial multiplexing can be regarded as forms of precoding.

Finally, in Section 3.7, we describe the application of these techniques, and especially precoding, in current and emerging wireless standards, especially the 3GPP LTE and the WiMAX, or IEEE 802.16 families of standards.

3.2 DIVERSITY AND MULTIPLEXING GAIN

3.2.1 Diversity

The idea of diversity is very simple and intuitive, and can be summed up in the phrase, "Don't put all your eggs in one basket!" In communication terms, if multiple and independent routes, called *diversity branches*, are provided for the same information, the probability that the information is lost due to fading is much reduced, since it would require all branches to fade simultaneously. If the branches are indeed independent then the probability of this is reduced according to the number of branches. Quantitatively, if each branch has an outage probability P_{out}, and the same information is transmitted over n branches, then the overall outage probability is the probability that all branches fade, namely P^n_{out}. n is then called the *diversity order*. This suggests that on the logarithmic scale commonly used to plot outage probabilities, the outage probability with n^{th} order diversity drops n times as fast with increasing Signal-to-Noise Ratio (SNR) as with a single branch (no diversity). Several books include more detailed descriptions of diversity as conventionally applied, for example [B01], pp. 255–258, [SS99], Chapter 15, or [BVJY03], pp. 53–60. Here we concentrate on the concept of diversity and the performance benefit it provides.

If the diversity branches are provided by spatially separated antennas, as in a MIMO system, this is known as *space diversity*. Let us consider the simplest possible case: a system with one transmit antenna and n_R receive antennas. If the receive antennas are subject to independent Rayleigh fading, then the channel can be described by a $1 \times n_R$ vector $\mathbf{h}$, whose entries are zero mean, circularly-symmetric complex Gaussian random variables. We further assume that a sequence $\mathbf{s}$ of symbols $s_l, l = 1 \ldots L$ is transmitted, and that the noise at the receive antennas in the lth symbol period can be represented by the $1 \times n_R$ vector $\mathbf{n}_l$. Further we suppose that the channel is *quasi-stationary*, that $\mathbf{h}$ can be assumed constant for the whole transmission period of sequence $\mathbf{s}$, although it may vary at random over a longer time period. Then the received signals in the lth period can be written as:

$$\mathbf{r}_l = \mathbf{h}s_l + \mathbf{n}_l. \qquad (3.1)$$

The entries of $\mathbf{n}_l$ are also circularly-symmetric complex Gaussian random variables, with power N. It can be shown [P01] that the optimum receiver, that maximizes the received signal-to-noise ratio, employs *maximum ratio combining* (MRC), in which a decision variable is formed by weighting the received signals by a factor proportional to the complex conjugate of the corresponding channel gain, and adding:

$$x_l = \mathbf{h}^H \mathbf{r}_l = \mathbf{h}^H \mathbf{h} s_l + \mathbf{h}^H \mathbf{n}_l = \|\mathbf{h}\|^2 s_l + \mathbf{h}^H \mathbf{n}_l. \tag{3.2}$$

Since the noise and the channel are uncorrelated, the noise term in this expression is complex Gaussian with mean power given by $\|\mathbf{h}\|^2 \overline{\|\mathbf{n}\|^2} = n_R N \|\mathbf{h}\|^2$. The signal power is $\|\mathbf{h}\|^4 \overline{|s_l|^2} = S \|\mathbf{h}\|^4$; hence, the SNR of the decision variable is given by:

$$\rho = S \|\mathbf{h}\|^2 / N n_R. \tag{3.3}$$

If $\mathbf{s}$ is long enough, and an appropriate Forward Error Correction (FEC) code is used, the sequence can be decoded without error provided the SNR is sufficient that the Shannon capacity [Sha48] of the equivalent channel is greater than is the information rate being transmitted, R bits/symbol, i.e., $\rho \geq \rho_0$ where $log_2(1+\rho_0) = R$. Otherwise an outage will occur for that sequence, or data frame. Thus, the frame error probability, P_f is the same as the outage probability for this channel, assuming that over a long time period the channel varies randomly. Since all other terms in Equation (1.3) are constant, this depends on $\|\mathbf{h}\|^2$; hence, we can write:

$$P_f = \Pr\left[\|\mathbf{h}\|^2 < \rho_0 N n_R / S\right] \tag{3.4}$$

Now $\|\mathbf{h}\|^2$ is the sum of the squared magnitude of n_R complex Gaussian random variables; therefore, (assuming that the mean square value of all the channel elements is the same, $\overline{|h|^2}$) it has the Chi-square distribution with $2n_R$ degrees of freedom, scaled by $\overline{|h|^2}/2$. The Cumulative Distribution Function (CDF) is [EWJ10]:

$$P_{\chi^2}\left(2z/\overline{|h|^2}, 2n_R\right) = \frac{\gamma\left(n_R, z/\overline{|h|^2}\right)}{\Gamma(n_R)} \tag{3.5}$$

where $\gamma(k,x)$ denotes the incomplete gamma function [EWA10a], and $\Gamma(k)$ the standard Gamma function [EWA10b], equal to $(k-1)!$ for integer argument. This can further be expressed in series form:

$$P_{\chi^2}\left(2z/\overline{|h|^2}, 2n_R\right) = \frac{1}{n_R!}\left(\frac{z}{\overline{|h|^2}}\right)^{n_R} + o\left(z^{n_R+1}\right). \tag{3.6}$$

Since the CDF denotes the probability that the random variable is less than its argument, the frame error probability from (1.4) is:

$$P_f = \frac{1}{(n_R-1)!}\gamma\left(n_R, \frac{\rho_0 n_R}{\overline{|h|^2} S/N}\right) \approx \frac{1}{n_R!}\left(\frac{\rho_0 n_R}{\overline{|h|^2} S/N}\right)^{n_R} \tag{3.7}$$

The approximation is asymptotically close at high SNR.

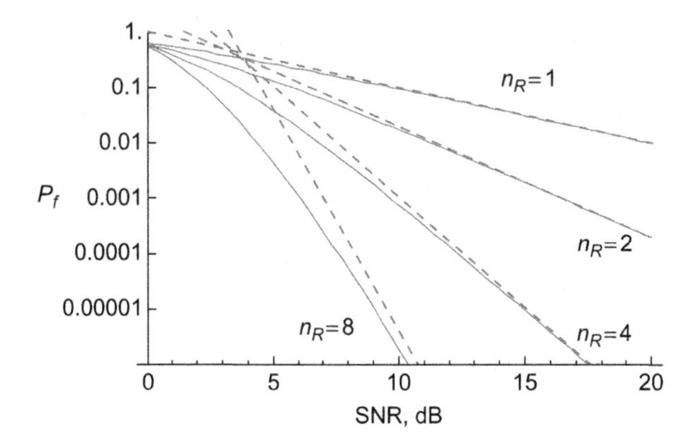

FIGURE 3.1

Frame error probability P_f versus SNR for diversity order 1, 2, 4, 8 (solid line) compared to asymptotic approximation (dashed line).

Figure 3.1 shows both the exact expression for P_f and the approximation, plotted on a logarithmic scale against SNR in dB. We observe that asymptotically at high SNR this becomes a straight line of slope $-n_R$ orders of magnitude per 10 dB. This agrees with the intuition we have already noted that the slope of P_f on a logarithmic scale is proportional to the diversity order.

Note that this derivation is obtained by information theoretic analysis of the channel. The diversity order is a property of the channel itself, not of any particular transmission technique we have assumed. However, as we will see the transmission technique may or may not allow the underlying diversity of the channel to be exploited.

Also note that in recent years the property illustrated by Figure 3.1 has become a definition of diversity order, in place of the intuitive definition based on the number of diversity branches available. Since the probability of bit error, or Bit Error Ratio (BER), is in general proportional to frame error probability, the BER also behaves in the same way. Thus, diversity order can be defined as the asymptotic slope of the logarithmic plot of BER against SNR (in dB) [ZT03]:

$$n_d = -\lim_{S\to\infty}\left[\frac{\log_{10} P_b}{\log_{10} S}\right] \tag{3.8}$$

where S denotes SNR, and P_b is BER. This definition has the advantage of being more flexible, since it applies to cases where the slope, and hence the diversity order, is not an integer. The relationship with the number of physical diversity branches is not however lost, since it becomes a property of the channel that its diversity order is equal to the number of diversity branches.

It is easy to see that in a MIMO system with n_T transmit and n_R receive antennas the diversity of the channel is limited to $n_T \times n_R$, the number of discrete paths

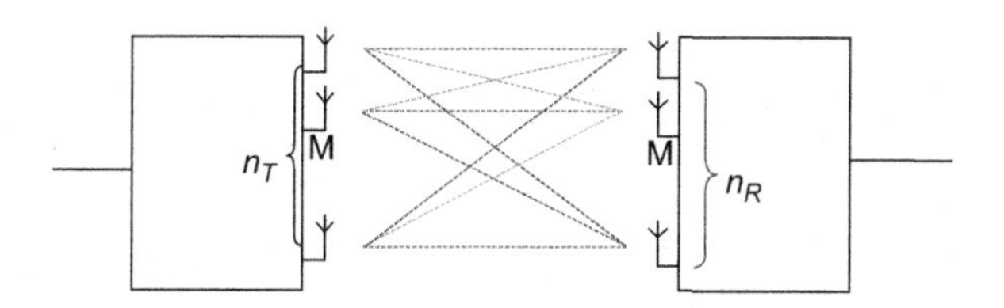

FIGURE 3.2

Diversity branches in MIMO channel.

between pairs of transmitting and receiving antennas (Figure 3.2). This multiplication effect means that large diversity orders can be achieved with relatively small numbers of antennas at each end of the link, which gives MIMO a substantial benefit in this respect compared to diversity reception (which can be characterized as *Single-input and Multiple-Output* (SIMO)). However, note that this assumes independent fading between all antennas. If there is limited multipath scattering in the channel, this results in correlation of the fading, which will limit the diversity order.

3.2.2 Multiplexing

However, the benefit of MIMO that is in many respects most significant is the increase in link capacity that it can provide. This can be characterized as *multiplexing gain*, and arises because the MIMO channel is able to transmit multiple data streams simultaneously over the same bandwidth, separated spatially; hence, the term *spatial multiplexing*. We will refer to these spatially separated channels as *subchannels*. The convexity of the graph of Shannon capacity against SNR (on a linear scale) means that the capacity of n_m subchannels with SNR equal to $(S/N)/n_m$ is much greater than that of a single channel with SNR S/N.

The MIMO channel matrix can be decomposed using the *singular value decomposition* (SVD) [GVL96], pp. 70–71:

$$\mathbf{H} = \mathbf{V}\Sigma\mathbf{U}^H \tag{3.9}$$

Therefore, the received signal can be written as:

$$\mathbf{r} = \mathbf{Hs} + \mathbf{n} = \mathbf{V}\Sigma\mathbf{U}^H\mathbf{s} + \mathbf{n} \tag{3.10}$$

where $\mathbf{V}$ and $\mathbf{U}$ are singular vector matrices, and $\Sigma = \sqrt{\Lambda}$ is the singular value matrix: a diagonal matrix whose elements are the square roots of the eigenvalues of $\mathbf{HH}^H$, $\lambda_i, i = 1 \ldots n_m$. $\mathbf{n}$ represents the received noise. We can then relate the transmitted signal to a matrix of data symbols $\mathbf{d}$ by $\mathbf{s} = \mathbf{Ud}$, and estimate the data from the received signal using $\hat{\mathbf{d}} = \mathbf{V^H r}$. Then:

$$\hat{\mathbf{d}} = \mathbf{V^H HUd} + \mathbf{V^H n} = \Sigma\mathbf{d} + \mathbf{V^H n}. \tag{3.11}$$

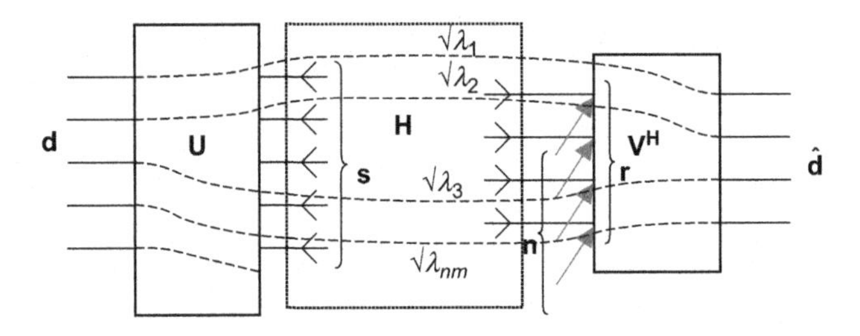

FIGURE 3.3

Subchannel model of MIMO channel.

This in effect creates a set of n_m noninterfering *subchannels*, whose power gains are given by the eigenvalues, as illustrated in Figure 3.3. Note that n_m is given by the rank of $\mathbf{HH}^H$, and is upper bounded $n_m \leq \min(n_T, n_R)$. Figure 3.3 shows how the number of usable subchannels is limited by both the number of transmit and receive antennas.

The capacity of the MIMO channel can be written [Fos96]:

$$C_{MIMO} = W \sum_{i=1}^{n_m} \log_2 \left(1 + \frac{S}{N} \frac{\lambda_i}{n_T}\right) \tag{3.12}$$

where W denotes the channel bandwidth. At very high SNR (as SNR tends to infinity):

$$\begin{aligned}\frac{C_{MIMO}}{W} &\to \sum_{i=1}^{n_m} \log_2 \left(\frac{S}{N} \frac{\lambda_i}{n_T}\right) = \sum_{i=1}^{n_m} \left(log_2 \left(\frac{S}{N}\right) + \log_2(\lambda_i) - \log_2(n_T)\right) \\ &= n_m \log_2 \left(\frac{S}{N}\right) - n_m \log_2(n_T) + \log_2 \left(\prod_{i=1}^{n_m} \lambda_i\right) \\ &= n_m \log_2 \left(\frac{S}{N}\right) - n_m \log_2(n_T) + \log_2 \det\left(\mathbf{HH}^H\right)\end{aligned} \tag{3.13}$$

Asymptotically the slope of the plot of this function against SNR in dB is given by $n_m \log_2(10)/10 = 0.332 n_m$ bits/s/Hz/dB, which (it is easy to see) is n_m times steeper than the same curve for a SISO system. This factor n_m is known as the *multiplexing gain*. It reflects the availability of n_m independent subchannels.

However the asymptotic curve is offset from the term $n_m \log_2(S/N)$ by the constant term $\log_2 \det\left(\mathbf{HH}^H\right) - n_m \log_2(n_T)$. Since $\det\left(\mathbf{HH}^H\right) = \prod_{i=1}^{n_m} \lambda_i$ and $\sum_{i=1}^{n_m} \lambda_i = \mathrm{trace}\left(\mathbf{HH}^H\right) = \|\mathbf{H}\|^2$, and $\|\mathbf{H}\|^2 = n_R n_T$ on average, according to the most commonly used normalizing assumption [Fos96], we can show that:

$$\log_2 \det\left(\mathbf{HH}^H\right) \leq n_m \log_2\left(n_T n_R / n_m\right) \leq n_m \log_2(n_T) \tag{3.14}$$

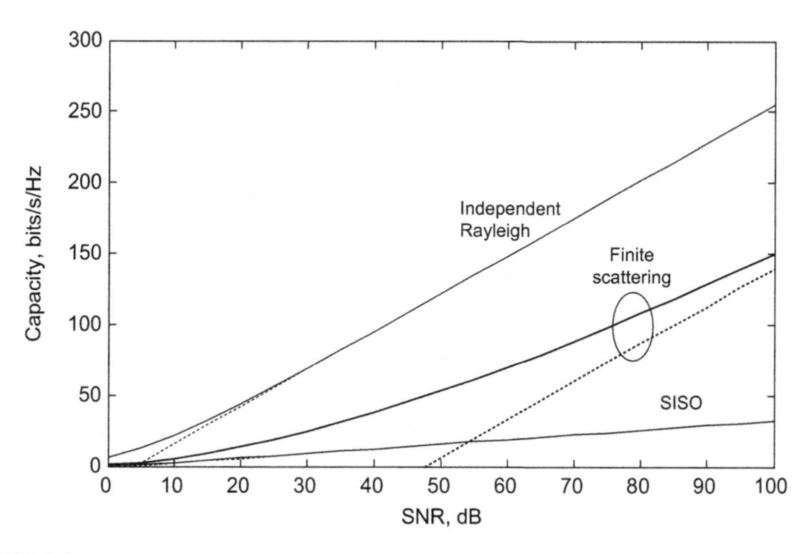

FIGURE 3.4

Capacity gain in 8×8 MIMO, showing capacity bounds for SISO and MIMO with i.i.d. Rayleigh and correlated fading channels, and asymptotes at high SNR (taken from [B09]).

provided $n_R \leq n_T$. Under these conditions, $n_m \log_2(S/N)$ provides an upper bound on the bandwidth efficiency.

The rank of $\mathbf{HH}^H$ is also limited by the number of discrete multipath components, n_S. It may readily be shown [Bur03] that it is also upper bounded by $n_S : n_m \leq \min(n_T, n_R, n_S)$. In general, correlation among the MIMO antennas will significantly reduce $\det\left(\mathbf{HH}^H\right)$, as illustrated in Figure 3.4, even if the rank is not reduced. The figure shows the capacity of an 8×8 MIMO uncorrelated Rayleigh channel, compared to a correlated MIMO and a SISO channel. The correlated MIMO channel is generated from a finite scattering channel with 16 scatterers and a total angular spread at the receiver of $\pi/6$. The effect of the correlation is to increase the spread of eigenvalues of the channel matrix, so that the smallest eigenvalue, and hence the determinant, is typically much smaller than for the uncorrelated case. The dashed lines show the asymptotes to the curves for the two MIMO channels. We note first that both show a multiplexing gain of $8\times$ compared to the SISO channel, demonstrated by the asymptotic slope of the curves. However, we also note that the asymptote for the correlated channel is much lower (some 100 bits/s/Hz lower) than the uncorrelated, and that the asymptote is only approached at around 100 dB SNR.

Observe that multiplexing gain, like diversity, is a property of the underlying channel, which can be determined by the application of information theory. It can also be related to physical features of the channel, in this case the number of independent subchannels that can be formed exploiting the transmit and receive antennas and the multiple transmission paths provided by the physical channel.

3.2.3 Diversity-multiplexing Trade-off

We have examined the fundamental benefits available from a MIMO channel—diversity and multiplexing—and shown that they are fundamentally properties of the underlying channel. However, it requires an appropriate transmission and detection scheme to realize these benefits in a communication system—such a scheme can in a general sense be described as a *space-time code,* as we will discuss in Section 3.3 below.

However, it is clear that the full benefits of diversity and multiplexing cannot be realized simultaneously. The diversity benefit assumes that the data rate is held constant and BER decreases as SNR increases, while multiplexing assumes that BER is held constant and data rate increases with SNR, by the use of increasingly bandwidth efficient modulation and coding schemes.

Zheng and Tse [ZT03] showed accordingly that it is in fact possible to trade one of these benefits against the other, and provide transmission schemes in which data rate increases as well as BER decreasing with SNR, albeit not as fast as in pure diversity or pure multiplexing schemes. They give an upper bound on the achievable diversity order n_d in terms of the multiplexing gain n_m provided by the transmission scheme:

$$n_d \leq (n_T - n_m)(n_R - n_m) \tag{3.15}$$

which applies provided the data is transmitted in blocks at least $n_T + n_R - 1$ symbols long, and providing $n_m \leq \min(n_R, n_T)$. As they suggest, it is as if n_m antennas at each end of the link must be used to provide the multiplexing gain, leaving the remaining antennas to provide diversity at each end. Figure 3.5 shows the trade-off for various numbers of transmit and receive antennas. It is important to realize that

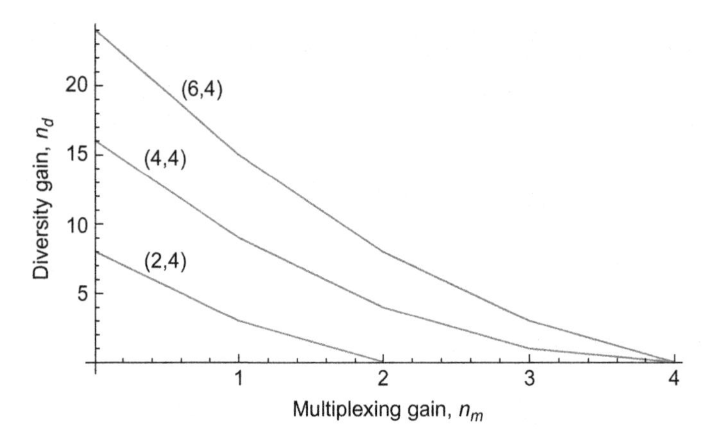

FIGURE 3.5

Diversity-multiplexing trade-off for various numbers of transmit and receive antennas (labeled in the form (n_T, n_R)).

zero multiplexing gain does not mean zero data rate, but rather that the rate does not change with SNR, because no adaptation is provided. Conversely, full multiplexing, with zero diversity, implies that an adaptive modulation and coding scheme is applied, so that BER remains constant with SNR while data throughput rate varies.

This concept of a trade-off is very simple and appealing, though its practical relevance is perhaps questionable for two reasons. The first is that it assumes both very high SNR, and ideal independent Rayleigh fading channels. The concepts of diversity and multiplexing both apply asymptotically as SNR tends to infinity, and as we have seen in practical channels the multiplexing gain especially may not reach its full asymptotic value until SNR is very high indeed.

Moreover, in such channels the full diversity and/or multiplexing order may not be available at all. Figure 3.6 shows the diversity-multiplexing trade-off extracted from measured channels, as described in [WC07a]. The dotted lines show the theoretical asymptotic trade-off described by [ZT03], the dashed lines the trade-off for i.i.d. Rayleigh channels at finite SNR, and the solid lines the trade-off applied to measured channels. The plots show that while the trade-off is close to the theory for 2×2 channels, for larger numbers of antennas there is a large deviation: in particular for the measured channels at low diversity order the theoretical multiplexing gain

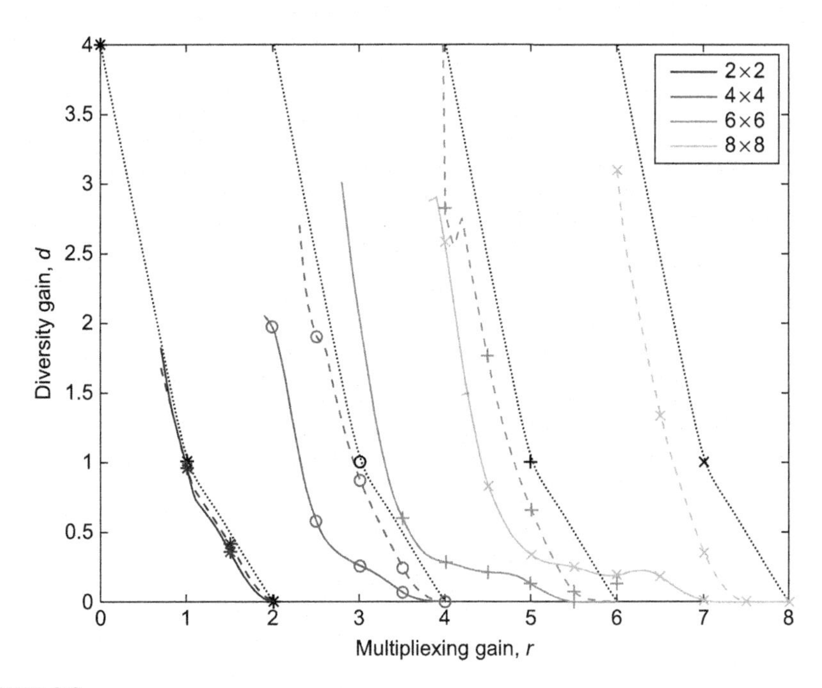

FIGURE 3.6

Div-mux trade-off for measured data (solid lines) and simulated i.i.d. complex Gaussian model (dashed lines) for SNR 30 dB, taken from [WC07b].

is not reached. At low diversity, even the i.i.d. channels show a significant loss of multiplexing gain.

The approach may also in any case have limited usefulness for system designers. Most systems are either required to transmit a fixed data rate at the lowest possible BER, which is usually obtained using full diversity, or at a certain maximum BER and the highest possible data rate, which is usually provided by a fully adaptive system, therefore which achieves maximum multiplexing gain. In most cases, there is little advantage to be had by a compromise between these cases, and space-time codes should be designed either to achieve full diversity or maximum multiplexing gain. We will consider next the criteria for code design in order to achieve full diversity.

3.3 THEORY OF SPACE-TIME CODING

3.3.1 Space-time Codes as a General Model of MIMO Transmission

The term *space-time code* is most commonly used to refer to a particular type of MIMO transmission scheme whose objective is to maximize diversity gain. However, in this section we will use the concept in a much more general sense, to apply to any MIMO transmission scheme. This will lead to criteria for the design of space-time codes in order to maximize diversity. These criteria have most commonly been applied to space-time codes in the more restrictive sense, but they also apply to MIMO schemes such as spatial multiplexing and precoding.

In the generalized sense, a space-time code is the generalization of a conventional code such as a Forward Error Correcting (FEC) code ([B01], Chapter 5). Such a code is defined as a set or *codebook* C of M distinct sequences of symbols, often binary but in wireless communication often drawn from a complex *constellation* representing modulated signals ([B01], Chapter 2). For a space-time code these symbols become length n_T complex vectors, whose elements represent the modulated signals transmitted on each transmit antenna. We will call these vector symbols *space-time symbols*, to distinguish them from the *modulation symbols* transmitted on each antenna, which are the elements of the vectors. The code sequences become arrays or matrices of complex modulation symbols. We can, as we will see later, define both block and convolutional space-time codes, but for simplicity in this section we will restrict our attention to block codes, consisting of n_w space-time symbols, and represented as a $n_T \times n_w$ array or matrix of modulation symbols.

In many cases, the constellation from which the modulation symbols are drawn is not defined, since the scheme will operate with any constellation. In such cases the code defines how the transmitted symbols are mapped to space-time codewords, rather than listing all possible codewords containing all possible constellation points. We can then represent a space-time encoder as shown in Figure 3.7, in which a block of k_{bin} bits is first modulated (and possibly also encoded using an FEC code or coded modulation scheme) to k_{sym} modulation symbols, then applied to a space-time mapper that maps these to a space-time codeword, represented as an $n_T \times n_w$ array of the modulation symbols.

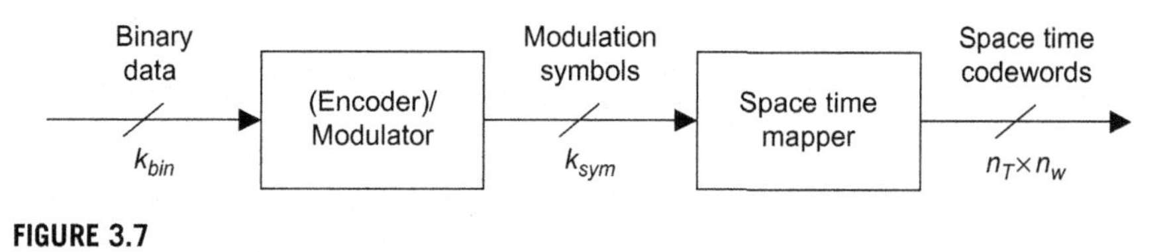

FIGURE 3.7

Structure of typical space-time encoder.

To achieve maximum multiplexing gain with such a space-time encoder, the encoder/modulator should be adaptive, choosing a modulation and coding scheme according to the channel SNR, such that the BER target is achieved at the receiver. (This of course requires a feedback channel from receiver to transmitter to provide channel state information (CSI) to the transmitter). Under these conditions, it is easy to see that the maximum multiplexing gain the scheme will support (assuming that the channel can support it) is k_{sym}/n_w, since the maximum information rate per codeword is proportional to k_{sym}, and there are n_w space time symbols per codeword. The data throughput rate in bits per space-time symbol is k_{bin}/n_w.

The diversity order achieved by a space-time code is however less straightforwardly determined. Here we will draw on a theoretical analysis based on our generalized concept of a space-time code to calculate the expected BER, and hence the diversity order. This will then lead to criteria for the design of space-time codes in order to maximize diversity.

3.3.2 Diversity of Space-time Codes

BER performance of a code depends on the distance properties of the code, as described in numerous texts on coding theory (see, for example [B01], Chapter 5). The same applies to space-time codes, with two differences: first, that the symbols are again the vector-valued space-time symbols referred to above (although their length is now the number of receive antennas, n_R); and second, that these parameters must be evaluated at the receiver rather than the transmitter, and thus are mediated by the MIMO channel.

Since the MIMO channel is random, so are the distance properties, which means that the BER is a function of the random channel, as well as being itself a statistical property due to the random noise at the receiver. In this chapter, we will assume a *quasi-stationary fading channel*; that is, that the channel can be treated as constant for the duration of at least one space-time codeword. However, we assume it varies randomly over a longer timescale, and analyze the statistics of the BER over this variation. In particular we will consider the average BER (sometimes called the *ergodic* BER), averaged over all random channels. What we may call the *outage BER*, that is, the BER exceeded only with a given probability—the outage probability—would also be of interest. The analysis below largely follows that of [TSC98].

To predict the BER performance of a code, we begin by considering the *Pairwise Error Probability* (PEP) $P_{ep}(\mathbf{B}|\mathbf{A})$: the probability that one given specific codeword

A will be decoded in error as another particular word, **B**. Knowing this for all pairs of codewords in the code allows us (in principle) to upper bound the average word error probability ([B01], Chapter 5):

$$P_{ew} = \frac{1}{M}\sum_{\mathbf{A}\in C} P\left(\bigcup_{\mathbf{B}\in C, \mathbf{B}\neq\mathbf{A}} \mathbf{A}\rightarrow\mathbf{B}\right) \leq \frac{1}{M}\sum_{\mathbf{A}\in C}\sum_{\mathbf{B}\in C, \mathbf{B}\neq\mathbf{A}} P_{ep}(\mathbf{B}|\mathbf{A}) \tag{3.16}$$

where $P\left(\bigcup_B \mathbf{A}\rightarrow\mathbf{B}\right)$ denotes the probability of the union of the events that codeword **A** is mis-decoded as any given **B**. Note that for most codes the inner summation is the same for all $\mathbf{A}\in C$, which greatly simplifies (3.16). The bound of (3.16) is called the *union bound*, and for most codes is reasonably tight to the actual BER, at least when the BER is low.

If the channel noise is Gaussian, the PEP in turn depends on the Euclidean distance $d_R(\mathbf{A},\mathbf{B})$ between the received words corresponding to **A** and **B**, according to:

$$P_{ep}(\mathbf{B}|\mathbf{A}) = Q\left(\frac{d_R(\mathbf{A},\mathbf{B})}{2\sigma}\right) \tag{3.17}$$

where σ is the standard deviation of the noise per dimension (i.e. real/imaginary part) on each antenna element, and Q is called the Q-function, defined as:

$$Q(z) = \frac{1}{\sqrt{2\pi}}\int_z^\infty \exp\left(-\frac{x^2}{2}\right) = \frac{1}{2}\text{erfc}\left(\frac{z}{\sqrt{2}}\right). \tag{3.18}$$

The Euclidean distance at the receiver corresponding to the two general codewords $\mathbf{A},\mathbf{B}$:

$$\begin{aligned} d_R^2(\mathbf{A},\mathbf{B}) &= \|\mathbf{HA}-\mathbf{HB}\|^2 = \|\mathbf{H}(\mathbf{A}-\mathbf{B})\|^2 = \|\mathbf{HD}\|^2 \\ &= \sum_{i=1}^{n_R}\sum_{m=1}^{n_w}\left|\sum_{k=1}^{n_T} H_{ik}D_{km}\right|^2 = \text{trace}\left(\mathbf{\Phi_{HH}\Phi_{DD}}\right) \\ &= \text{trace}\left(\mathbf{HH}^H\mathbf{V_D}^H\mathbf{\Lambda_D V}\right) = \text{trace}\left(\mathbf{VHH}^H\mathbf{V_D}^H\mathbf{\Lambda_D}\right) \\ &= \text{trace}\left(\mathbf{\Phi_{H'H'}\Lambda_D}\right) = \sum_{i=1}^{r}(\mathbf{\Phi_{H'H'}})_{ii}\lambda_{D,i} \end{aligned} \tag{3.19}$$

where $\mathbf{D}=\mathbf{A}-\mathbf{B}$ is the *codeword difference matrix*, and $\mathbf{\Phi_{DD}}=\mathbf{DD}^H$ is its correlation matrix, while $\mathbf{\Phi_{HH}}=\mathbf{H}^H\mathbf{H}$ is the transmit end channel correlation matrix. $\mathbf{V_D}^H\mathbf{\Lambda_D V}$ denotes the singular value decomposition of $\mathbf{\Phi_{DD}}$, where $\mathbf{V_D}$ is a unitary matrix and $\mathbf{\Lambda_D}$ is the diagonal eigenvalue matrix, with diagonal values $\lambda_{\mathbf{D},i}, i=1\ldots r$ being the eigenvalues of $\mathbf{\Phi_{DD}}$, and r its rank. Note that $\mathbf{H}'=\mathbf{V_D H}$ has the same statistics as **H**. The diagonal elements of $\mathbf{\Phi_{H'H'}}$ are given by:

$$(\mathbf{\Phi_{H'H'}})_{ii} = \sum_{i=1}^{n_R} H'^{*}_{ki} H'_{ki} = \sum_{i=1}^{n_R} |H'_{ki}|^2, \quad i = 1 \ldots r \tag{3.20}$$

Assuming **H** is independent Rayleigh fading, the distribution of these diagonal elements is chi-square with $2n_R$ degrees of freedom. Assuming further that the mean square magnitude of the elements of **H** is unity, the distribution should be scaled by 1/2 to maintain the mean, which is n_R.

We now return to calculating the PEP, from (3.17). Since the Q function is analytically intractable, and following [TSC98], we use the Chernoff bound ([P01], p. 55). (Note that the Chernoff bound is only one of several bounds that may be used; others are available which might give a tighter bound, but the Chernoff bound is straightforward to use.) It may be shown that:

$$Q(x) \leq \frac{1}{2} \exp\left(-\frac{x^2}{2}\right) \tag{3.21}$$

(Note that the factor 1/2 included here applies when $x \geq 0$). Then:

$$P_{ep}(\mathbf{B}|\mathbf{A}) \leq \frac{1}{2} \exp\left(-\frac{d_R^2(\mathbf{A},\mathbf{B})}{8\sigma^2}\right) = \frac{1}{2} \exp\left(-\frac{1}{8\sigma^2} \sum_{i=1}^{r} (\mathbf{\Phi_{H'H'}})_{ii} \lambda_{\mathbf{D},i}\right) = P_{ep}(\mathbf{D}) \tag{3.22}$$

For brevity of notation, we write the ith diagonal element of $\mathbf{\Phi_{HH}}$ as $x_i/2$, such that x_i has the Chi-square distribution. Then the mean PEP:

$$\begin{aligned} \overline{P_{ep}(\mathbf{D})} &\leq \int_0^\infty \ldots \int_0^\infty \exp\left(-\frac{1}{8\sigma^2} \sum_{i=1}^{r} \frac{\lambda_{D,i} x_i}{2}\right) p(x_1) \ldots p(x_r)\, d_{x_r} \ldots dx_1 \\ &= \int_0^\infty \ldots \int_0^\infty \prod_{i=1}^{r} \exp\left(-\frac{\lambda_{D,i} x_i}{16\sigma^2}\right) p(x_1) \ldots p(x_r)\, d_{x_r} \ldots dx_1 \\ &= \prod_{i=1}^{r} \int_0^\infty \exp\left(-\frac{\lambda_{D,i} x_i}{16\sigma^2}\right) p(x_i)\, dx_i = \prod_{i=1}^{r} \overline{\exp\left(-\frac{\lambda_{D,i} x_i}{16\sigma^2}\right)} \end{aligned} \tag{3.23}$$

where the overbar denotes mean. It can be shown that if x has the Chi-square distribution with ν degrees of freedom then $\overline{\exp(-\alpha x)} = (2\alpha + 1)^{-\nu/2}$. Then:

$$\overline{P_{ep}(\mathbf{D})} \leq \prod_{i=1}^{r} \left(1 + \frac{\lambda_{\mathbf{D},i}}{8\sigma^2}\right)^{-n_R}. \tag{3.24}$$

Asymptotically, as SNR $\rightarrow \infty$ this becomes:

$$\overline{P_{ep}(\mathbf{D})} \leq \prod_{i=1}^{r} \left(\frac{\lambda_{\mathbf{D},i}}{8\sigma^2} \right)^{-n_R} = \frac{\left(8\sigma^2\right)^{n_R r}}{\left(\det\left(\mathbf{D}\mathbf{D}^H\right)\right)^{n_R}} \tag{3.25}$$

since the determinant is the product of the eigenvalues of a matrix.

The exponent of σ^2 in (3.25) shows that the diversity order achieved, at least in respect of one pairwise error probability, is $n_R r$. Asymptotically at high SNR the BER will certainly be dominated by the PEP corresponding to the smallest diversity order, that is to the codeword difference matrix with the smallest rank over all possible pairs of codewords.

Second, for codeword difference matrices with the same rank those with the smallest determinant will tend to dominate the BER. Since the determinant is the product of the eigenvalues, if one eigenvalue is small, the determinant will also be small. Since the sum of the eigenvalues is given by the sum of squared magnitudes in the codeword difference matrix, for a given number of entries in the matrix (which we refer to as its *Hamming weight*) the determinant will be maximized if the eigenvalues are as nearly as possible the same, a condition we refer to as *balanced eigenvalues*. Note also that difference matrices with minimum determinant are likely to correspond to those with relatively small Hamming weight, though not always the minimum Hamming weight.

However, we should remember that (3.25) applies only in the limit of high SNR. In (3.24) we observe that terms corresponding to small eigenvalues approximate to unity for $\sigma^2 \gg \lambda_{\mathbf{D},i}/8$. Hence the asymptotic diversity order is reached only above the SNR corresponding to $\sigma^2 \gg \min_i(\lambda_{\mathbf{D},i})/8$, which for unbalanced eigenvalues may be relatively high.

Finally, referring to (3.16), we note that the overall word error probability depends also on the number of codeword pairs having the minimum rank and determinant, known as the *multiplicity*.

3.3.3 Design Criteria

These factors determining the BER of a space-time code also provide criteria for the design of codes. They should be applied in order, from the most important (in the sense that they have most effect on the BER) to the least.

We have noted that, at least at high SNR, the most important factor is the diversity order, which is given by the minimum rank of codeword difference matrix corresponding to any pair of codewords. This is known as the *rank criterion* [TSC98], and should normally be applied first.

However, at lower SNR the effective diversity is determined by the number of eigenvalues of $\mathbf{\Phi}_{\mathbf{DD}}$ greater than $8\sigma^2$. Under these conditions, the rank criterion should be replaced by what we may call a *significant eigenvalues criterion*. The number of eigenvalues greater than this threshold.

The second most important factor, for high SNR, is the determinant of $\mathbf{\Phi_{DD}}$, so the second criterion to apply is the *determinant criterion*: select the code whose minimum rank codeword difference has maximum determinant [TSC98]. At lower SNR, the determinant should be replaced by the product of the eigenvalues greater than the threshold $8\sigma^2$: This could be called the *product of significant eigenvalues criterion*.

The determinant criterion and the significant eigenvalues criterion both depend on avoiding small eigenvalues of $\mathbf{\Phi_{DD}}$. Hence, they are encompassed by the *balanced eigenvalues criterion*: codes should be selected so that $\mathbf{\Phi_{DD}}$ corresponding to all pairs of codewords has eigenvalues as nearly equal as possible. However, as a means of selecting codes this is less precise than the other criteria we have described.

Finally, the multiplicity should be minimized for codeword pairs with minimum rank and determinant. This may be called the *multiplicity criterion*.

A common goal for space-time codes is to achieve *full diversity*; that is, the maximum diversity available from the channel, namely $n_T n_R$ (although in the presence of restricted multipath the actual diversity available may be less than this, as we have seen). The rank criterion shows that this requires that the codeword difference matrix has full rank, n_T. Note that it also requires that the codeword length n_w be at least n_T space-time symbols. Full diversity is in particular the main objective of conventional space-time codes, which we consider next.

3.4 SPACE-TIME CODES

Historically, space-time codes as usually described arose from attempts to achieve *transmit diversity*; that is, to exploit multiple antennas at the transmit end of a link, as opposed to the receive end of the link, which was well known. This is of course of particular importance on the downlink of mobile radio systems, since it is easier to provide multiple antennas at the base station than on a mobile handset. Some early approaches to the provision of transmit diversity were described in the early 1990s by Wittneben [W93] and by Seshadri and Winters [SW94], which depend on transmitting different versions of the data (in the latter case a delayed version) from different antennas, and using signal processing possibly with the aid of FEC coding to separate the signals at the receiver. Subsequently, Tarokh [TSC98] formalized the concept of a space-time code, describing the criteria we have considered above, and introduced space-time trellis codes. At the same time, Alamouti [Ala98] described what was in fact the first space-time block code, after which Tarokh [TJC99] formalized and extended this concept also, as we will see below. In all this work, the primary goal was to achieve the maximum diversity available, with a secondary goal to achieve what in this context is described as *full rate*; that is, one data symbol per channel use, corresponding to unit multiplexing gain. As we have already seen, higher rates and higher multiplexing gains are in fact possible, so that the term full rate is in a sense misleading.

3.4.1 STTC

The first space-time codes to be developed based on the code design criteria outlined in Section 3.3 were in fact trellis codes. The criteria, although they are couched in terms of block codes, can readily be applied also to trellis codes, as we will see. They were introduced in the same paper [TSC98] that described the rank and determinant criteria.

Trellis codes (or *convolutional codes* as they are known when they are simple binary codes) are based on an encoder having memory [B01], Chapter 7. The current output block (for STTC usually one symbol long) depends not only on the current data block (length k symbols), but also on several previous blocks (the total number of blocks, including the current one, being the *constraint length*, ν). Typically, the blocks are very short, often including only one or two data bits. The encoder is implemented as a shift register. An example is shown in Figure 3.8, which shows the encoder for an example STTC with constraint length $\nu = 2$ and input block length $k = 2$. Each output space-time symbol depends on the current input block (two bits, entering the shift register in parallel), plus the two previous bits from the previous symbol period, stored in the two parallel shift register elements. These are fed to a space-time modulator, which maps the four resulting bits to two modulated symbols to be transmitted on two antennas.

The space-time modulator can be described by means of a *generator matrix* like that for conventional FEC codes. If we construct a vector of data bits in the form:

$$\mathbf{d} = \left[d_i^0, d_i^1 \ldots d_i^{k-1}, d_{i-1}^0 \ldots d_{i-1}^{k-1}, \ldots d_{i-\nu+1}^0, \ldots d_{i-\nu+1}^{k-1} \right] \quad (3.26)$$

then the transmitted symbols are given by:

$$\mathbf{s} = \text{Mod}(\mathbf{dG})^T \quad (3.27)$$

where Mod(.) here denotes the modulation and $\mathbf{G}$ is the generator matrix. The elements of $\mathbf{G}$ are M_c-ary, where M_c is the number of points in the modulation constellation, and the matrix multiplication in (3.27) is carried out in modulo-M_c

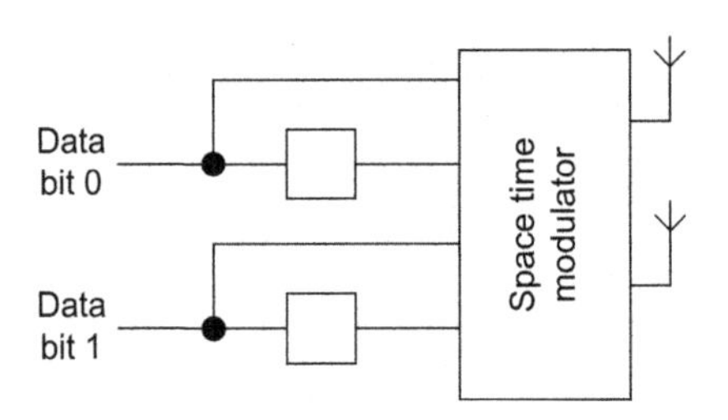

FIGURE 3.8

Example STTC encoder: data block length $k = 2$; constraint length $\nu = 2$.

arithmetic. In our example code, which is in fact one due to Tarokh in [TSC98], the generator matrix:

$$\mathbf{G} = \begin{bmatrix} 2 & 0 \\ 1 & 0 \\ 0 & 2 \\ 0 & 1 \end{bmatrix} \tag{3.28}$$

and the modulation is QPSK, with constellation points numbered in natural order, so that the point numbers correspond to complex modulated signals as shown in Figure 3.9.

A trellis code, as the name suggests, can be represented by a *trellis diagram*. The trellis diagram for our example code is given in Figure 3.10, which shows all the possible states of the encoder and all the possible state transitions as a function of time. Each column of the diagram represents a block (input bit) period, while each node represents the state of the encoder at the end of the input block period, and the branches show the permitted transitions between them. The branches are labeled according to the constellation point to be transmitted from each transmit antenna. Note that to avoid difficulty labeling the diagram the branch labels have been placed on the left of the diagram, adjacent to the state labels. They take the form of a list

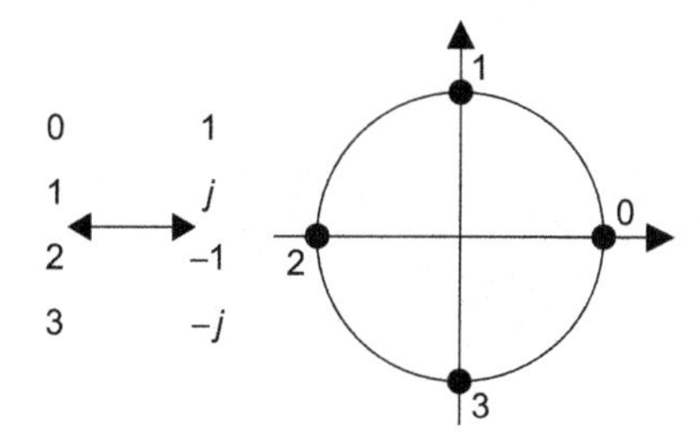

FIGURE 3.9

QPSK constellation mapping for example code.

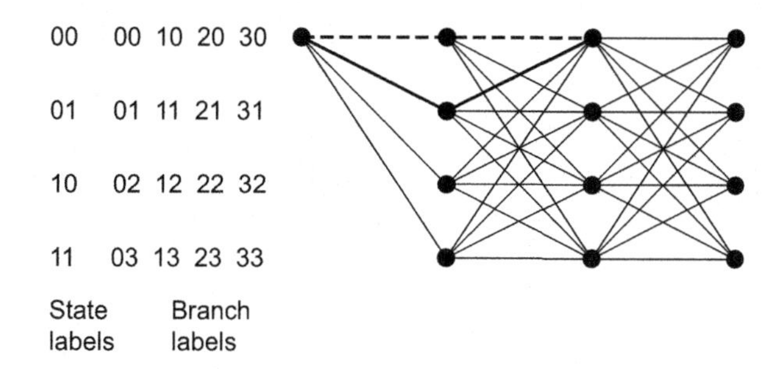

FIGURE 3.10

Trellis for example space-time trellis code.

of the labels, which apply to the branches diverging from that state, from top to bottom. For the correspondence of input data to branches diverging from a state, we assume natural numbering from top to bottom: that is, the branches correspond to data 00, 01, 10 11.

The value of the trellis diagram lies in the fact that every valid code sequence corresponds to a path through the trellis. Unlike block codes, the output of a trellis encoder cannot be divided into independent blocks: we must deal with potentially infinite code sequences instead. The value of the trellis diagram lies in the fact that every valid code sequence corresponds to a path through the trellis. The decoder must find the sequence or path closest to the received signal, which is most likely to be the correct transmitted sequence. This is known as *maximum likelihood* (ML) decoding, since it returns the code sequence most likely to have been transmitted, given the received signal. An *error event* occurs when the decoder compares two candidate sequences corresponding to a pair of paths that begin and end on the same state, and selects the wrong one. The *error event probability* is the equivalent of the pairwise error probability in block codes, and can be bounded in the same way as described in Section 3.2. The code design criteria from Section 3.3 also can be applied to all such pairs of sequences in the trellis, in the same way as to all pairs of codewords in a block code.

Consider, for example, the pair of paths in the trellis of Figure 3.10 marked with dashed and heavy lines, respectively. The dashed line is the all zeros sequence, corresponding to data 00, 00..., while the heavy line corresponds to data 01, 00... The dashed line also generates a string of 0 constellation symbols, i.e., signals +1, +1,... The data vectors **d** and modulated signals **s** corresponding to the heavy line are:

$$\mathbf{d} = \begin{bmatrix} 0 & 1 & 0 & 0 \\ 0 & 0 & 0 & 1 \end{bmatrix}; \ \mathbf{dG} = \begin{bmatrix} 1 & 0 \\ 0 & 1 \end{bmatrix}; \ \mathbf{s} = \begin{bmatrix} j & 1 \\ 1 & j \end{bmatrix} \tag{3.29}$$

which correspond also to the branch labels in Figure 3.10. The codeword difference matrix corresponding to this pair of sequences, and its correlation matrix, is:

$$\mathbf{D} = \begin{bmatrix} 1-j & 0 \\ 0 & 1-j \end{bmatrix}; \ \mathbf{DD}^H = \begin{bmatrix} 2 & 0 \\ 0 & 2 \end{bmatrix} \tag{3.30}$$

This has rank 2 and determinant 4, which corresponds to full diversity. From Figure 3.10 we note that on the path marked constellation symbol 1 (corresponding to data 01) is transmitted twice, once from the first antenna and once from the second. An error event is likely to occur only if the channels from both antennas are faded, which indicates that second order diversity should indeed be achieved.

We should use the same approach to search for the minimum rank and determinant of all codeword difference matrices corresponding to such pairs of paths. Because of the structure of the encoder and the linear generator matrix approach used to encode it, in fact we need only compare the all zero path (the dashed line and its extension) with all other paths which branch from it and return to it later in the trellis. It is quite easy to establish that all such paths give rank 2, and none gives a determinant less than 4. All in effect involve repeating a data symbol on each antenna. Hence, this code does achieve full diversity, 2.

The drawback of space-time trellis codes is that for larger numbers of antennas to achieve full diversity requires the encoder to have longer memory (a longer shift register), and the trellis to contain exponentially more states. Since decoding complexity is proportional to the number of trellis states and branches, this tends to become excessively computationally expensive.

3.4.2 STBC

Space-time block codes, as the name suggests, are block rather than trellis-based. In their best known form they avoid the complexity problem mentioned above, since they have a maximum likelihood decoder which is very simple. They take the approach illustrated in Figure 3.7, and consist of a mapping scheme that maps modulated symbols to transmit antennas.

As mentioned above, the earliest space-time block code (though not called by that name) was described by Alamouti [Ala98], and is commonly called the *Alamouti scheme*. It provided full (second order) diversity over two transmit antennas, and was also able to transmit at what has been described as "full rate," that is, one modulation symbol per space time symbol. (Of course, "full rate" is not the maximum throughput rate possible in a MIMO system, as we shall see—in fact, it corresponds to a multiplexing gain of 1.) In addition, the scheme allowed the simple ML decoder previously mentioned.

The Alamouti scheme was subsequently generalized by Tarokh [TJC99], who coined the term *space-time block code*. He showed that the desirable properties of the Alamouti scheme arose because the transmit signal matrix is an orthogonal matrix, regardless of the modulated data symbols being transmitted. His generalization extended this by means of generalized *complex orthogonal designs*, which map a set of arbitrary complex numbers representing the modulated data to an orthogonal matrix. He also showed that unfortunately such complex orthogonal designs provide full rate only for two antennas: the maximum rate for more than this is less than one data symbol per space-time symbol.

For simplicity, we will illustrate the scheme with reference to the Alamouti scheme. The transmitted signal matrix for a code block transmitting two modulated data symbols, s_1 and s_2 is:

$$\mathbf{S}_A = \begin{bmatrix} s_1 & s_2^* \\ s_2 & -s_1^* \end{bmatrix} \tag{3.31}$$

(Notice this matrix is the transpose of that given by Alamouti and Tarokh—the rows correspond to space-time symbol periods and the columns to antennas. Here, for consistency, we use the more common notation for MIMO in general.) We note that:

$$\mathbf{S}_A\mathbf{S}_A^H = \begin{bmatrix} s_1 & s_2^* \\ s_2 & -s_1^* \end{bmatrix}\begin{bmatrix} s_1^* & s_2^* \\ s_2 & -s_1 \end{bmatrix} = \begin{bmatrix} |s_1|^2+|s_2|^2 & 0 \\ 0 & |s_1|^2+|s_2|^2 \end{bmatrix} \tag{3.32}$$

that is, $\mathbf{S}_A$ is orthogonal, independent of s_1 and s_2, Hence, the codeword difference matrix $\mathbf{D}=\mathbf{S}_A-\mathbf{S}_A'$ is also orthogonal for all possible codeword pairs $\mathbf{S}_A,\mathbf{S}_A'$. Also, $\mathbf{DD}^H$ is diagonal, with diagonal entries $|s_1-s_1'|^2+|s_2-s_2'|^2$, which is full rank, for all $\mathbf{S}_A,\mathbf{S}_A',\mathbf{S}_A\neq\mathbf{S}_A'$. The minimum determinant clearly arises if one of the two modulated symbols is the same in $\mathbf{S}_A$ and $\mathbf{S}_A'$, and the other has the minimum distance in the constellation $d_{\min}$ between them, in which case the determinant is $d_{\min}^4$. Hence the scheme has full diversity. We also note that the transmission matrix ensures that each modulated symbol, s_1, s_2 is transmitted on each antenna, and hence we can expect it to be received provided not more than one of the antennas is subject to fading—this also suggests second order diversity.

It is convenient to note that when this transmit matrix is sent over a 2×1 MISO channel with channel matrix $\mathbf{H}=\begin{bmatrix}h_1 & h_2\end{bmatrix}$, the received signal is:

$$\begin{aligned}\mathbf{R}=\begin{bmatrix}r_1 & r_2\end{bmatrix}=\mathbf{HS}_A+\mathbf{N}&=\begin{bmatrix}h_1 & h_2\end{bmatrix}\begin{bmatrix} s_1 & s_2^* \\ s_2 & -s_1^* \end{bmatrix}+\begin{bmatrix}n_1 & n_2\end{bmatrix}\\ &=\begin{bmatrix}h_1s_1+h_2s_2 & h_1s_2^*-h_2s_1^*\end{bmatrix}+\begin{bmatrix}n_1 & n_2\end{bmatrix}\end{aligned} \tag{3.33}$$

Defining $\mathbf{R}'=\begin{bmatrix} r_1 \\ r_2^* \end{bmatrix}$ we have:

$$\mathbf{R}'=\begin{bmatrix} h_1s_1+h_2s_2 \\ h_1^*s_2-h_2^*s_1 \end{bmatrix}+\begin{bmatrix} n_1 \\ n_2^* \end{bmatrix}=\begin{bmatrix} h_1 & h_2 \\ h_1^* & -h_2^* \end{bmatrix}\begin{bmatrix} s_1 \\ s_2 \end{bmatrix}+\begin{bmatrix} n_1 \\ n_2^* \end{bmatrix}=\mathbf{H}'\mathbf{s}+\mathbf{N}' \tag{3.34}$$

That is, the combination of the space time encoder and the channel can be treated as an equivalent channel matrix $\mathbf{H}'$. Note in particular that this matrix is also orthogonal:

$$\mathbf{H}'^H\mathbf{H}'=\begin{bmatrix} |h_1|^2+|h_2|^2 & 0 \\ 0 & |h_1|^2+|h_2|^2 \end{bmatrix} \tag{3.35}$$

The ML receiver finds the data symbol vector $\hat{\mathbf{s}}$ that minimizes the total squared distance between the corresponding received signal $\mathbf{H}'\hat{\mathbf{s}}$ and the actual received signal matrix $\mathbf{R}'$. Formally:

$$\hat{\mathbf{s}}=\arg\min_{\tilde{\mathbf{s}}}\left(\|\mathbf{H}'\tilde{\mathbf{s}}-\mathbf{R}'\|^2\right). \tag{3.36}$$

The argument can be expanded:

$$\begin{aligned}\|\mathbf{H}'\tilde{\mathbf{s}}-\mathbf{R}'\|^2 &= (\mathbf{H}'\tilde{\mathbf{s}}-\mathbf{R}')^H(\mathbf{H}'\tilde{\mathbf{s}}-\mathbf{R}')\\ &= \tilde{\mathbf{s}}^H\mathbf{H}'^H\mathbf{H}'\tilde{\mathbf{s}}+\mathbf{R}'^H\mathbf{R}'-\mathbf{R}'^H\mathbf{H}'\tilde{\mathbf{s}}-\tilde{\mathbf{s}}^H\mathbf{H}'^H\mathbf{R}'\\ &= \left(|r_1|^2+|r_2|^2\right)+\left(|\tilde{s}_1|^2+|\tilde{s}_2|^2\right)\left(|h_1|^2+|h_2|^2\right)-2\mathrm{Re}\left[\tilde{\mathbf{s}}^H\mathbf{H}'^H\mathbf{R}'\right]\end{aligned} \tag{3.37}$$

Writing $\mathbf{H}'^H\mathbf{R}'=\mathbf{x}=\begin{bmatrix}x_1 & x_2\end{bmatrix}^T$

$$\mathrm{Re}\left[\tilde{\mathbf{s}}^H\mathbf{H}'^H\mathbf{R}'\right]=\mathrm{Re}\left[\tilde{s}_1^*x_1\right]+\mathrm{Re}\left[\tilde{s}_2^*x_2\right] \tag{3.38}$$

Then:

$$\begin{aligned}\hat{\mathbf{s}} &= \arg\min_{\tilde{\mathbf{s}}}\left(\|\mathbf{H}'\tilde{\mathbf{s}}-\mathbf{R}'\|^2\right)\\ &= \arg\min_{\tilde{\mathbf{s}}}\left(\left(|r_1|^2+|r_2|^2\right)+\left(|\tilde{s}_1|^2+|\tilde{s}_2|^2\right)\left(|h_1|^2+|h_2|^2\right)-\mathrm{Re}\left[\tilde{s}_1^*x_1+\tilde{s}_2^*x_2\right]\right)\\ &= \arg\min_{\tilde{\mathbf{s}}}\begin{pmatrix}\left(|r_1|^2+|r_2|^2\right)\\ +\,|\tilde{s}_1|^2\left(|h_1|^2+|h_2|^2\right)-2\mathrm{Re}\left[\tilde{s}_1^*x_1\right]\\ +\,|\tilde{s}_2|^2\left(|h_1|^2+|h_2|^2\right)-2\mathrm{Re}\left[\tilde{s}_2^*x_2\right]\end{pmatrix}\end{aligned} \tag{3.39}$$

Since this argument consists of a term which is independent of $\tilde{s}_1$ and $\tilde{s}_2$, a term dependent only on $\tilde{s}_1$ and another dependent only on $\tilde{s}_2$, we can minimize it for $\tilde{s}_1$ and $\tilde{s}_2$ separately, requiring only $2M$ comparisons instead of M^2. Effectively the operation is reduced to the equivalent of two simple demodulation operations.

Moreover, if we expand the variable $\mathbf{x}$ defined above:

$$\mathbf{x}=\mathbf{H}'^H\mathbf{R}'=\mathbf{H}'^H\left(\mathbf{H}'\mathbf{s}+\mathbf{N}'\right)=\left(|h_1|^2+|h_2|^2\right)\mathbf{s}+\mathbf{H}'^H\mathbf{N}'. \tag{3.40}$$

Thus, $\mathbf{x}$ provides a decision variable for $\mathbf{s}$, corrupted by Gaussian noise of power $\left(|h_1|^2+|h_2|^2\right)N$, where N is the received noise power per antenna. It can be shown [TJC99] that the signal-to-noise ratio is not affected by the transformation. Hence, the data can be recovered without loss by forming the decision variable $\mathbf{x}$ by a simple linear transformation, then applying a conventional demodulator to x_1 and x_2.

As already mentioned, Tarokh et al. [TJC99] generalized the Alamouti scheme to more than two antennas, in the sense of providing larger matrices with the same orthogonality properties. Full rate designs do not however exist for this case, but Tarokh found rate 1/2 codes for three and four antennas based on the following designs:

$$\mathbf{S}_{T3}=\begin{bmatrix}s_1 & -s_2 & -s_3 & -s_4 & s_1^* & -s_2^* & -s_3^* & -s_4^*\\ s_2 & s_1 & s_4 & -s_3 & s_2^* & s_1^* & s_4^* & -s_3^*\\ s_3 & -s_4 & s_1 & s_2 & s_3^* & -s_4^* & s_1^* & s_2^*\end{bmatrix} \tag{3.41}$$

$$\mathbf{S}_{T4} = \begin{bmatrix} s_1 & -s_2 & -s_3 & -s_4 & s_1^* & -s_2^* & -s_3^* & -s_4^* \\ s_2 & s_1 & s_4 & -s_3 & s_2^* & s_1^* & s_4^* & -s_3^* \\ s_3 & -s_4 & s_1 & s_2 & s_3^* & -s_4^* & s_1^* & s_2^* \\ s_4 & s_3 & -s_2 & s_1 & s_4^* & s_3^* & -s_2^* & s_1^* \end{bmatrix} \tag{3.42}$$

However, these have rate only 1/2. They also described codes with rate 3/4, but Ganesan and Stoica [GS01] found simpler codes:

$$\mathbf{S}_{G4} = \begin{bmatrix} s_1 & 0 & s_2 & -s_3 \\ 0 & s_1 & s_3^* & s_2^* \\ -s_2^* & -s_3 & s_1^* & 0 \\ s_3^* & -s_2 & 0 & s_1^* \end{bmatrix} \tag{3.43}$$

$$\mathbf{S}_{G3} = \begin{bmatrix} s_1 & 0 & s_2 & -s_3 \\ 0 & s_1 & s_3^* & s_2^* \\ -s_2^* & -s_3 & s_1^* & 0 \end{bmatrix} \tag{3.44}$$

Tarokh also shows that rate 1/2 codes exist for any number of antennas.

All these codes are capable of providing full diversity, of order $n_T n_R$, since their orthogonality means that the argument of (3.32) applies equally here. Orthogonality also means that ML detection can be implemented in the same simple linear form as for the Alamouti scheme. Specifically we can define an equivalent channel matrix $\mathbf{H}'$ as in (3.34), which can then be used for linear detection followed by separate demodulation of the modulation symbols. We will show in Section 3.6.3 how this may be done in general.

Their multiplexing gain, however, is given by the number of distinct modulation symbols transmitted per symbol period, which is the same as the rate of the code in the sense we have used the term in this section, and is therefore never greater than 1, and in fact is less than 1 for all codes except the Alamouti scheme. Since Tarokh's results have shown that even rate 1 cannot be achieved using strictly orthogonal transmission matrices, subsequent researchers have considered relaxing the orthogonality constraint.

In particular Jafarkhani [J01] proposed *quasi-orthogonal space-time block codes* (QO-STBC). He describes a rate 1 code for 4 transmit antennas whose transmission matrix is:

$$\mathbf{S}_{J4} = \begin{bmatrix} s_1 & -s_2^* & -s_3^* & s_4 \\ s_2 & s_1^* & -s_4^* & -s_3 \\ s_3 & -s_4^* & s_1^* & -s_2 \\ s_4 & s_3^* & s_2^* & s_1 \end{bmatrix} \tag{3.45}$$

Then:

$$\mathbf{S}_{J4}\mathbf{S}_{J4}^H = \begin{bmatrix} \begin{matrix}|s_1|^2+|s_2|^2\\+|s_3|^2+|s_4|^2\end{matrix} & 0 & 0 & \begin{matrix}2\mathrm{Re}\left[s_1^*s_4\right]\\-2\mathrm{Re}\left[s_2s_3^*\right]\end{matrix} \\ 0 & \begin{matrix}|s_1|^2+|s_2|^2\\+|s_3|^2+|s_4|^2\end{matrix} & \begin{matrix}2\mathrm{Re}\left[s_2^*s_3\right]\\-2\mathrm{Re}\left[s_1s_4^*\right]\end{matrix} & 0 \\ 0 & \begin{matrix}2\mathrm{Re}\left[s_2^*s_3\right]\\-2\mathrm{Re}\left[s_1s_4^*\right]\end{matrix} & \begin{matrix}|s_1|^2+|s_2|^2\\+|s_3|^2+|s_4|^2\end{matrix} & 0 \\ \begin{matrix}2\mathrm{Re}\left[s_1^*s_4\right]\\-2\mathrm{Re}\left[s_2s_3^*\right]\end{matrix} & 0 & 0 & \begin{matrix}|s_1|^2+|s_2|^2\\+|s_3|^2+|s_4|^2\end{matrix} \end{bmatrix} \quad (3.46)$$

The matrix is not unconditionally orthogonal: for some codewords, there are nonzero values on the anti-diagonal. For example if $s_1 = s_4$ and $s_2 = -s_3$, then the matrix of (3.16) becomes:

$$4\begin{bmatrix} 1 & 0 & 0 & 1 \\ 0 & 1 & -1 & 0 \\ 0 & -1 & 1 & 0 \\ 1 & 0 & 0 & 1 \end{bmatrix}$$

which has rank 2, and hence the diversity order of the scheme is $2n_R$: full diversity is not achieved. Moreover, the simple separable ML decoder described by (3.16) is also not available, because the equivalent channel matrix $\mathbf{H}'$ resulting from this scheme is also not orthogonal. However, it is possible [J01] to separate the ML detection into two terms to be separately minimized, in which each depend on two of the symbols, which greatly reduces complexity compared to full ML decoding.

It is, however, possible to design space-time block codes using nonorthogonal designs which do achieve full diversity, albeit not unconditionally for any constellation. One example is obtained by using two different constellations in the QO-STBC of (3.45), one for symbols s_1 and s_2, and the other for s_3 and s_4. Under these circumstances, the matrix of (3.46) will maintain full rank, and hence full diversity, although the eigenvalues may still be somewhat unbalanced.

3.5 SPATIAL MULTIPLEXING

Conversely to space-time codes, spatial multiplexing arose from the attempt to achieve maximum multiplexing gain, and hence maximum capacity. It was originally described by Foschini et al. [Fos96] at Bell Labs, and was termed BLAST (for Bell Labs Layered Space Time). Two years later Bell Labs issued a press release entitled "Bell Labs Scientists Shatter Limit on Fixed Wireless Transmission" [BL10]. Of course, the advertised work did not breach the Shannon bound to which it referred—as we have already seen, the Shannon bound is fundamental to determining the capacity of MIMO systems—and of course the reference is somewhat ironical, since Shannon was working at Bell Labs when he described his bound.

The fundamental concept of BLAST is to demultiplex the data stream into substreams, referred to as *layers*, one per transmit antenna. There are several flavors of BLAST depending on exactly how these are mapped to the antennas. Also the layers may be encoded with an FEC code per layer before the mapping. The mapping is conveniently described with reference to an array the rows of which represent the symbols transmitted on the same antenna, while the columns contain symbols transmitted at the same time, on different antennas.

3.5.1 V-BLAST

The simplest scheme, though historically not the first to be described, is known as V-BLAST [GFVW99], the V denoting "vertical." Figure 3.11 shows the structure of the transmitter and receiver, and Figure 3.12 shows the array representation mentioned above.

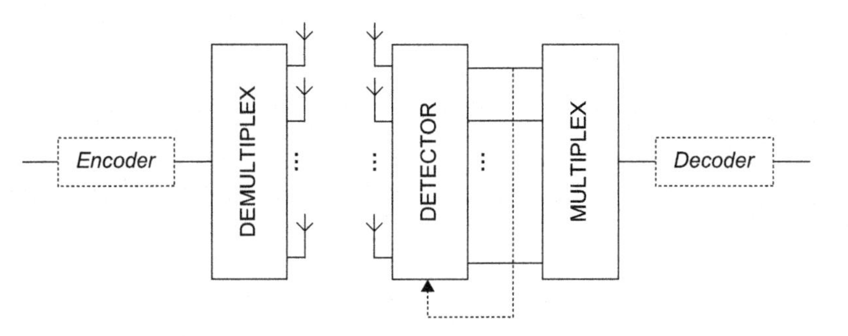

FIGURE 3.11

Structure of V-BLAST transmitter and receiver.

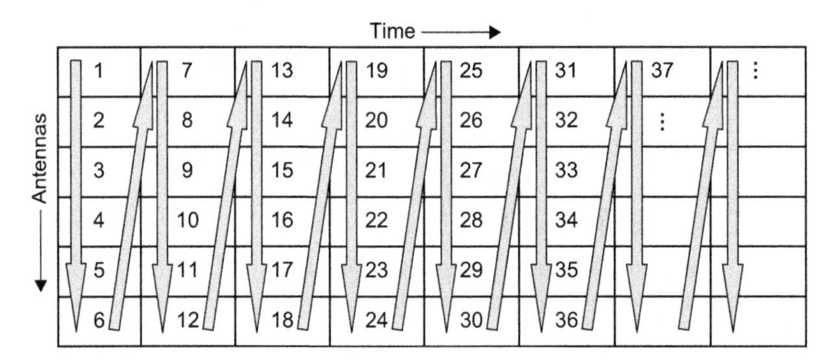

FIGURE 3.12

Array description of V-BLAST scheme.

The numbers in the array of Figure 3.12 show the order of data symbols in the original encoded stream. It will be observed that the stream is mapped vertically onto successive columns of the transmission array: hence the term "vertical." However, the scheme is also the most straightforward way of multiplexing a single stream in space over multiple antennas; hence, the term *spatial multiplexing*.

3.5.2 D-BLAST

D-BLAST (D for "diagonal") was in fact the earliest form of spatial multiplexing proposed, being the form referred to in the Bell Labs press release mentioned above, and described in [Fos96]. D-BLAST uses the layering concept as described above, with independent FEC codes applied to each layer, as illustrated in Figure 3.13. There are as many layers as transmit antennas; however, the layers are not directly transmitted one per antenna, but are cyclically reordered so that only a small number of successive bits are transmitted on a given antenna. The mapping of the layers (here labeled 1-6) to the transmit array mentioned above is shown in Figure 3.14. The advantage of this mapping compared to the simpler V-BLAST mapping will become clearer in Section 3.5.4. However, Figure 3.13, compared to Figure 3.11, shows that the decoder required is more complex than for V-BLAST, requiring the layer decoders to

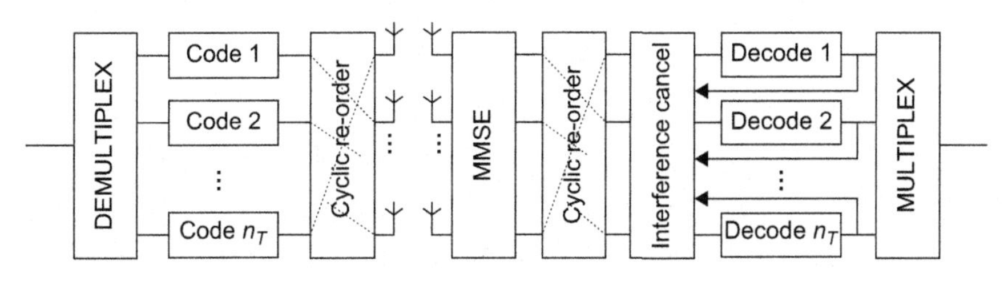

FIGURE 3.13

D-BLAST encoder and decoder (taken from [BZH04]).

Time →

Antennas ↓

1	6	5	4	3	2	1	6
2	1	6	5	4	3	2	1
3	2	1	6	5	4	3	2
4	3	2	1	6	5	4	3
5	4	3	2	1	6	5	4
6	5	4	3	2	1	6	5

FIGURE 3.14

Transmission of D-BLAST layers 1-6 in transmit array (taken from [BZH04]).

Time →

Antennas ↓

1	13	25	37	49	61	7	19	31	43	55	67
68	2	14	26	38	50	62	8	20	32	44	56
57	69	3	15	27	39	51	63	9	21	33	45
46	58	70	4	16	28	40	52	64	10	22	34
35	47	59	71	5	17	29	41	53	65	11	23
24	36	48	60	72	6	18	30	42	54	66	12

FIGURE 3.15

"D-BLAST like" mapping order of modulated symbols onto transmission array (numbers are symbol orders in the encoded sequence) (taken from [BZH04]).

be integrated into the detector, which motivated the later development of V-BLAST as a simplified alternative.

As we will see in the following, most of the advantages of D-BLAST can be retained while using a simpler system requiring only one encoder and decoder. Figure 3.15 shows an example mapping order for the encoded and modulated symbols from the encoder onto the transmission array. The important point here is that the diagonal structure of the D-BLAST scheme is retained. We will refer to this scheme in the sequel as a "D-BLAST-like" scheme.

3.5.3 Detection of Spatial Multiplexing

While the transmission of spatial multiplexing, especially V-BLAST, is very simple and straightforward, detection is more difficult since the signals from each antenna (each layer) are in principle received by all the receive antennas, and there is no inherent orthogonality to assist with separating them, as there is for O-STBC. However a range of signal processing techniques exist for separating such signals, especially those used in multiuser detection for multiuser wireless systems such as CDMA. These can be classified as Maximum Likelihood (ML), linear, and nonlinear.

The concept of ML detection we have considered above for STBC. Here it can be implemented per space-time symbol by applying a version of (3.36) directly:

$$\hat{\mathbf{s}} = \arg\min_{\tilde{\mathbf{s}}} \left(\left\| \mathbf{H}\tilde{\mathbf{s}} - \mathbf{r} \right\|^2 \right) \tag{3.47}$$

where $\mathbf{s}$ represents the space-time symbol (one column of the array of Figure 3.12), $\mathbf{H}$ the channel matrix and $\mathbf{r}$ the corresponding received signal vector. The search implicit in (3.47) runs over all possible transmitted space time symbols, and hence its complexity is proportional to $M_c^{n_T}$., where M_c is the number of symbols in the modulation constellation. Reduced complexity methods such as *sphere detection* (described elsewhere in this book) can reduce this complexity significantly.

Linear detection is based on multiplying by the inverse of the channel matrix, to reconstruct the transmitted signal. In *Zero Forcing* (ZF) we use the pseudo-inverse of the channel directly:

$$\hat{\mathbf{s}} = \left(\mathbf{H}^H\mathbf{H}\right)^{-1}\mathbf{H}^H\mathbf{r} \tag{3.48}$$

The complexity of this method per space-time symbol is much less than for ML, since it involves simply a linear transformation including $n_T n_R$ complex multiplications. It does involve calculating the inverse $\left(\mathbf{H}^H\mathbf{H}\right)^{-1}$, but on relatively slowly fading channels this needs to be calculated only once per frame, not per space-time symbol.

However, ZF detection may give rise to noise enhancement, since:

$$\hat{\mathbf{s}} = \left(\mathbf{H}^H\mathbf{H}\right)^{-1}\mathbf{H}^H\left(\mathbf{H}\mathbf{s}+\mathbf{n}\right) = \mathbf{s} + \left(\mathbf{H}^H\mathbf{H}\right)^{-1}\mathbf{H}^H\mathbf{n} \tag{3.49}$$

Writing $\mathbf{H}$ as $\mathbf{V}\Sigma\mathbf{U}^H$ using the singular value decomposition, as in (3.10) (where $\Sigma^2 = \Lambda$, a diagonal matrix of the eigenvalues), the noise term becomes:

$$\left(\mathbf{H}^H\mathbf{H}\right)^{-1}\mathbf{H}^H\mathbf{n} = \left(\mathbf{U}\Lambda\mathbf{U}^H\right)^{-1}\mathbf{U}\Sigma\mathbf{V}^H\mathbf{n} = \mathbf{U}\Lambda^{-1}\mathbf{U}^H\mathbf{U}\Sigma\mathbf{V}^H\mathbf{n} = \mathbf{U}\Lambda^{-1}\mathbf{n}' \tag{3.50}$$

where $\mathbf{n}' = \mathbf{V}^H\mathbf{n}$, remembering that $\mathbf{U}$ and $\mathbf{V}$ are unitary matrices. Hence, if any of the eigenvalues of $\mathbf{H}^H\mathbf{H}$ are small, the inversion will mean that the noise is amplified. It can be shown that if $n_R = n_T$ the distribution of the smallest eigenvalue is approximately exponential, the same as the power gain of a Rayleigh fading channel. Since a fade will affect the signal to noise ratio in the same way as for a Rayleigh fading channel, the resulting effective diversity order is unity, and hence not even the available receive diversity is provided. However, if $n_R > n_T$ the diversity order is $n_R - n_T + 1$.

Linear detection can be improved by using the *minimum mean square error* (MMSE) criterion, which minimizes the total error due to the combination of noise and distortion. In this case, we use the detection transformation:

$$\hat{\mathbf{s}} = \left(\mathbf{H}^H\mathbf{H} + \sigma^2\mathbf{I}\right)^{-1}\mathbf{H}^H\mathbf{r} \tag{3.51}$$

where σ^2 is the variance of the noise. However, at high SNR, the term $\sigma^2\mathbf{I}$ becomes negligible, and the asymptotic performance is the same as ZF. This also results in the same limitations on diversity order as for ZF.

Nonlinear methods involve decision feedback in some form, in order to remove the interference between the signals from each antenna. (The line in Figure 3.11 shown dotted around the detector for the V-BLAST scheme denotes this feedback.) In this context the most commonly used approach is called *Ordered Successive Interference Cancellation* (OSIC), and is the original algorithm proposed in [GFVW99] for detection of V-BLAST transmissions. The principle is to use linear detection to detect first the modulation symbol on the antenna least affected by noise enhancement, then subtract the interference due to that symbol on the other symbols, then detect the next strongest signal, until all symbols are detected. The algorithm can be described in pseudo-code as follows:

For each symbol period ($n = 1 \ldots N_f$)

$\widehat{\mathbf{H}} = \mathbf{H}$

$\widehat{\mathbf{r}} = \mathbf{r}$

For each antenna ($i = 1..n_T$):

Find next least faded antenna:

$\mathbf{g} = \text{diag}\left(\left(\widehat{\mathbf{H}}^H \widehat{\mathbf{H}}\right)^{-1}\right)$

$k_i = \arg\min_k (g_k)$

Obtain MMSE estimate of antenna signal s_{k_i}:

$\mathbf{X}^{(i)} = \left(\widehat{\mathbf{H}}^{\mathbf{H}} \widehat{\mathbf{H}} + \sigma^2 \mathbf{I}\right)^{-1} \widehat{\mathbf{H}}^H$

$s_{k_i} = \left[\mathbf{X}^{(i)} \widehat{\mathbf{r}}\right]_{k_i}$

Make decision on this symbol:

$\hat{s}_{k_i} = \mathbf{D}\left(s_{k_i}\right)$

Subtract interference due to this symbol from $\widehat{\mathbf{r}}$:

$\widehat{\mathbf{r}} = \widehat{\mathbf{r}} - \hat{s}_{k_i} [\mathbf{H}]_{k_i}$

Delete k_i^{th} column of $\widehat{\mathbf{H}}$ and k_i^{th} entry of $\widehat{\mathbf{r}}$:

$\widehat{\mathbf{H}} = [\widehat{\mathbf{H}}]_{1 \ldots k_i - 1, k_i + 1 \ldots n_T - i + 1}$

$\widehat{\mathbf{r}} = [\widehat{\mathbf{r}}]_{1 \ldots k_i - 1, k_i + 1 \ldots n_T - i + 1}$

In this description, $[\mathbf{x}]_l$ represents the lth element of a vector $\mathbf{x}$, while $[\mathbf{X}]_l$ represents the lth column of a matrix $\mathbf{X}$. For multiple subscripts the notation represents a vector (or matrix) containing the subset of elements (or columns) indicated by the subscript values. $\mathbf{D}(x)$ denotes the demodulator slicing function.

OSIC improves upon linear detection, since it applies selection diversity to the first symbol detected, and for subsequent symbols the interference cancellation increases the effective diversity. However, in practice the result is a long way from the maximum diversity available. Figure 3.16 shows the BER performance of a coded V-BLAST system with an OSIC detector compared to the MMSE detector and the Chernoff bound, which shows the diversity order achievable with an ML detector. While the MMSE detector performs a little better than first order diversity, the OSIC detector is a little better than that, close to second order in this range of SNR. However, the diversity achievable with ML detection is at least 4th order, as shown by the Chernoff bound.

Figure 3.17 shows an iterative detector and decoder for the "D-BLAST-like" system mentioned in Section 3.5.2. Following MMSE detection in a first iteration, the code sequence is fed to a soft-output decoder using the *maximum a posteriori* (MAP) algorithm [BCJR74], which is used to tentatively cancel interference in a second iteration. The decoding is then repeated until convergence is reached. We will show below that this detector can achieve most of the receive end diversity available on the channel.

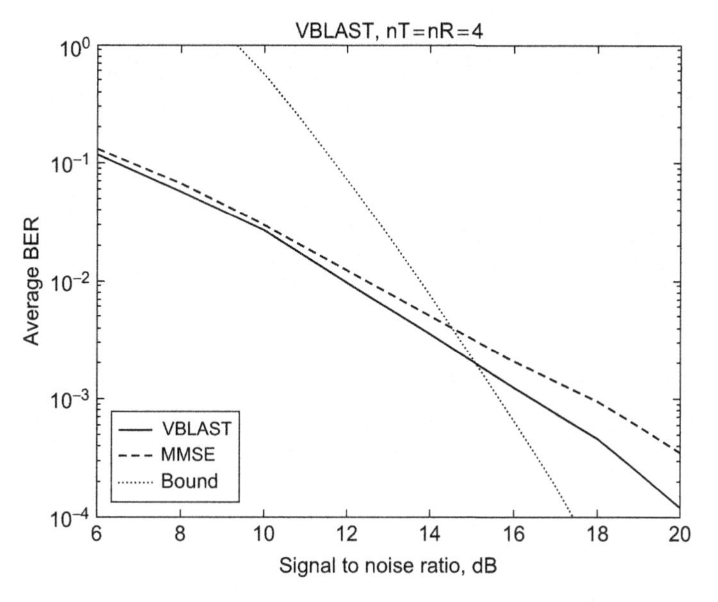

FIGURE 3.16

BER performance of 4 × 4 coded V-BLAST system with MMSE and OSIC (VBLAST) detection, compared with Chernoff bound.

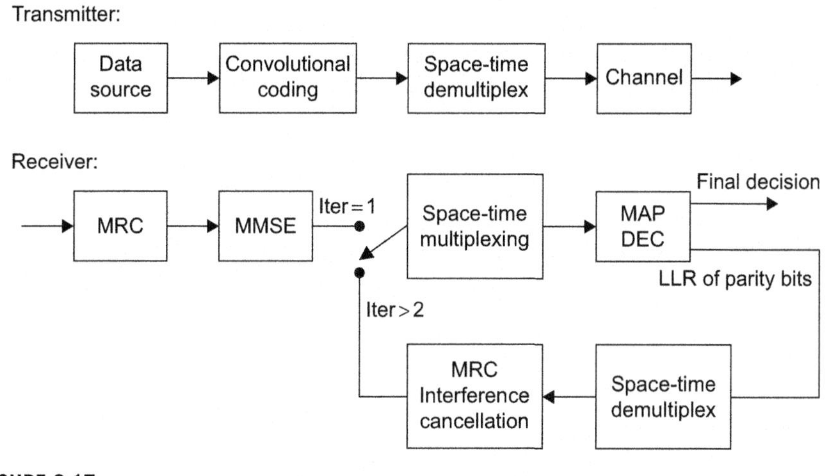

FIGURE 3.17

Iterative decoder for "D-BLAST-like" scheme (taken from [M04]).

3.5.4 Transmit Diversity of Coded Spatial Multiplexing

However, we will first review the application of the theory of space-time coding from Section 3.3 to coded spatial multiplexing, showing how the diversity criteria of Section 3.3.3 can also be applied here. The space-time codewords consist of permitted combinations of coded symbols mapped to a space-time array of the form of Figure 3.12. As in Section 3.3.2 we are interested in pairs of words with the minimum number of differences, which correspond to code sequences with the minimum number of nonzero symbols (which we will refer to as '1's). Typically in such sequences the '1's will tend to occur within "bursts," within which about half the symbols are '1', with '0's everywhere else in the array. Hence, in V-BLAST each column of a codeword will tend to contain a large number of '1's. This will give rise to nonzero off-diagonal terms in the matrix $\mathbf{\Phi_{DD}} = \mathbf{DD}^H$ (see (3.19)), which as discussed will tend to reduce the rank of the codeword.

We will illustrate this with reference to the constraint length 3 convolutional code with generators (7, 5) in octal, which has minimum Hamming distance 5, whose minimum weight code sequence contains ...000111011000.... This is used to encode a 4×4 V-BLAST system. The difference matrix and the correlation matrix might take the form:

$$\mathbf{D} = \begin{bmatrix} 1 & 1 & 0 & 0 & 0 \\ 1 & 1 & 0 & 0 & 0 \\ 1 & 0 & 0 & 0 & 0 \\ 0 & 0 & 0 & 0 & 0 \end{bmatrix}; \ \mathbf{\Phi_{DD}} = \begin{bmatrix} 2 & 2 & 1 & 0 \\ 2 & 2 & 1 & 0 \\ 1 & 1 & 1 & 0 \\ 0 & 0 & 0 & 0 \end{bmatrix} \tag{3.52}$$

The rank of this is however only 2, and the second eigenvalue turns out to be much smaller than the first, which means that the effective transmit end diversity at medium SNR is only first order.

On the other hand, in D-BLAST the "bursts" will tend to lie along a diagonal of the array. Under most conditions there can be no more than one '1' in each column of the array, and thus the rank of the matrix is the number of rows containing at least one '1'. Depending on the number of transmit antennas, this may approach the minimum Hamming distance of the code. Thus, coded D-BLAST can in principle provide a diversity gain as well as increasing capacity, while V-BLAST cannot guarantee to do this, even with coding.

With our example code the difference and correlation matrices become:

$$\mathbf{D} = \begin{bmatrix} 1 & 0 & 0 & 0 & 1 \\ 0 & 1 & 0 & 0 & 0 \\ 0 & 0 & 1 & 0 & 0 \\ 0 & 0 & 0 & 0 & 0 \end{bmatrix}; \ \mathbf{\Phi_{DD}} = \begin{bmatrix} 2 & 0 & 0 & 0 \\ 0 & 1 & 0 & 0 \\ 0 & 0 & 1 & 0 \\ 0 & 0 & 0 & 0 \end{bmatrix} \tag{3.53}$$

which has rank 3 and eigenvalues 2, 1, 1, which are much more balanced than for V-BLAST, achieving a significantly larger transmit end diversity. It is worth noting that the diagonal mapping in the D-BLAST-like scheme means that this can also provide the same diversity order.

Figure 3.18 compares the V-BLAST and D-BLAST-like schemes with OSIC detection, demonstrating the diversity order improvement that is achieved simply by changing the symbol mapping. The diversity order here is around 3, suggesting that the transmit end diversity predicted above is achieved. Figure 3.19 shows how the

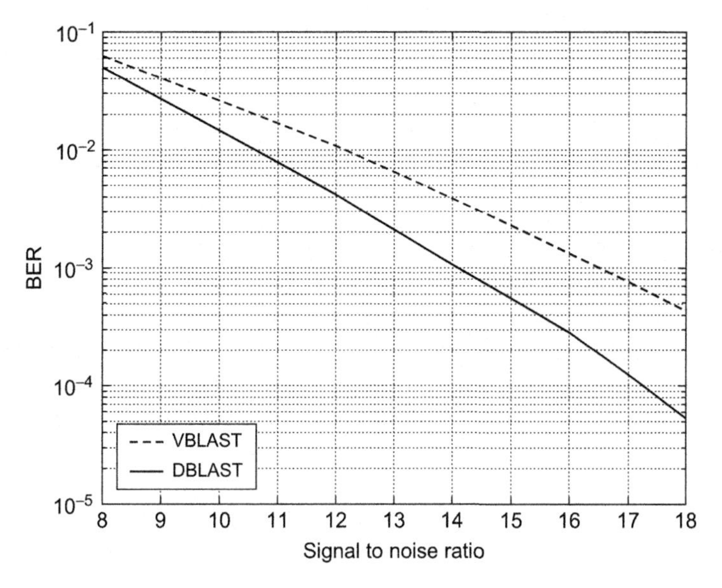

FIGURE 3.18

Comparison of V-BLAST and "D-BLAST-like" systems, 4 × 4 MIMO system, convolutional (7,5) code, constraint length 3, with OSIC detection (taken from [BZH04]).

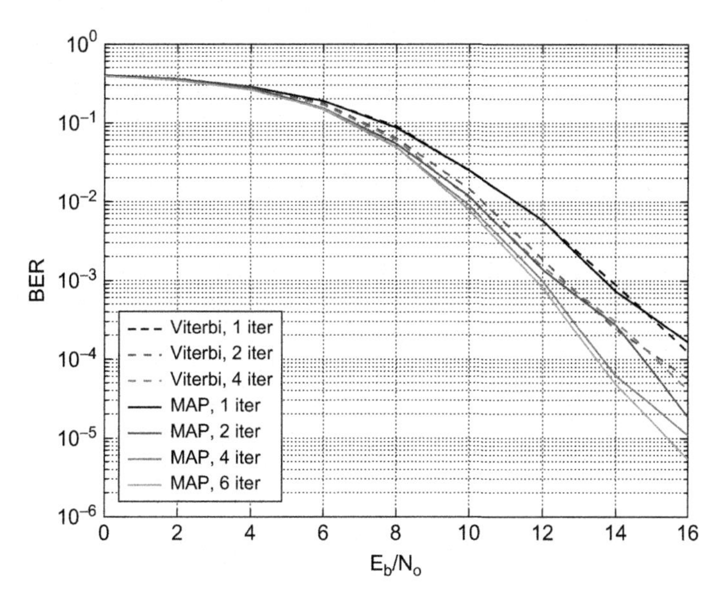

FIGURE 3.19

Performance of iterative decoding of D-BLAST-like scheme with hard-output and MAP decoding (taken from [BZH04]).

iterative detector improves upon this, as the interference cancellation causes the performance to tend to that of ML detection, so the receive end diversity is also partially achieved. Note that performance with the soft-output MAP decoder is significantly better than with the hard-output Viterbi decoder. Full diversity of this 4 × 4 system is still not achieved (the order is around 6), but the code has a relatively small Hamming distance.

Figure 3.20 shows the performance of a similar system with a turbo code, compared with the bound on outage capacity for a MIMO system. (Note that in quasi-static fading the outage probability for a given capacity target is a bound on the Frame Error Ratio (FER) of a wireless system. Also, note that the outage capacity for a given constellation—the *Constrained Modulation* (CM) capacity—is a few dB less than the unconstrained capacity which is usually plotted for MIMO systems.) This shows that for a 4 × 4 system the FER performance achieved is only about 1 dB poorer than the capacity bound, and the diversity order is the same as predicted by the bound, for practical purposes. Here a pseudo-random interleaver is used in place of the diagonal mapping, but the effect is very similar.

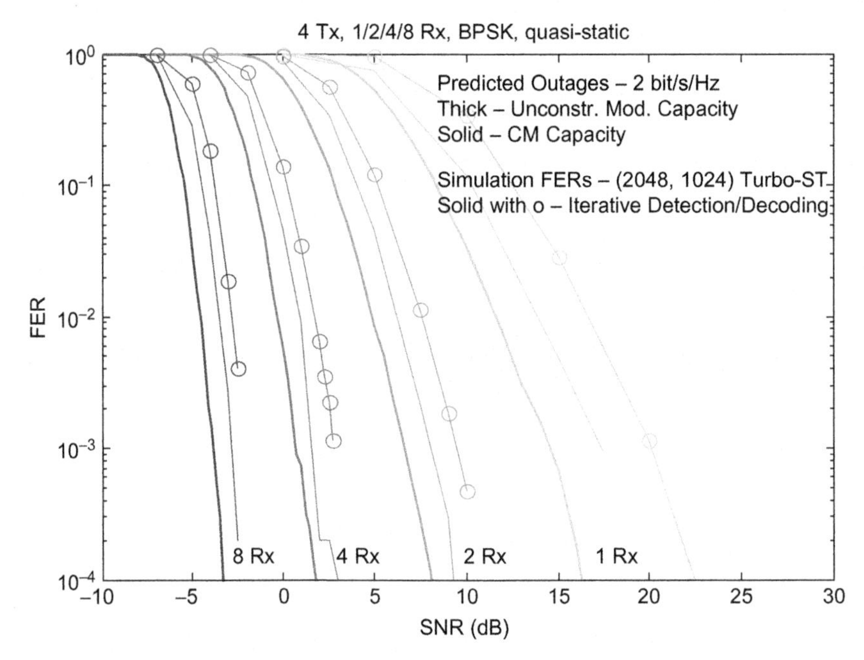

FIGURE 3.20

Frame Error Ratio (FER) performance of turbo coded spatial multiplexing with iterative receiver, compared to capacity bounds, for 4 transmit antennas and BPSK (taken from [BZH04]).

3.6 PRECODING

3.6.1 Adaptive Transmission using the SVD

Both space-time coding and spatial multiplexing are designed primarily for the case where the channel is unknown at the transmitter. Equation (3.11) suggests an alternative method if the channel is known: the data vector **d** is multiplied by a unitary matrix **U**, the right-hand singular vector matrix of the channel **H**, to generate the transmit vector **s** which will be sent over the channel. Then at the receiver, the received vector **r** is multiplied by another unitary matrix $\mathbf{V}^H$, where **V** is the left hand singular vector matrix of **H**. This explicitly creates the eigenvalue subchannels shown in Figure 3.3, and allows perfect separation of the data symbols at the receiver requiring only linear processing at the receiver and without noise enhancement, since $\mathbf{V}^H$ is unitary and hence does not affect the statistics of the noise. [LGF05] describes this approach and considers some practical issues surrounding its use.

In fact, it allows increased capacity because the power can be optimally distributed between the subchannels, using the principle of *water-filling* [S49]. Figure 3.21 illustrates this principle, showing how the optimum transmitted power is determined. If different subchannels have different gains (a), the received signal is first equalized (b), so that all subchannels have the same gain but different noise power. The principle of water-filling is that the signal power is then added (c), such that as far as possible the total of signal plus (equalized) noise is the same in all sub-channels. We observe that less power or none at all is transmitted in sub-channels with smaller gain (corresponding to smaller eigenvalues). Since this concentrates power where it will be most effective, it increases capacity: Equation (3.12) becomes:

$$C_{MIMO-WF} = W \sum_{i=1}^{n_m} \log_2 \left(1 + \frac{S_i}{N} \lambda_i \right) \tag{3.54}$$

where S_i is the signal power transmitted in the ith subchannel, with $\sum_{i=1}^{n_m} S_i = S$.

The effect is greatest on channels with reduced rank or very unbalanced eigenvalues, which typically occurs in relatively poor multipath environments, and can be

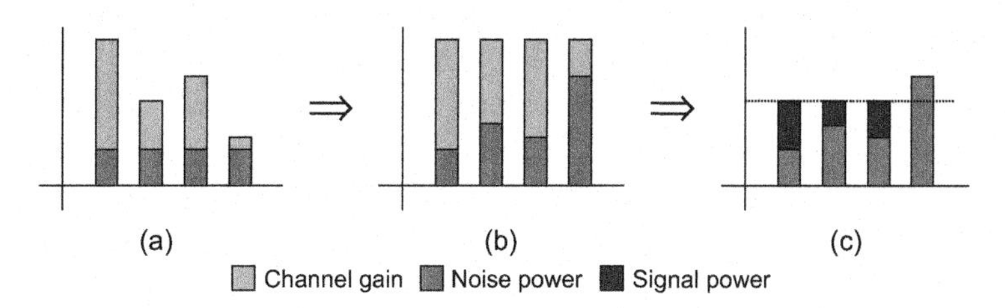

FIGURE 3.21

Principle of water-filling: (a) channel gains and noise power on channel; (b) channel gains and noise power after equalization; (c) determining signal power.

equivalent to an increase in SNR of a factor up to the number of transmit antennas. This approach can be described as *adaptive* because the transmission is adapted to the channel.

3.6.2 Codebook-based Precoding

The multiplication by the matrix **U** at the transmitter can be regarded as a form of *precoding*. The disadvantage of this form of precoding is that since the channel matrix **H** is in general estimated at the receiver, not at the transmitter, it requires the full matrix **H** to be transmitted at full precision back to the transmitter on some return channel. This would usually involve an unacceptable overhead, and so in practice the permissible set of precoding matrices is restricted to the entries of a *codebook* of matrices. The selection of a precoding matrix is then performed at the receiver following channel estimation, and the index of the optimum matrix within the codebook is then signaled back to the transmitter, incurring a much-reduced overhead.

This approach is used in most of the recent and emerging wireless standards which involve MIMO, especially WiMAX (IEEE 802.16e/m [IEE09a], [IEEE16mSDD]) and 3GPP-LTE [3GPP07] and LTE-Advanced [MB09]. The basic codebook used in 3GPP-LTE with two transmit antennas and one or two sub-channels ("layers" in LTE terminology) is given in (3.55) [3GPP07].

$$\mathbf{F}_{3gpp} \in \left\{ \begin{array}{c} \begin{bmatrix} 1 \\ 0 \end{bmatrix}, \begin{bmatrix} 0 \\ 1 \end{bmatrix}, \sqrt{\tfrac{1}{2}}\begin{bmatrix} 1 & 0 \\ 0 & 1 \end{bmatrix}, \\ \sqrt{\tfrac{1}{2}}\begin{bmatrix} 1 \\ 1 \end{bmatrix}, \sqrt{\tfrac{1}{2}}\begin{bmatrix} 1 \\ -1 \end{bmatrix}, \tfrac{1}{2}\begin{bmatrix} 1 & 1 \\ 1 & -1 \end{bmatrix}, \\ \sqrt{\tfrac{1}{2}}\begin{bmatrix} 1 \\ j \end{bmatrix}, \sqrt{\tfrac{1}{2}}\begin{bmatrix} 1 \\ -j \end{bmatrix}, \tfrac{1}{2}\begin{bmatrix} 1 & 1 \\ j & -j \end{bmatrix} \end{array} \right\}. \tag{3.55}$$

Note that each of these matrices corresponds to a form of MIMO transmission: those matrices with only one column correspond to a single layer, and transmit only one data symbol per channel use. Also in the first row of (3.55) the first two matrices correspond to antenna selection, since the single symbol is transmitted either on the first or the second antenna. The third provides two layers, and corresponds in fact to simple spatial multiplexing, since each of the layers is transmitted on one antenna. The remaining four matrices with a single column can be regarded as simple forms of beam-forming. For example, $[11]^T$ forms a beam in the broadside direction from the antenna array, while $[1-1]^T$ generates a null in the broadside direction. The remaining two 2×2 matrices transmit two layers, providing a different beam-pattern for each layer. The multiplying factors $1/\sqrt{2}$ and 1/2 ensure total transmit power is the same for all precoding matrices.

Figure 3.22 shows the BER performance of a simple MIMO system using this codebook (or subsets of it) for precoding. In this system we maintain a constant data throughput of 4 bits per channel use, using two streams with QPSK modulation when two layers are available, or one stream with 16-QAM when one is used. Detection is

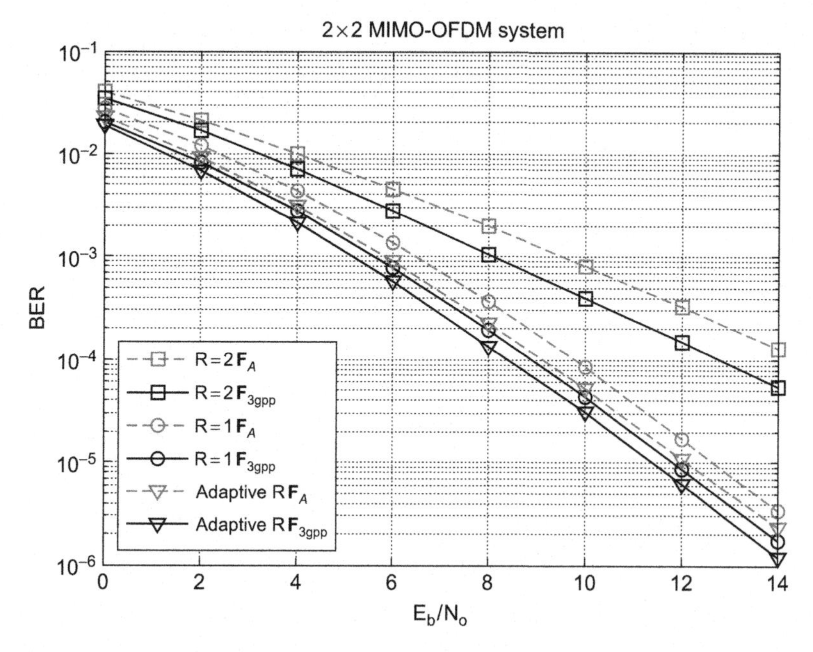

FIGURE 3.22

BER of precoded 2 × 2 MIMO scheme transmitting 4 bits per channel use (taken from [DCAB08]).

ML. The different curves correspond to different subsets of the codebook: "R = 2" in the legend means only the two layer precoders, "R = 1" only the single layer, and "Adaptive" includes both. "F_A" refers to just the first three matrices in (3.55).

We observe that when the use of two layers is enforced, only second order diversity is achieved; this is receive end diversity as a result of using ML detection. Allowing a single layer increases diversity order to 4, because transmit end diversity is then also available. A larger adaptive codebook results in an improvement in coding gain, because a precoder closer to the matrix **U** can be used.

3.6.3 Generalized Precoding

We have noted above that some of the precoding matrices in the codebook $\mathbf{F}_{3gpp}$ are equivalent to other forms of MIMO transmission, for example conventional spatial multiplexing (V-BLAST), antenna selection and beamforming. This suggests that it may be possible to treat a wider range of MIMO schemes using the notation of precoding. This leads to the concept of *generalized precoding* [AB08].

It is less straightforward; however, to treat schemes like D-BLAST and STBC in this way, since in these cases, data may be mapped to different space-time symbols in a different way. However, a generalization based on a "stacked" form of the transmitted space-time matrix can be used:

$$\mathrm{vec}(\mathbf{S}) = \mathbf{Fd} \tag{3.56}$$

where the function $\mathrm{vec}(\mathbf{X})$ has the effect of stacking the columns of the $(p \times q)$ matrix $\mathbf{X}$ to form a $(p \cdot q \times 1)$ column vector. Space-time block codes, however, raise a further problem because of the use of complex conjugates in the complex orthogonal designs. A solution to this has been known for some time: to split the modulated symbols into their real and imaginary parts [PTSL02]. On this basis, the Alamouti scheme can be represented as:

$$\mathrm{vec}(\mathbf{S}) = \left[\begin{array}{cc|cc} 1 & 0 & j & 0 \\ 0 & 1 & 0 & j \\ \hline 0 & 1 & 0 & -j \\ -1 & 0 & j & 0 \end{array}\right] \begin{bmatrix} \mathrm{Re}\{d_1\} \\ \mathrm{Re}\{d_2\} \\ \mathrm{Im}\{d_1\} \\ \mathrm{Im}\{d_2\} \end{bmatrix} = \mathbf{F}_A \mathbf{d}_{ri} \tag{3.57}$$

where $\mathbf{d}_{ri}$ represents the data symbol vector in which the real and imaginary part are separated as shown. This form is used in the 3GPP LTE standard [3GPP07].

However, this "stacked" form of $\mathbf{S}$ has disadvantages, especially that we cannot apply it directly to the channel matrix $\mathbf{H}$ in its usual form. An alternative approach, explored in [AB08] is to use *tensor* notation for the precoder. For our purposes, a tensor can be regarded simply as a multidimensional array of (possibly complex) data elements, the generalization in multiple dimensions of the vector and the matrix. We will use upper-case bold italics to refer to tensors (whose elements will in general be complex), and will also denote them using subscripted elements within square brackets:

$$\boldsymbol{X} \equiv \left[X_{i_1 \cdots i_N}\right] \in \mathbb{C}^{I_1 \times I_2 \cdots \times I_N} \tag{3.58}$$

where N is the *order* of the tensor, i.e., the number of dimensions, and I_1 to I_N denote the size of the tensor in each dimension. We will use the symbol "$\otimes$" to denote tensor multiplication as a generalization of matrix multiplication, with a superscript or superscripts on the left to denote which dimensions of the left hand multiplicand are to be summed over, and a subscript on the right similarly to indicate the dimensions of the right hand multiplicand summed over. For example:

$$\begin{aligned} \boldsymbol{Z} &= \boldsymbol{X}^1 \otimes_1 \boldsymbol{Y} \equiv \left[\sum_k X_{klm} Y_{kn}\right] \\ \boldsymbol{Z} &= \boldsymbol{X}^{2,3} \otimes_{1,4} \boldsymbol{Y} \equiv \left[\sum_m \sum_l X_{klm} Y_{lnpm}\right] \end{aligned} \tag{3.59}$$

In this notation ordinary matrix multiplication is equivalent to the ${}^2\otimes_1$ product of two second-order tensors.

Then the transmitted signal matrix can be written:

$$\mathbf{S} = \boldsymbol{F}^2 \otimes_1 \mathbf{d} \tag{3.60}$$

where $\boldsymbol{F}$ is the precoding tensor. Now we can identify the dimensions of the array with signal domains: the first dimension (row number) represents space, the second (column number) is the index of the modulated data symbol, and the third represents time.

Moreover the received signal matrix can be written:

$$\begin{aligned}\mathbf{R} &= \mathbf{HS} = \boldsymbol{H}^2 \otimes_1 \left(\boldsymbol{F}^2 \otimes_1 \mathbf{d}\right) \\ &= \left(\boldsymbol{H}^2 \otimes_1 \boldsymbol{F}\right)^2 \otimes_1 \mathbf{d} = \boldsymbol{H}'^2 \otimes_1 \mathbf{d}\end{aligned} \tag{3.61}$$

where $\boldsymbol{H}'$ is an equivalent channel tensor, generalizing the equivalent channel matrix of (3.34).

As an example, consider the Alamouti scheme. The "stacked" precoding matrix $\mathbf{F}_A$ of (3.57) can be expressed in tensor form as in Figure 3.23. Here we have added a fourth dimension, of size 2, separating the real and imaginary parts of the data, and resulting in a fourth order tensor. In the figure, the fourth dimension is shown schematically as the dashed arrow.

The data then becomes a matrix $\mathbf{D}' = \left[\mathrm{Re}\{\mathbf{d}\} \quad \mathrm{Im}\{\mathbf{d}\}\right]$ containing the real and imaginary parts of the data symbols in separate columns. The transmitted signal matrix can be written:

$$\mathbf{S} = \boldsymbol{F}_A^{2,4} \otimes_{1,2} \mathbf{D}' \tag{3.62}$$

and the received matrix:

$$\begin{aligned}\mathbf{R} &= \mathbf{HS} = \boldsymbol{H}^2 \otimes_1 \left(\boldsymbol{F}^{2,4} \otimes_{1,2} \mathbf{D}'\right) \\ &= \left(\boldsymbol{H}^2 \otimes_1 \boldsymbol{F}\right)^{2,4} \otimes_{1,2} \mathbf{D}'\end{aligned} \tag{3.63}$$

Figure 3.24 shows the resulting fourth order tensor representing the combined channel and precoder. Note that if we redefine a received signal matrix $\mathbf{R}'$ in which the second MIMO symbol of the block is conjugated then we can combine the real and

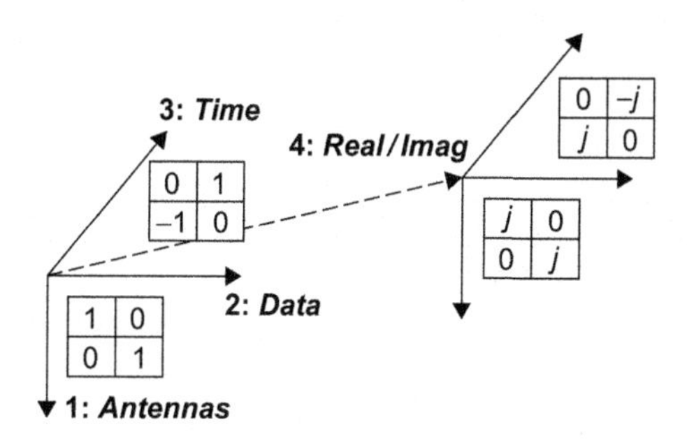

FIGURE 3.23

Precoding tensor for Alamouti scheme (taken from [B08]).

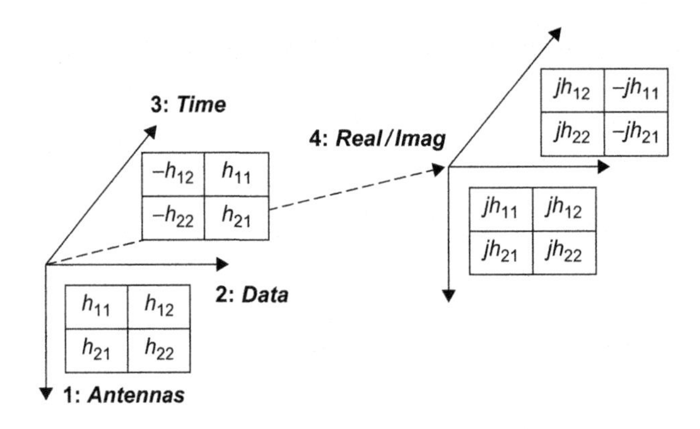

FIGURE 3.24

Combined precoding and channel tensor for the Alamouti scheme (taken from [B08]).

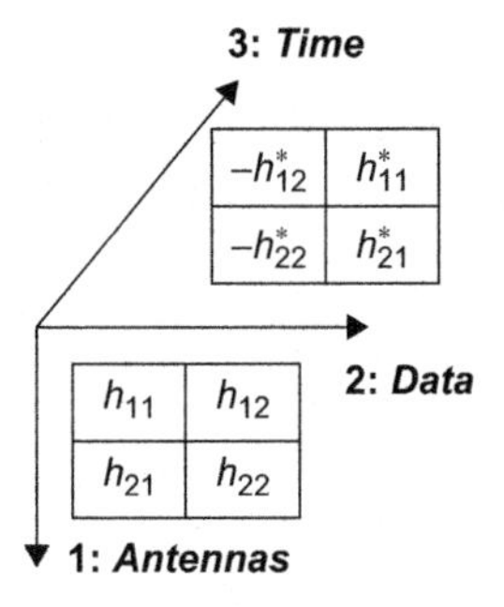

FIGURE 3.25

Combined Alamouti precoding and channel tensor with conjugate second space time symbol (taken from [B08]).

imaginary parts of $\mathbf{d}$ and rewrite the combined precoder as a third order tensor, as shown in Figure 3.25. The slices of this tensor in the time/data plane will be recognized as the matrix commonly used to represent the scheme in 2×1 Alamouti-coded systems, but the tensor representation allows this to be extended to multiple receive antennas.

More details of this tensor-based precoding framework are given in [B08], from which the notation and the diagrams above are taken. Note that it also subsumes, or provides an alternative to, *linear dispersion codes* [HH02] and *linear constellation precoding* [WXG03], as well as the forms of MIMO transmission we have reviewed above.

The framework also allows new MIMO transmission schemes to be developed, both adaptive and nonadaptive. For example, in [LB10] the authors describe a synthesis technique for precoding tensors based on the PARAFAC decomposition [H70], according to which the precoding tensor can be written:

$$\boldsymbol{F} = \left[\sum_{f} \boldsymbol{\Omega}_{if} \boldsymbol{\Psi}_{kf} C_{lf} \right] \tag{3.64}$$

where $\boldsymbol{\Omega}_{if}, \boldsymbol{\Psi}_{kf}, C_{lf}$ are elements of the antenna allocation matrix $\boldsymbol{\Omega}$, the data allocation matrix $\boldsymbol{\Psi}$, and the spreading code matrix $\mathbf{C}$, respectively. It can be shown that the rank criterion for maximum diversity is fulfilled provided all of the following matrices have rank at least n_T:

$$\boldsymbol{\Omega}, \mathbf{C}, \mathrm{diag}\left(\boldsymbol{\Psi}\Delta\mathbf{d}\right), \boldsymbol{\Omega}\,\mathrm{diag}\left(\boldsymbol{\Psi}\Delta\mathbf{d}\right), \mathrm{diag}\left(\boldsymbol{\Psi}\Delta\mathbf{d}\right) C^T, \boldsymbol{\Omega}\,\mathrm{diag}\left(\boldsymbol{\Psi}\Delta\mathbf{d}\right) C^T \tag{3.65}$$

for all pairs of modulated data vectors, where $\Delta\mathbf{d}$ denotes the difference vector between these pairs. This allows schemes to be defined such that these conditions are fulfilled. Note that unlike O-STBC, the design of which ensures orthogonality and hence full rank for any data symbols, the rank of these schemes may be dependent on the modulation constellation. As a consequence, however, they are not restricted in available rate by the requirements of orthogonality, as we have seen for O-STBC, and yet they can provide full rank for some constellations, unlike QO-STBC.

3.7 MIMO IN CURRENT AND EMERGING STANDARDS

Because of the remarkable capacity and diversity improvements it provides, MIMO has become essential in the most recent standards for broadband wireless communication, and especially in what has become known as the "fourth generation" of wireless communications, or 4G. The accepted definition of 4G is that given by the International Telecommunications Union – Radio (IMT-R) under the name International Mobile Telecommunications (IMT) Advanced [ITU-R08]. The leading standards that are now accepted within this definition are WiMAX (defined as IEEE 802.16 [IEE09a]) and 3GPP LTE-Advanced [MB09]. In this section, we briefly review the MIMO techniques used in these standards. A good overview is given in [LLL+10].

The techniques used may be grouped in several ways. First, we distinguish single-user (SU) MIMO from multi-user (MU). In multiuser MIMO multiple terminals are served from the same frequency subband and time-slot, being separated by a form of beamforming or precoding—as we have already seen, the two are equivalent. However, the details are beyond the scope of this chapter, and here we focus on SU-MIMO techniques. Multiple users can of course still be served by SU-MIMO, using Frequency or Time-Division Multiple Access (F/TDMA).

Second, we distinguish between Closed-Loop (CL) and Open-Loop (OL) techniques. Closed-loop techniques are used when CSI is available at the transmitter; otherwise open-loop techniques must be used. Note however that these standards employ adaptive modulation and coding, which implies that at least a channel quality indication (CQI) should be available at the transmitter to allow different modulation and code rates to be selected; however, the CSI required for CL-MIMO must fully describe the channel matrix to some degree of accuracy.

Third, in general different techniques are used on the downlink and the uplink. In particular, in LTE and LTE-Advanced single-carrier FDMA (SC-FDMA) is used on the uplink for multiple access, while OFDMA (Orthogonal Frequency Division Multiple Access) is used on the downlink. SC-FDMA can be described as OFDMA pre-coded by multiplication by a Discrete Fourier Transformation (DFT) matrix, which has the effect of regenerating an effective single carrier signal, with improved peak to average power ratio (PAPR) compared to OFDMA [MLG06]. WiMAX uses OFDMA on both links [IEE09a]. Generally, OFDMA allows more flexibility in the MIMO techniques that can be employed, but the techniques used on the downlink of LTE and both links of WiMAX are quite similar.

For open loop, both allow the use of STBC in the form of Alamouti for two transmit antennas. In LTE the Ganesan and Stoica rate 3/4 code [GS01] is used for four antennas. Both are expressed in the standard as precoding matrices in the "stacked" form with separated real and imaginary parts, as in (3.57). Both also allow simple spatial multiplexing, again expressed in simple matrix form. They also use closed-loop precoding using predefined codebooks of precoding matrices. However, more advanced precoding methods are being developed for the next generation of the standards, 802.16m and LTE-Advanced: adaptive codebooks in the former and user-specified codebooks in the latter. These allow more precision in the description of the channel, while exploiting compression techniques to minimize the feedback overhead.

Perhaps most significant difference between the standards [LLL+10] concerns the way FEC coding is applied to precoding involving multiple layers. In LTE there are usually separate codes applied to each layer, rate-adapted to the SNIR provided by the layer. In many cases, this can maximize the link capacity. In WiMAX a single stream is encoded and multiplexed across the layers, which has the effect of improving diversity, as discussed previously.

3.8 SUMMARY

In this chapter, we have considered the techniques available for achieving in practice the significant capacity increases that are in principle available on the MIMO wireless channel. We have reviewed the main performance benefits of MIMO, namely multiplexing gain and diversity gain, which provide respectively increased throughput and increased robustness, and showed how these are in fact properties of the underlying channel, but require care in the design of transmission schemes if they are to be realized in MIMO systems. We consider also the extent to which they can be traded off against one another, though noting that the idealized diversity-multiplexing trade-off may not be available under more realistic conditions.

The theory of space-time coding provides a useful basis for understanding how the diversity and multiplexing benefits can be achieved. It leads to criteria for the transmitted signal, especially for achieving maximum diversity gain. The most important criterion is that the rank of the transmitted signal matrix should be maximized. This has allowed us to consider the design of MIMO schemes for achieving maximum diversity, leading to what are commonly called space-time codes, and maximum multiplexing gain, leading to spatial multiplexing. However, the theory of space-time codes can in fact also be applied to spatial multiplexing in order to optimize diversity here also. A third transmission technique, precoding, which is widely used in current and developing wireless standards, was also described, and we have shown that the other transmission schemes can in fact conveniently be described in terms of precoding.

Finally, we have briefly reviewed the application of MIMO techniques in these emerging standards, especially 3GPP LTE/LTE-Advanced, and WiMAX (IEEE 802.16). In both the formalism of precoding is widely used, especially for closed-loop MIMO systems, where channel state information is exploited at the transmitter to adapt the transmitted signal to the MIMO channel.

CHAPTER

Interference Functions – A Mathematical Framework for MIMO Interference Networks

4

Martin Schubert

The author wishes to acknowledge Holger Boche
for his contributions to this work.

Future wireless networks will need to support high data rates in order to meet the requirements of multimedia services. Also the user density will be much higher than today. In addition to the communication between people, there will be an increasing amount of data communication between devices.

In this context, interference emerges as the key performance limiting factor. Interference determines how many users/terminals per area can be served at a certain data rate. Assigning each user a separate resource is not always an efficient way of organizing the system. If the number of users is high then each user only gets a small fraction of the overall resource. Shortages occur when many users have high capacity requirements. Hence, the classical design paradigm of independent point-to-point communication links is gradually being replaced by a new network-centric point of view, where users interact and compete for the limited system resources.

Dynamic interference management and resource allocation will play a crucial role in fully exploiting the available system capacity. The goal is to optimize the system performance by assigning the available resources in a flexible way, taking into account the current interference situation.

MIMO offers another degree of freedom to avoid and mitigate interference. The multiuser case is quite different from the single-user case, because the users compete for the available resources. This typically means that there is no single optimum strategy. The performance of some users can be increased at the cost of decreasing the performance of other users. The chosen operating point is often a compromise between fairness and efficiency.

There are still many open questions about how to optimize MIMO in a multiuser network. The main difficulty lies in the interdependencies caused by mutual interference. In order to fully exploit the available resources, a joint optimization of all MIMO links and the resource allocation is required. This often results in cross-layer optimization problems that are difficult to handle.

This chapter discusses a general mathematical framework, namely interference functions [Yat95, SB06], that can be used to solve different optimization problems

MIMO. DOI: 10.1016/B978-0-12-382194-2.00004-6

in MIMO interference networks, with a special emphasis on power control (QoS balancing).

We begin with a general review of multiuser MIMO channels in Section 4.1. This section provides a brief overview on some key concepts for multiuser MIMO systems, with a special emphasis on beamforming. We discuss some practical examples of interference functions. These are special cases of the more general mathematical framework of interference functions which will be introduced later in Section 4.2. Readers who are familiar with these concepts can skip this section and begin directly with Section 4.2.

In Section 4.2 we introduce and discuss the axiomatic framework of interference functions. The framework of *interference functions* provides a unifying model that captures core properties of typical interference scenarios.

Finally, in Section 4.3, we show how properties of interference functions can be exploited for the development of iterative algorithms.

4.1 MULTIUSER CHANNELS

Figure 4.1 illustrates three fundamental multiuser channels that are often encountered in wireless networks. We assume that there is no cooperation between the transmitting or receiving terminals. Independent information is transmitted to each user. Each terminal is possibly equipped with multiple antennas, which can be used for *spatial filtering*, provided that some knowledge about the channel is available. If the channel matrix of a transmitter-receiver pair has rank greater than one, then it is possible to spatially multiplex several data streams on a single communication link. In the following, however, we will discuss the simpler case where there is just a single data stream, as illustrated in Figure 4.1. This is sometimes referred to as *beamforming*.

The *multiple access channel* (MACh) is a channel in which there are several transmitters and a single receiver. A typical example is the *uplink* of a cellular or satellite wireless network. If the terminals transmit independent information on the same resource, then there is mutual interference.

The *broadcast channel* (BC) is a channel in which there is one sender and several receivers. Here, we are not interested in the case where the same information is broadcasted to all terminals, we rather assume that independent information

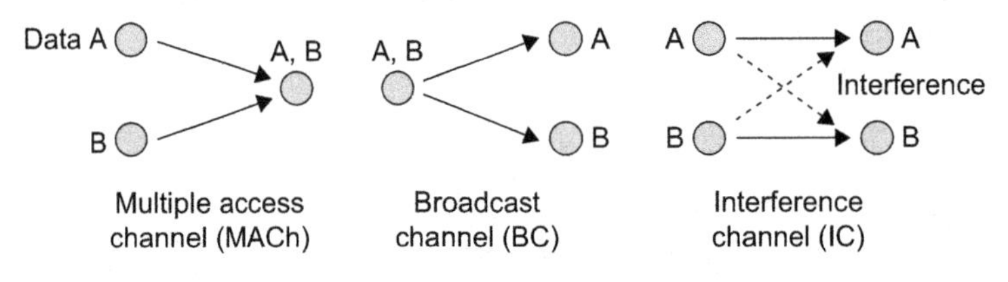

FIGURE 4.1

Multiuser channels.

is transmitted to selected receivers. Interference between the data streams can be avoided by properly pre-processing the signals. The receivers cannot cooperate (e.g., because of different geographical locations). An example is the *downlink* of a cellular or satellite communication system.

Another interesting case is *multicasting*, where a single data stream is transmitted to selected terminals. Multicasting is gaining interest in the context of cellular wireless communication. For example, if several users subscribe to the same live video stream, then the stream can be transmitted over a single multicast link, thus avoiding the bandwidth-costly usage of several independent links.

Most concepts for multicasting have been developed in the context of network protocols. However, in a wireless network with MIMO links, the physical layer greatly affects the achievable performance. For example, the spatial radiation pattern of the transmitting antenna array can be optimized for individual users [SDL06]. Thereby, the SNR can be improved and interference can be avoided. There resulting performance crucially depends on how the radiation pattern (the "beam") is adjusted.

A cellular network with multiple noncooperating base stations and multiple users can be modeled as an *interference channel* (IC). The IC is more difficult to handle than MACh and BC. While the capacity regions of the Gaussian MACh and BC are known, even for the MIMO case [CS00, CS03, VT03, VJG03, WSS04], the capacity region of the IC is still unknown, except for special cases (see, e.g., [ETW08]).

All these channels have an upper performance limit, which is achieved by allowing ideal cooperation between transmitters and receivers. In this case, the terminals on either side can be regarded as an antenna array whose signals can be combined coherently. The resulting channel is the conventional point-to-point MIMO link. Consequently, the discussed multiuser channels can be regarded as a MIMO channel with restricted cooperation between the input or output terminals.

Consequently, one important way to increase the network throughput is the development of strategies that enable cooperation between terminals. Here, "cooperation" can mean many things, involving different degrees of complexity. For example, terminals can monitor their environment (interference sensing), they can exchange channel state information, or they can even jointly transmit coherent signals from multiple base stations acting as a virtual antenna array. The latter case is also known as *network MIMO* (cf. Section 4.1.4).

An axiomatic approach for interference modeling that is applicable to all the discussed channels, will be discussed later in Section 4.2. To this end, we will briefly review some fundamentals of spatial filtering. This leads to a particular nonlinear interference model which is a special case of the more general framework that will be discussed in Section 4.2.

4.1.1 Spatial Interference Filtering

We begin by focusing on a multiple access channel (uplink), which is easier to handle analytically. The results can be transferred to broadcast (downlink) channels by exploiting the *uplink/downlink duality* which will be discussed in Section 4.1.3.

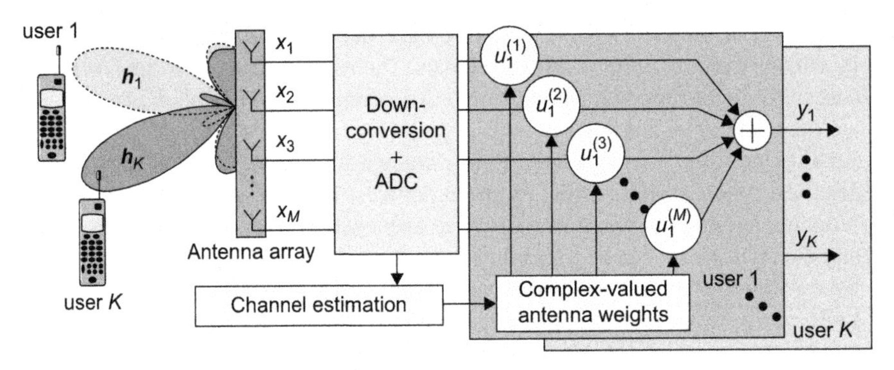

FIGURE 4.2

Spatial filtering: superimposed signals are separated by a bank of spatial filters (beamformers).

To some extent, this framework can also be applied to interference channels. However, many issues, like duality or capacity remain an open issue for the IC.

Consider an uplink system with K communication links (users) $\mathcal{K} = \{1,2,\ldots,K\}$ and an M-element antenna array at the receiver, as illustrated in Figure 4.2. The following discussion is confined to the case where each user transmits a single data stream. The K signals are modeled as random variables $S_1,\ldots,S_K$ with $p_k = \mathrm{E}[|S_k|^2]$.

The following signal model can easily be extended to the case where each user has multiple data streams (multiplexing). For example, data can be multiplexed over different beams (spatial multiplexing) or carriers (OFDM). This adds additional degrees of freedom but it also complicates the optimization of the communication links.

Let $\boldsymbol{t}_l$, with $\|\boldsymbol{t}_l\|_2 = 1$, be a complex-valued vector whose length is equal to the number of transmit antennas of user l. The vector $\boldsymbol{t}_l$ maps the signal S_1 to the transmit antennas. The resulting signal $\boldsymbol{t}_l S_1$ is transmitted over the propagation channel, which is characterized by a matrix $\boldsymbol{H}_l$, which is assumed to be a random variable. User l is transmitting over an *effective channel* $\boldsymbol{h}_l = \boldsymbol{H}_l \boldsymbol{t}_l$, which is random as well.

The matrix $\boldsymbol{H}_l$ contains the channel coefficients between all transmit and receive antennas of user l. The number of columns equals the number of transmit antennas, and the number of rows equals the number of receive antennas. Such a multiplicative channel model can be used if the signal bandwidth is relatively small in comparison to the coherence bandwidth of the propagation channel. This is the case, for example, in an OFDM system with narrowband carriers.

At the receiver, the resulting ouput of the antennas is $\boldsymbol{h}_l S_l$. The overall array output vector $\boldsymbol{x}$ is the superposition of the signals of all K users plus a random noise signal $\boldsymbol{n}$.

$$\boldsymbol{x} = \sum_{l \in \mathcal{K}} \boldsymbol{h}_l S_l + \boldsymbol{n}.$$

Assuming M receive antennas, we have $\boldsymbol{x} \in \mathbb{C}^M$. The transmitted signals $S_1,\ldots,S_K$ can be recovered by a bank of linear filters $\boldsymbol{w}_1,\ldots,\boldsymbol{w}_K \in \mathbb{C}^M$ ("beamformers"). The output of the kth filter is

$$y_k = \boldsymbol{w}_k^H \left(\sum_{l \in \mathcal{K}} \boldsymbol{h}_l S_l + \boldsymbol{n} \right) = \underbrace{\boldsymbol{w}_k^H \boldsymbol{h}_k S_k}_{\text{desired signal}} + \underbrace{\sum_{l \in \mathcal{K};\, l \neq k} \boldsymbol{w}_k^H \boldsymbol{h}_l S_l}_{\text{interference}} + \underbrace{\boldsymbol{w}_k^H \boldsymbol{n}}_{\text{effective noise}} . \tag{4.1}$$

The desired signal is corrupted by interference and noise. An important performance measure is the *signal-to-interference-plus-noise ratio* (SINR). Many other performance measures, like bit error rate, or capacity can be linked to the SINR (see Subsection 4.1.2).

The SINR of each user depends on the transmission powers $\boldsymbol{p} = [p_1, \ldots, p_K]^T$ and the noise variance σ_n^2, which can be collected in a single variable

$$\underline{\boldsymbol{p}} = [p_1, \ldots, p_K, \sigma_n^2]^T. \tag{4.2}$$

Now, we introduce $\boldsymbol{R}_k = \mathrm{E}[\boldsymbol{h}_k \boldsymbol{h}_k^H]$, which is the spatial covariance matrix of the channel of user l. Exploiting that the signals are uncorrelated and $\mathrm{E}[\boldsymbol{n}\boldsymbol{n}^H] = \sigma^2 \boldsymbol{I}$, the SINR of user k is

$$\mathrm{SINR}_k(\underline{\boldsymbol{p}}, \boldsymbol{w}_k) = \frac{\mathrm{E}\left[\left| \boldsymbol{w}_k^H \boldsymbol{h}_k S_k \right|^2 \right]}{\mathrm{E}\left[\left| \sum_{l \neq k} \boldsymbol{w}_k^H \boldsymbol{h}_l S_l + \boldsymbol{w}_k^H \boldsymbol{n} \right|^2 \right]} = \frac{p_k \boldsymbol{w}_k^H \boldsymbol{R}_k \boldsymbol{w}_k}{\boldsymbol{w}_k^H \left(\sum_{l \neq k} p_l \boldsymbol{R}_l + \sigma_n^2 \boldsymbol{I} \right) \boldsymbol{w}_k}. \tag{4.3}$$

The beamformer that maximizes the SINR can be found efficiently via eigenvalue decomposition [MM80]. The SINR is maximized by any $\boldsymbol{w}_k$ fulfilling

$$\left(\sum_{l \neq k} p_l \boldsymbol{R}_l + \sigma_n^2 \boldsymbol{I} \right)^{-1} \boldsymbol{R}_k \cdot \boldsymbol{w}_k = \lambda_{\max} \cdot \boldsymbol{w}_k \tag{4.4}$$

where $\lambda_{\max}$ is the maximum eigenvalue of the matrix $\left(\sum_{l \neq k} p_l \boldsymbol{R}_l + \sigma_n^2 \boldsymbol{I} \right)^{-1} \boldsymbol{R}_k$. The resulting interference-plus-noise power of user k can be written as

$$\begin{aligned} \mathcal{I}_k(\underline{\boldsymbol{p}}) &= \frac{p_k}{\max_{\|\boldsymbol{w}_k\|=1} \mathrm{SINR}_k(\underline{\boldsymbol{p}}, \boldsymbol{w}_k)} = \min_{\|\boldsymbol{w}_k\|=1} \frac{p_k}{\mathrm{SINR}_k(\underline{\boldsymbol{p}}, \boldsymbol{w}_k)} \\ &= \min_{\|\boldsymbol{w}_k\|_2=1} \frac{\sum_{l \neq k} p_l \boldsymbol{w}_k^H \boldsymbol{R}_l \boldsymbol{w}_k + \|\boldsymbol{w}_k\|^2 \sigma_n^2}{\boldsymbol{w}_k^H \boldsymbol{R}_k \boldsymbol{w}_k} = \min_{\|\boldsymbol{w}_k\|_2=1} \underline{\boldsymbol{p}}^T \boldsymbol{v}_k \end{aligned} \tag{4.5}$$

where $\boldsymbol{v}_k$ is a vector of coupling coefficients, whose lth element is defined as follows.

$$[\boldsymbol{v}_k]_l = \begin{cases} \frac{\boldsymbol{w}_k^H \boldsymbol{R}_l \boldsymbol{w}_k}{\boldsymbol{w}_k^H \boldsymbol{R}_k \boldsymbol{w}_k} & 1 \leq l \leq K,\ l \neq k \\ \frac{\|\boldsymbol{w}_k\|^2}{\boldsymbol{w}_k^H \boldsymbol{R}_k \boldsymbol{w}_k} & l = K+1, \\ 0 & l = k. \end{cases} \tag{4.6}$$

The notation in (4.5) might appear unnecessarily complicated, but it will turn out to be useful later in Section 4.2, where it will be shown that the interference function (4.5) is a special case of a more general framework.

The vector $v_k(w_k)$ determines how user k is interfered by other users. For any given $p > 0$, the beamformer w_k can adapt to the current interference situation.

One possible way of choosing w_k is to enforce $w_k^H R_l w_k = 0$. This approach, known as *zeroforcing*, is suboptimal in the presence of power constraints since it neglects the noise enhancement factor $\|w_k\|^2$ in (4.5). Another strategy is to minimize $\|w_k\|^2$. This maximizes the signal-to-noise ratio (SNR) and is known as the *spatial matched filter*. Clearly, this approach has the disadvantage of not completely eliminating the interference. The aforementioned SINR maximization strategy (4.5) provides a compromise between eliminating interference and minimizing $\|w_k\|^2$.

A special case occurs if the channels h_k are deterministic, then $R_k = h_k h_k^H$. Such a deterministic model is usually assumed if the channel h_k is constant within the time scale of interest, so that it can be estimated. Replacing R_k in (4.4) by $h_k h_k^H$, it can be observed that the SINR is maximized by scalar multiples of the vector

$$w_k^{\text{opt}} = \left(\sum_{l \neq k} p_l R_l + \sigma_n^2 I \right)^{-1} h_k. \tag{4.7}$$

Plugging w_k^{opt} in the SINR (4.3), we obtain a closed-form expression of the interference.

$$\mathcal{I}_k(\underline{p}) = \frac{1}{h_k^H \left(\sigma_n^2 I + \sum_{l \neq k} p_l R_l \right)^{-1} h_k}. \tag{4.8}$$

Note, that the function (4.8) is concave in $\underline{p}$. This is a consequence of (4.8) being a special case of (4.5). Recall that the minimum of linear functions is concave. Later, in Section 4.3 it will be discussed how concavity can be exploited for the design of resource allocation algorithms.

4.1.2 Capacity Region

Consider a Gaussian multiple access channel (MACh) with single-antenna transmitters and an antenna array at the receiver. If the channels $h_1, \ldots, h_K$ are known at the receiver, then minimum mean square error (MMSE) filtering with *successive interference cancellation* (SIC) is a capacity achieving strategy [CT91, VG97, VT03, VJG03, YC04]. The main idea of SIC is to decode the users successively. In each step, the already decoded part of the signal is subtracted off. Assuming a decoding order $K \ldots 1$, the resulting interference-plus-noise covariance at user k is

$$Z_k = \sigma_n^2 I + \sum_{l=1}^{k-1} p_l h_l h_l^H. \tag{4.9}$$

For any given p, the MMSE beamformer of user k is $w_k^{\text{MMSE}} = p_k Z_k^{-1} h_k$. Note, that any MMSE beamformer maximizes the SINR. Conversely, not every beamformer that maximizes the SINR also minimizes the MSE.

An interesting question is how to adjust the powers $\boldsymbol{p}$ so as to achieve a certain point from the capacity region. With (4.8), the capacity of user k is

$$C_k(\boldsymbol{p}) = \log\big(1+\mathrm{SINR}_k(\boldsymbol{p})\big) = \log\big(1+p_k\boldsymbol{h}_k^H\boldsymbol{Z}_k^{-1}\boldsymbol{h}_k\big). \tag{4.10}$$

Because of the monotonic relationship between capacity and SINR, the problem of achieving certain capacity targets can be transferred to the problem of achieving SINR targets $\gamma_1,\ldots,\gamma_K$.

This can be achieved by a simple algorithm presented in [SB05]. It exploits the recursive structure of $\boldsymbol{Z}_k$. Because of SIC, the first user "sees" no interference, so $\boldsymbol{Z}_1 = \sigma_n^2\boldsymbol{I}$. Given $\boldsymbol{Z}_1$, the covariances of all other users can be computed recursively.

$$\boldsymbol{Z}_{k+1} = \boldsymbol{Z}_k + p_k\boldsymbol{h}_k\boldsymbol{h}_k^H. \tag{4.11}$$

Applying the *matrix inversion lemma* (Sherman-Morrison formula [Mey00]), we obtain

$$\boldsymbol{Z}_{k+1}^{-1} = \boldsymbol{Z}_k^{-1} - \frac{p_k\boldsymbol{Z}_k^{-1}\boldsymbol{h}_k\boldsymbol{h}_k^H\boldsymbol{Z}_k^{-1}}{1+p_k\boldsymbol{h}_k^H\boldsymbol{Z}_k^{-1}\boldsymbol{h}_k}.$$

This recursive interference structure facilitates an efficient computation of the optimal transmission powers for achieving arbitrary SINR targets $\gamma_1,\ldots,\gamma_K > 0$. Given the initialization $\boldsymbol{Z}_1^{-1} := \boldsymbol{I}/\sigma_n^2$, the following K steps are carried out.

1: **for** $k = 1$ to K **do**

2: $\quad p_k^{opt} := \dfrac{\gamma_k}{\boldsymbol{h}_k^H\boldsymbol{Z}_k^{-1}\boldsymbol{h}_k}$

3: $\quad \boldsymbol{Z}_{k+1}^{-1} := \boldsymbol{Z}_k^{-1} - \dfrac{p_k^{opt}\boldsymbol{Z}_k^{-1}\boldsymbol{h}_k\boldsymbol{h}_k^H\boldsymbol{Z}_k^{-1}}{1+\gamma_k}$

4: **end for**

The resulting power vector $\boldsymbol{p}^{opt}$ achieves the SINR targets $\boldsymbol{\gamma} = [\gamma_1,\ldots,\gamma_K]$. Assume that there exists another vector $\boldsymbol{p}_2$ which also achieves these targets, then $\boldsymbol{p}^{opt} \le \boldsymbol{p}_2$ (component-wise). That is, $\boldsymbol{p}^{opt}$ is optimal in terms of power expenditure. Associated with $\boldsymbol{p}^{opt}$ is a set of optimal beamformers (4.7).

It is perhaps interesting to note that arbitrary targets can be achieved. In the absence of power constraints, the capacity region is the non-negative orthant $\mathbb{R}_+^K$, no matter how many users there are. When increasing the number of users, then also the interference increases. However, this can always be outweighed by increasing the power. This means that the system performance is not interference-limited, but only limited by the combined effect of noise and power constraints. This behavior is a consequence of the assumed SIC. Without SIC the system might very well be interference-limited. This case will be discussed in the following sections, where a more general interference model will be assumed. In particular, the problem of minimizing the total transmission power subject to SINR constraints are addressed in Subsection 4.2.3.

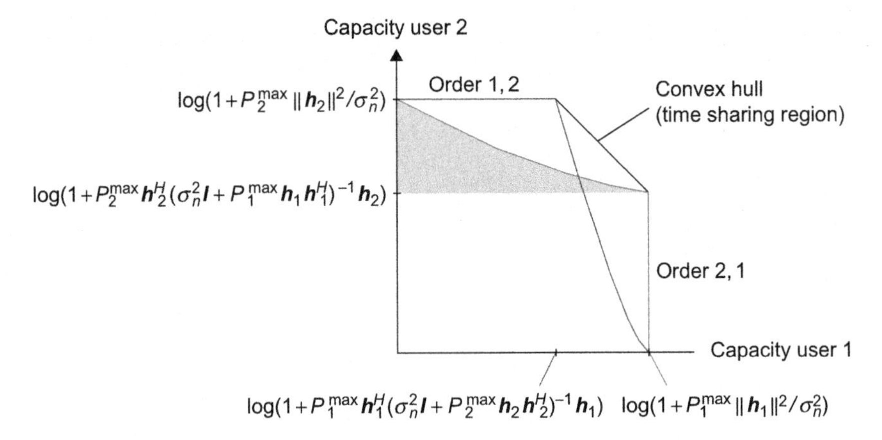

FIGURE 4.3

Capacity region of the Gaussian MACh with individual power constraints $P_1 \leq P_1^{\max}$ and $P_2 \leq P_2^{\max}$.

The solution discussed here can be regarded as a special case of this generalized framework. That is, the power vector $\boldsymbol{p}^{opt}$ minimizes the total power subject to SINR constraints. This shows that under certain assumption on the interference model (here: SIC), the solution of the power minimization problem becomes simple. Other interference scenarios will be discussed later.

The capacity region for individual power constraints $P_1^{\max}, P_2^{\max}$ and $K = 2$ is illustrated in Figure 4.3.

For explanation, consider the decoding order 2,1, which is associated with the region outlined in blue. User 1 sees no interference from user 2. Thus, user 1 can always achieve a capacity $\log\left(1 + P_1^{\max}\|\boldsymbol{h}_1\|^2/\sigma_n^2\right)$, no matter how large the power of user 1 is. User 2, however, is interfered by user 1. If user 1 transmits at maximum power $P_1^{\max}$, then with (4.10) the capacity of user 2 is $\log\left(1 + P_2^{\max}\boldsymbol{h}_2^H\left(\sigma_n^2\boldsymbol{I} + P_1^{\max}\boldsymbol{h}_1\boldsymbol{h}_1^H\right)^{-1}\boldsymbol{h}_2\right)$. User 2 can only achieve a higher capacity if user 1 reduces its interference to user 2. This corresponds to the shaded region above the dashed line.

The sum capacity as a function of the transmission powers is

$$C_{sum}(\underline{p}) = \sum_{k=1}^{K} \log\left(1 + \mathrm{SINR}_k(\underline{p})\right). \tag{4.12}$$

Using the determinant $|\cdot|$, we have

$$\begin{aligned}\log(1 + \mathrm{SINR}_k(\underline{p})) &= \log(1 + p_k\boldsymbol{h}_k^H\boldsymbol{Z}_k^{-1}\boldsymbol{h}_k) \\ &= \log|\boldsymbol{I} + p_k\boldsymbol{h}_k\boldsymbol{h}_k^H\boldsymbol{Z}_k^{-1}| \\ &= \log|p_k\boldsymbol{h}_k\boldsymbol{h}_k^H + \boldsymbol{Z}_k| - \log|\boldsymbol{Z}_k|.\end{aligned} \tag{4.13}$$

Exploiting that the interference-plus-noise spatial covariance Z_k has the special recursive structure (4.11), we can write

$$\begin{aligned}
C_{sum}(\underline{p}) &= \sum_{k=1}^{K} \log|Z_{k+1}| - \sum_{k=1}^{K} \log|Z_k| \\
&= \log|Z_{K+1}| - \log|Z_1| \\
&= \log\left|\sigma_n^2 I + \sum_{k=1}^{K} p_k h_k h_k^H\right| - \log|\sigma_n^2 I| \\
&= \log\left|I + \frac{1}{\sigma_n^2}\sum_{k=1}^{K} p_k h_k h_k^H\right| \\
&= \log\left|I + \frac{1}{\sigma_n^2} H P H^H\right| \qquad (4.14)
\end{aligned}$$

where $P = \mathrm{diag}\{p_1, \ldots, p_K\}$ and $H = [h_1, \ldots, h_K]$.

Note, that (4.13) has the same form as the point-to-point MIMO capacity

$$\log\left|I + \frac{1}{\sigma_n^2} H Q H^H\right| \qquad (4.15)$$

where Q is a positive-semidefinite transmit covariance. However, in (4.13), the transmit covariance P is restricted to be diagonal. From a practical point of view, this means that the transmitting antennas are uncorrelated. That is, there is no cooperation between the terminals. The capacity of the fully cooperative system (4.14) provides an upper bound for the capacity region. For further reading on this aspect, we refer to [VJG03, JB04].

The throughput-wise optimal power allocation is obtained by solving

$$\max_{p \in \mathcal{P}} C_{sum}(\underline{p}) \qquad (4.16)$$

where $\mathcal{P}$ is a compact set of possible transmission powers, resulting from power constraints (e.g., total power constraint or per-user power constraints). The function $C_{sum}(p)$ is concave and can be efficiently optimized (see e.g., [VBW98]).

4.1.3 Duality

Throughout the literature, it was observed that the feasible utility region of certain multiuser channels is preserved when reversing the role of transmit and receive strategies. Examples are SIR balancing [ZF94], beamforming [RLT98, VT03, BS02, SB04], MMSE optimization [SSB07, HJU09]. Related results were observed in an information theoretic context, where it was shown that the Gaussian MIMO MACh has the same capacity region as the Gaussian broadcast channel [CS00, CS03, VT03, VJG03, WSS04]. This reciprocity between certain multiuser channels is sometimes referred to as *duality*. Here, the term duality differs from the conventional definition used in optimization theory, although there are some interesting parallels, as shown in [WES06].

Multiuser duality is useful in optimizing the communication over certain channels that are too difficult to handle directly. As an example, we will briefly discuss the SINR duality observed in [ZF94, RLT98, SB04, BS02, VT03]. To this end, we use the SINR model developed in Section 4.1.1. Since the SINR is invariant with respect to a scaling of $\boldsymbol{w}_k$, we can assume $\|\boldsymbol{w}_k\| = 1$ without loss of generality. The interference coupling coefficients are collected in a matrix $\boldsymbol{\Psi} \in \mathbb{R}_+^{K \times K}$, defined as

$$[\boldsymbol{\Psi}]_{kl} = \begin{cases} \boldsymbol{w}_k^H \boldsymbol{R}_l \boldsymbol{w}_k, & l \neq k \\ 0, & l = k. \end{cases} \tag{4.17}$$

The resulting SINR can be written as

$$\mathrm{SINR}_k(\underline{\boldsymbol{p}}, \boldsymbol{w}_k) = \frac{p_k \cdot \boldsymbol{w}_k^H \boldsymbol{R}_l \boldsymbol{w}_k}{\left[\boldsymbol{\Psi}\boldsymbol{p} + \sigma_n^2 \boldsymbol{1}\right]_k}, \quad \forall k \in \mathcal{K}. \tag{4.18}$$

Assume that we want to achieve certain SINR values $\boldsymbol{\gamma} = [\gamma_1, \ldots, \gamma_K] > 0$. We say that $\boldsymbol{\gamma}$ is *feasible* if we can find $\underline{\boldsymbol{p}}$ and $\boldsymbol{w}_k$ such that the following inequality is satisfied.

$$\mathrm{SINR}_k(\underline{\boldsymbol{p}}, \boldsymbol{w}_k) \geq \gamma_k, \quad \forall k \in \mathcal{K}. \tag{4.19}$$

Defining the diagonal matrix $\boldsymbol{D} = \mathrm{diag}(\gamma_1/\boldsymbol{w}_1^H \boldsymbol{R}_1 \boldsymbol{w}_1, \ldots, \gamma_K/\boldsymbol{w}_K^H \boldsymbol{R}_K \boldsymbol{w}_K)$, the feasibility condition (4.18) can be rewritten in matrix notation as $\boldsymbol{p} \geq \boldsymbol{D}[\boldsymbol{\Psi}\boldsymbol{p} + \sigma_n^2 \boldsymbol{1}]$, where the inequality is meant component-wise. Equivalently, we have

$$\left[\boldsymbol{I} - \boldsymbol{D}\boldsymbol{\Psi}\right]\boldsymbol{p} \geq \boldsymbol{D}\boldsymbol{1}\sigma_n^2. \tag{4.20}$$

Whether or not there exists a $\boldsymbol{p} > 0$ that fulfills (4.18), respectively (4.19), depends on the choice of the spatial filters $\boldsymbol{w}_1, \ldots, \boldsymbol{w}_K$. A positive vector $\boldsymbol{p} > 0$ fulfilling (4.19) exists if and only if the *spectral radius* $\lambda_{\max}(\boldsymbol{D}\boldsymbol{\Psi})$ (which in this case is the same as the maximum eigenvalue) fulfills $\lambda_{\max}(\boldsymbol{D}\boldsymbol{\Psi}) \leq 1$. This is a consequence of the convergence properties of the Neumann series [Mey00, SB04]. Thus, the SINR region is

$$\mathcal{S} = \{\boldsymbol{\gamma} \geq 0 : \min_{\boldsymbol{w}_1, \ldots, \boldsymbol{w}_K} \lambda_{\max}(\boldsymbol{D}\boldsymbol{\Psi}) \leq 1\}. \tag{4.21}$$

The minimization of the maximum eigenvalue (4.21) can be carried out efficiently by the algorithm [BS09].

For any given $\boldsymbol{w}_1, \ldots, \boldsymbol{w}_K$, we have $\lambda_{\max}(\boldsymbol{D}\boldsymbol{\Psi}) = \lambda_{\max}(\boldsymbol{D}\boldsymbol{\Psi}^T)$, so $\mathcal{S}$ can be equivalently expressed in terms of the transpose coupling matrix $\boldsymbol{\Psi}^T$. This transpose operation corresponds to reversing the role of transmitters and receivers. If $\boldsymbol{\Psi}$ models an uplink channel, as above, then $\boldsymbol{\Psi}^T$ models the "dual" downlink channel. As a consequence, the SINR region resulting from condition (4.18) is the same as the region resulting from the following downlink SINRs.

$$\mathrm{SINR}_k^{DL}(\underline{\boldsymbol{q}}, \boldsymbol{w}_1, \ldots, \boldsymbol{w}_K) = \frac{q_k \cdot \boldsymbol{w}_k^H \boldsymbol{R}_k \boldsymbol{w}_k}{\left[\boldsymbol{\Psi}^T \boldsymbol{q} + \boldsymbol{1}\sigma_n^2\right]_k}, \quad \forall k. \tag{4.22}$$

Here, we use a different variable $\boldsymbol{q}$ instead of $\boldsymbol{p}$ to emphasize that generally different "power vectors" are required to achieve the same SINR point in both uplink and downlink. This means that duality does not necessarily hold under additional power constraints, as discussed later.

Unlike the uplink SINR (4.17), the downlink SINR (4.21) depends on *all* spatial filters $\boldsymbol{w}_1,\ldots,\boldsymbol{w}_K$. This is a consequence of using $\boldsymbol{\Psi}^T$ instead of $\boldsymbol{\Psi}$. This means that a joint approach is required for optimizing the downlink (transmit) beamformers in (4.21), whereas uplink (receive) beamformers can be optimized independently for a given power vector $\boldsymbol{p}$.

In some cases, the downlink can be optimized directly, as shown in [BO01, WES06]. But we can also exploit the duality between uplink and downlink channels, which states that the optimal downlink beamformers can be found indirectly, by optimizing a "virtual" uplink channel. This sometimes leads to efficient algorithmic solutions, as shown in [RLT98, SB04].

Consider SINR targets $\boldsymbol{\gamma} > 0$ with a corresponding matrix $\boldsymbol{D}$. If $\lambda_{\max}(\boldsymbol{D\Psi}) < 1$, then it can be shown with (4.19) that the optimum vector achieving $\boldsymbol{\gamma}$ is

$$\boldsymbol{p} = \left[\boldsymbol{D}^{-1} - \boldsymbol{\Psi}\right]^{-1}\sigma_n^2\mathbf{1}. \tag{4.23}$$

Here, "optimum" means that for any vector $\hat{\boldsymbol{p}}$ fulfilling the inequality (4.19), we have $\boldsymbol{p} \le \hat{\boldsymbol{p}}$ (component-wise). Because of $\lambda_{\max}(\boldsymbol{D\Psi}) = \lambda_{\max}(\boldsymbol{D\Psi}^T)$, we know that $\boldsymbol{\gamma}$ is also achievable in the dual "downlink" channel. The optimum downlink solution is

$$\boldsymbol{q} = \left[\boldsymbol{D}^{-1} - \boldsymbol{\Psi}^T\right]^{-1}\sigma_n^2\mathbf{1}. \tag{4.24}$$

Although $\boldsymbol{p}$ and $\boldsymbol{q}$ achieve the same SINR targets in uplink and downlink, respectively, they are different. Thus, if additional power constraints are imposed, uplink and downlink can have different SINR regions. An exception is the total power constraint $\sum_k p \le P_{\max}$, as observed in [VT03, SB04, BS02, YL07]. This follows from (4.19). Defining $\mathbf{1} = [1,\ldots,1]^T$, the minimum sum power for achieving uplink and downlink targets $\boldsymbol{\gamma}$ is

$$\mathbf{1}^T\boldsymbol{p} = \mathbf{1}^T\left[\boldsymbol{D}^{-1} - \boldsymbol{\Psi}\right]^{-1}\mathbf{1}\cdot\sigma_n^2 = \mathbf{1}^T\left[\boldsymbol{D}^{-1} - \boldsymbol{\Psi}^T\right]^{-1}\mathbf{1}\cdot\sigma_n^2 = \mathbf{1}^T\boldsymbol{q}. \tag{4.25}$$

That is, if a point is feasible in the uplink, then it also feasible in the downlink, and the same sum power is needed in order to achieve this point [VT03, SB04, YL07]. Moreover, this point is achieved by the same beamformers. Hence, for a downlink channel with a sum power constraint or with no power constraints, the optimum transmit beamformers can be obtained indirectly by optimizing a virtual uplink channel, which is obtained by using the transpose of the coupling matrix.

Another form of duality was presented in [YL07], where *per-antenna power constraints* were investigated. Finding a duality for per-user power constraints is still an open problem.

4.1.4 Network MIMO

Over the last decades, cellular wireless networks have witnessed significant throughput gains. Most of these gains are due to the ability to reuse the spectrum geographically, meaning an increased density of base stations in cellular networks in the most populated areas (micro-, pico- and soon femto-cellularization) [CA08]. Because of this aggressive reuse of spectrum, interference is becoming more and more problematic.

The detrimental effect of interference is particularly severe at the cell edge, where interference can be very strong, and at the same time the signal energy is weak. Therefore, strategies for improving the cell-edge performance are of particular interest in current standardization activities. For example, a quote from [3GP09] says

> *"A more homogeneous distribution of the user experience over the coverage area is highly desirable and therefore a special focus should be put on improving the cell edge performance."*

The downlink of the multicellular system can be regarded as a MIMO system with geographically distributed transmit antennas (the base stations) and distributed receive antennas (the user equipment). An upper limit on the achievable throughput is obtained by letting the transmitting base stations cooperate (see Figure 4.4).

Here, cooperation means that the same signal is transmitted from different base stations. Assuming that the different propagation delays are compensated prior to transmission, the signals coherently combine at the receiver. By properly adjusting the phases of the signals, one can achieve a constructive combination of the signals at

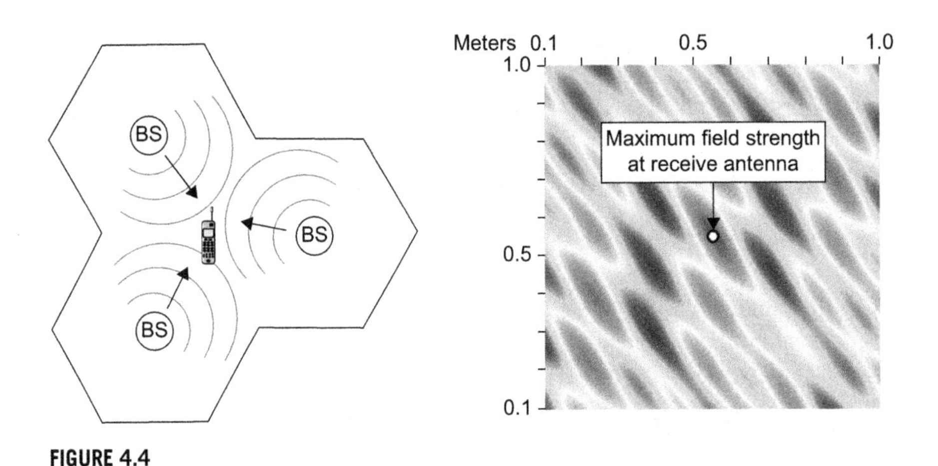

FIGURE 4.4

Cooperative transmission: The SNR is improved by coherently combining the signals from multiple base stations (BS). The resulting field strength pattern is computed for 2 Ghz carrier frequency and an antenna spacing of $\approx$ 100 m.

the desired receiver, while steering "nulls" towards other points. As a consequence, the SNR of the desired link is improved and interference between users can be avoided. This technique is closely related to beamforming (see Section 4.1.1). It differs from beamforming in that the distance of the antennas is rather large, compared with the typical half-wavelength distance of an antenna array. This leads to strongly varying field strength pattern (see Figure 4.4), which makes network MIMO quite sensitive to errors (synchronization, channel estimation, etc.).

An extreme case is cooperation based on the knowledge of perfect channel state information. This provides a theoretic limit for the achievable throughput, which can serve as a benchmark. From a mathematical point of view, this corresponds to the case of a downlink channel with a single virtual antenna array, consisting of the base stations in the system. All the techniques discussed in the previous section apply to this scenario.

However, there are additional practical constraints, like the power limitations of the amplifiers. In this case, the optimization of the powers resources and beamformers needs to be carried out under the assumption of per-antenna power constraints. This case is more difficult to handle than the basic beamforming scenario discussed in the previous sections. Besides, coherent transmission from geographically distributed antennas is very demanding in terms of required synchronization between the base stations. In order to put network MIMO into practice, a high-capacity backbone (e.g., optical fiber line) is probably required.

Despite these difficulties, network MIMO is a promising concept, since it offers a fundamentally new method to overcome the interference-limitedness of cellular system. It is quite likely that the current hierarchical structure of cellular systems will be gradually replaced by more flexible system architectures, which offer a better spectral efficiency. Network MIMO is one step towards this direction. It offers additional degrees of freedom, which can be exploited for improving the cell edge SNR or for mitigating interference. A challenging task is the combination of network MIMO with dynamic resource allocation. The ultimate goal is to deliver the resources to the antennas where they are needed and where they cause as little interference as possible.

4.2 A GENERAL FRAMEWORK FOR OPTIMIZING INTERFERENCE NETWORKS

In the previous section we have discussed some examples where joint spatial processing and power allocation can be optimized efficiently. However, many open problems remain. For example, the joint optimization of transmit and receiver beamformers is difficult, except for a few special scenarios (for example, sum-rate maximization [VJG03, YC04, VT03, WSS04] or sum-MMSE minimization with perfect channel state information [SSB07]). A major challenge is finding good communication strategies for the interference channel, whose capacity region is still unknown.

These problems cannot be conquered without a thorough mathematical modeling. In this section we will discuss an abstract mathematical approach for modeling interference. The main idea is to model interference by a framework of axioms, defining some core properties. The framework is analytically appealing. It offers enough structure to allow for efficient algorithmic solutions. On the other hand, the model is quite general, so it includes many practical interference scenarios as special case. A simple example that has been discussed in the previous Section 4.1 is the interference (4.5) resulting from adaptive beamforming.

4.2.1 Interference Functions

We begin by introducing a basic interference model. Consider a coupled multiuser system with L resources $\boldsymbol{r} = [r_1, r_2, \ldots, r_L]^T \in \mathbb{R}^L_+$. For example, think of a system with L communication links, where $\boldsymbol{r}$ contains the transmission powers assigned to these links. Different links can belong to the same user, e.g., when data streams are multiplexed over frequencies, or spatial beams. Inactive links correspond to zero entries of the vector $\boldsymbol{r}$.

But $\boldsymbol{r}$ is not confined to powers. There are other system variables that can be measured in terms of non-negative real values (e.g., SINR, as discussed in the following Section 4.2.2).

The system performance depends on the choice of the resource vector $\boldsymbol{r}$. In some cases, this dependency can be modeled using the theory of *interference functions* [Yat95, SB06]. Let $\mathcal{I} : \mathbb{R}^L_+ \mapsto \mathbb{R}_+$. We say that $\mathcal{I}$ is a general interference function (or simply *interference function*) if the following axioms are fulfilled:

A1 (positivity)	There exists an $\boldsymbol{r} > 0$ such that $\mathcal{I}(\boldsymbol{r}) > 0$,
A2 scale invariance)	$\mathcal{I}(\alpha\boldsymbol{r}) = \alpha\mathcal{I}(\boldsymbol{r})$ for all $\alpha > 0$,
A3 (monotonicity)	$\mathcal{I}(\boldsymbol{r}) \geq \mathcal{I}(\boldsymbol{r}')$ if $\boldsymbol{r} \geq \boldsymbol{r}'$.

Throughout this chapter, all vector inequalities are element-wise, i.e., $\boldsymbol{r} \geq \boldsymbol{r}'$ means that $r_l \geq r'_l$ for all elements $l = 1, 2, \ldots, L$.

Property A3 is quite intuitive. If we increase the amount of expended resources $\boldsymbol{r}$, then the resulting quantity $\mathcal{I}$ will increase (or at least not decrease). An immediate example is power control, where $\boldsymbol{r}$ is a vector of transmission powers, and $\mathcal{I}(\boldsymbol{r})$ is the resulting interference at some other user. Hence, the name "interference function."

The axioms A1, A2, A3 were proposed and studied in [SB06], with extensions in [BS08c, BS08b, BS08a]. The connection with Yates' well-known framework of *standard interference functions* [Yat95] will be discussed later in Section 4.2.3.

We will begin with an example where interference functions are used as an indicator for feasibility of Quality of Service (QoS) values in a network.

4.2.2 Quality of Service Region

In this chapter, *Quality of Service* (QoS) stands for an arbitrary performance measure, which depends on the SINR by a strictly monotonic and continuous function ϕ

defined on $\mathbb{R}_+$. The QoS of user k is

$$u_k(\boldsymbol{p}) = \phi_k\big(\mathrm{SINR}_k(\boldsymbol{p})\big), \quad k \in \mathcal{K}. \tag{4.26}$$

Many performance measures depend on the SINR in this way. Examples are SINR, logarithmic SINR, MMSE, bit error rate, etc.

Let γ_k be the inverse function of ϕ_k, then $\gamma_k(u_k)$ is the minimum SINR level needed by the kth user to satisfy some QoS target u_k. Assume that the QoS is defined on some domain $\mathbb{Q}$, and the K-dimensional domain is denoted by $\mathbb{Q}^K$. Let $\boldsymbol{u} \in \mathbb{Q}^K$ be a vector of QoS values, then the corresponding SINR vector is

$$\boldsymbol{\gamma}(\boldsymbol{u}) = [\gamma_1(u_1), \ldots, \gamma_K(u_K)]^T. \tag{4.27}$$

That is, $\boldsymbol{u}$ is feasible if and only if the SINR targets $\boldsymbol{\gamma} := \boldsymbol{\gamma}(\boldsymbol{u})$ are feasible. Thus, we can focus on the analysis of the SINR region.

We say that $\boldsymbol{\gamma}$ is feasible if for any $\epsilon > 0$ there exists a $\boldsymbol{p}_\epsilon > 0$ such that

$$\mathrm{SINR}_k(\boldsymbol{p}_\epsilon) \geq \boldsymbol{\gamma}_k - \epsilon, \quad \forall k \in \mathcal{K}. \tag{4.28}$$

If this is fulfilled for $\epsilon = 0$, then the targets $\boldsymbol{\gamma}$ can actually be attained, otherwise they are achieved in an asymptotic sense. For practical scenarios, the former case is mostly ensured by noise and power constraints. A necessary and sufficient condition for feasibility is $f(\boldsymbol{\gamma}) \leq 1$, where we define the *indicator function*

$$f(\boldsymbol{\gamma}) = \inf_{\boldsymbol{p}>0}\left(\max_{k\in\mathcal{K}} \frac{\gamma_k}{\mathrm{SINR}_k(\boldsymbol{p})}\right). \tag{4.29}$$

The resulting SINR region is

$$\mathcal{S} = \{\boldsymbol{\gamma} \geq 0 : f(\boldsymbol{\gamma}) \leq 1\}. \tag{4.30}$$

This definition of an SINR (resp. QoS) region is quite general, and not tied to any particular channel or interference mitigation strategy. The definition (4.29) contains other known SINR regions as special cases.

For example, if we consider an uplink channel with adaptive beamforming at the receiver, then we obtain the previously discussed interference functions (4.5) with the corresponding SINR region (4.20). For this particular case, we have [SB06]

$$f(\boldsymbol{\gamma}) = \min_{\boldsymbol{w}_1,\ldots,\boldsymbol{w}_K} \lambda_{\max}(\boldsymbol{D}(\boldsymbol{\gamma})\boldsymbol{\Psi}). \tag{4.31}$$

The indicator function $f(\boldsymbol{\gamma})$ fulfills the axioms A1, A2, A3. Thus, $\mathcal{S}$ is a subset of an interference function. This has some interesting consequences: Certain properties of $\mathcal{S}$ directly correspond with the properties of $f(\boldsymbol{\gamma})$. By analyzing $f(\boldsymbol{\gamma})$ we obtain valuable information about the structure of QoS regions [BS08c].

For example, the property A3 translates into *comprehensiveness*. A set $\mathcal{V} \subset \mathbb{R}^K_{++}$ (strictly positive reals) is said to be *upward-comprehensive* if for all $\boldsymbol{w} \in \mathcal{V}$ and $\boldsymbol{w}' \in \mathbb{R}^K_{++}$, the inequality $\boldsymbol{w}' \geq \boldsymbol{w}$ implies $\boldsymbol{w}' \in \mathcal{V}$. If the inequality is reversed, then $\mathcal{V}$ is said to be *downward-comprehensive*. In the context of game theory, comprehensiveness is interpreted as *free disposability of utility* [Pet92]. If a user can achieve a certain utility, then it can freely dispose of all utilities below.

It was shown in [BS08c] that any set from $\mathbb{R}^K_{++}$ is closed downward-comprehensive if and only if it is a sublevel set of an interference function. Likewise, it is closed upward-comprehensive if and only if it is a superlevel set of an interference function. This shows a one-to-one correspondence between interference functions and comprehensive sets. Moreover, it was shown that such a comprehensive set is convex if and only if the corresponding interference function is convex.

Convexity of the QoS region is a desirable property which often facilitates efficient algorithmic solutions. However, many QoS regions are comprehensive but not convex. A standard approach is to "convexify" the utility set by randomization techniques (see e.g., [Zho97, SDHM07]), or by resource sharing. However, such a strategy may not always be possible or relevant in all situations.

Another, more elegant way is to exploit "*hidden convexity*," if available. Sometimes, a given problem is not convex but there exists an equivalent convex problem formulation. That is, the original nonconvex problem can be solved indirectly by solving the equivalent problem instead.

An example is *logarithmic convexity* (log-convexity). A function $f(x)$ is said to be log-convex on its domain if $\log f(x)$ is convex. Log-convexity was already exploited in the context of power control [Sun02, CIM04, BS05, BS04, SWB06, CTP+07, TPC07]. Assume that $\text{SINR}_k(\boldsymbol{p}) = p_k/\mathcal{I}_k(\boldsymbol{p})$, and $\mathcal{I}_1,\ldots,\mathcal{I}_K$ are log-convex interference functions after a change of variable $\boldsymbol{s} = \log \boldsymbol{p}$. That is, $\mathcal{I}(\exp(\boldsymbol{s}))$ is log-convex with respect to $\boldsymbol{s}$. For example, this is fulfilled, for all linear interference functions, and also for certain worst-case designs [BS08a].

Under the assumption of SINRs defined by log-convex interference functions, the resulting indicator function $f(\boldsymbol{\gamma})$ itself is a log-convex interference function after another change of variable $q_k = \log \gamma_k$ [BS08a]. That is, $f(\exp(\boldsymbol{q}))$ is log-convex with respect to $\boldsymbol{q} = (q_1,\ldots,q_K)$. Every log-convex function is convex, and a sublevel set of a convex function is convex. Thus, the set $\mathcal{S}$ is convex on a logarithmic scale. This can be exploited for the development of algorithms that operate on the boundary of the region [SWB06].

4.2.3 Power Control

Interference functions were originally proposed for power control problems. Consider K communication links with powers $\boldsymbol{p} \in \mathbb{R}^K_{++}$. Here, K can stand for the number of users, but we can also model the case where each user has multiple links, as discussed at the beginning of Section 4.2.1. Optimization strategies are mostly based on the SIR or the SINR, depending on whether the model includes noise or not.

The case with no noise corresponds to a system without power constraints. If the transmission powers can be arbitrarily large, then the impact of noise is negligible. This case is interesting from a theoretical point of view, since it allows to focus on the impact of interference coupling. Noiseless scenarios have been addressed in the early literature (e.g., [Zan92]) and later in the context of beamforming [GP96]. More recent extensions of these results can be found in [BS08a, BS09, SWB08]. Studying the noiseless case helps to obtain a thorough understanding of the underlying structure of power control problems. Two mathematical theories have been proved useful in this context. First, the Perron-Frobenius theory of non-negative matrices. Second, the theory of interference functions [Yat95, SB06].

In the following we will discuss a system with receiver noise power $\sigma_n^2 > 0$. In order to incorporate noise in the framework A1, A2, A3, we introduce an extended power resource vector

$$\underline{p} = \begin{bmatrix} \boldsymbol{p} \\ \sigma_n^2 \end{bmatrix} = [p_1, \ldots, p_K, \sigma_n^2]^T. \tag{4.32}$$

Then $\mathcal{I}(\underline{p})$ stands for interference plus noise, as in the previous example (4.5). While the impact of noise is evident in (4.5), it is not so obvious when defining the interference via the axioms A1, A2, A3 alone. In order for the noise to have any impact, we need the following additional property.

A4 (strict monotonicity) $\mathcal{I}(\underline{p}) > \mathcal{I}(\underline{p}')$ if $\underline{p} \geq \underline{p}'$ and $\underline{p}_{K+1} > \underline{p}'_{K+1}$.

If $\underline{p}_{K+1} > 0$, then A4 ensures that $\mathcal{I}(\underline{p}) > 0$ for arbitrary $\boldsymbol{p} \geq 0$. This can be easily shown by contradiction: Suppose that $\mathcal{I}(\underline{p}) = 0$, then for any α with $0 < \alpha < 1$ we have

$$0 = \mathcal{I}(\underline{p}) > \mathcal{I}(\alpha\underline{p}) = \alpha\mathcal{I}(\underline{p}),$$

which would lead to the contradiction $0 > \lim_{\alpha\to 0} \alpha\mathcal{I}(\underline{p}) = 0$.

Axiom A4 connects the framework A1, A2, A3 with the framework of *standard interference functions* [Yat95]. In [Yat95] it was first proposed to model core properties of interference by an axiomatic approach. A function $Y(\boldsymbol{p})$ is called a standard interference function if the following axioms are fulfilled.

Y1 (positivity) $Y(\boldsymbol{p}) > 0$ for all $\boldsymbol{p} \in \mathbb{R}_+^K$,
Y2 (scalability) $\alpha Y(\boldsymbol{p}) > Y(\alpha\boldsymbol{p})$ for all $\alpha > 1$,
Y3 (monotonicity) $Y(\boldsymbol{p}) \geq Y(\boldsymbol{p}')$ if $\boldsymbol{p} \geq \boldsymbol{p}'$.

Properties of this framework were studied in [KYH95, HY98, LSWL04]. The connection between standard interference functions and the axiomatic framework A1, A2, A3 was studied in [BS10]. It was shown that any standard interference function Y can be expressed by a general interference function. To be precise, a function Y is a standard interference function if and only if the function

$$\mathcal{I}_Y(\underline{p}) := \underline{p}_{K+1} \cdot Y\left(\frac{p_1}{\underline{p}_{K+1}}, \ldots, \frac{p_K}{\underline{p}_{K+1}}\right) \tag{4.33}$$

fulfills A1, A2, A3 plus strict monotonicity A4. If we choose a constant value $\underline{p}_{K+1} = 1$, then we simply have $\mathcal{I}_Y(\underline{p}) = Y(\boldsymbol{p})$. That is, $\mathcal{I}_Y$ is a standard interference function with respect to the first K components of its argument.

This means that the axiomatic framework A1, A2, A3 is not confined to the analysis of QoS regions or power control with no noise. It is also suitable for a broad class of SINR-based power control problems, with interference that can be modeled by standard interference functions. Examples are beamforming

[FN98, RLT98, BO01, SB04, SB06, HBO06, WES06], CDMA [KYH95, UY98], base station assignment [Han95, YC95], and robust designs [BSG04, PPIL07].

In this context, a standard problem is to minimize the total power subject to QoS targets $\boldsymbol{u}^{\min} = [u_1^{\min}, \ldots, u_K^{\min}]^T > 0$, i.e.,

$$\min_{\boldsymbol{p} \in \mathcal{P}} \sum_{l \in \mathcal{K}} p_l \quad \text{s.t.} \quad u_k(\boldsymbol{p}) \geq u_k^{\min} \text{ for all } k \in \mathcal{K} \tag{4.34}$$

where $\mathcal{P}$ is some compact set of possible transmission powers, and $u_k(\boldsymbol{p})$ is defined by (4.25).

Problem (4.33) is closely connected with the following max-min problem.

$$\max_{\boldsymbol{p} \in \mathcal{P}} \left(\min_{k \in \mathcal{K}} \frac{u_k(\boldsymbol{p})}{u_k^{\min}} \right). \tag{4.35}$$

If the power set $\mathcal{P}$ is bounded, then both problems (4.33) and (4.34) are equivalent in a sense that the solution of one problem can be found indirectly via a bisection strategy based on solutions of the other problem. If $\boldsymbol{u}^{\min}$ is a point on the boundary of the feasible set, then both problems yield the same optimizer. Thus, both problems can be comprehended under the name "utility balancing."

We will now focus on problem (4.33) which is more convenient to analyze. Let $\phi^{[-1]}$ be the inverse function of ϕ, then $\gamma_k = \phi_k^{[-1]}(u_k^{\min})$ is the minimum SINR level needed by the kth user to achieve some feasible target $u_k^{\min}$. Therefore, the optimizer of (4.33) is obtained by solving the SINR balancing problem

$$\min_{\boldsymbol{p} \in \mathcal{P}} \sum_{l \in \mathcal{K}} p_l \quad \text{s.t.} \quad \frac{p_k}{\mathcal{I}_k(\underline{\boldsymbol{p}})} \geq \gamma_k, \text{ for all } k \in \mathcal{K}. \tag{4.36}$$

Let's assume that the power set $\mathcal{P}$ and the targets $\boldsymbol{\gamma}$ are such that the constraints are feasible. It was shown in [Yat95] that problem (4.35) has a unique optimizer, which is the exact point where all constraints are fulfilled with equality. This is the fixed-point obtained by the iteration

$$p_k^{(n+1)} = \gamma_k \mathcal{I}_k(\underline{\boldsymbol{p}}^{(n)}), \quad \forall k \in \mathcal{K}, \quad \boldsymbol{p}^{(0)} = \boldsymbol{0}. \tag{4.37}$$

This iteration is component-wise monotonic convergent to the global optimizer $\boldsymbol{p}^*$. This was shown for standard interference functions in [Yat95]. An alternative study based on the framework A1, A2, A3, A4 was presented in [SB06]. The iteration has linear convergence [HY98, BS08d], regardless of the actual choice of $\mathcal{I}_k$.

Convexity is commonly considered as the dividing line between "easy" and "difficult" problems. Many examples can be found in the context of multiuser MIMO [LDGW04, LY06, BO01, HBO06, WES06]. For example, equivalent convex reformulations exist for the well-known downlink beamforming problem, as observed in [BO01, HBO06, WES06]. When investigating a problem, a common approach is to first look whether the problem is convex or not. Yates' pioneering work on interference functions [Yat95] has shown that it is not just convexity that matters. The above problem (4.36) can be solved efficiently without exploiting convexity, by merely relying on monotonicity and scalability properties.

This underlines the importance of exploiting all available structure of the problem at hand. Standard approaches from convex optimization theory do not make sufficient use of the special structure offered by interference functions. On the other hand, Yates' framework of standard interference functions [Yat95] does not exploit convexity or concavity. As an example, consider the interference function (4.5), which results from optimum multiantenna combining. It fulfills the axioms A1, A2, A3, and it is concave in addition. However, concavity is not exploited by the fixed-point iteration (4.36).

In the remainder of this section, we discuss how concavity (resp. convexity) and monotonicity can be exploited jointly. We revisit the problem of QoS balancing (4.33), which was already solved for some special cases, like beamforming or robust signal processing. Here, we formulate the problem in its most general form, based on an axiomatic interference model. By only exploiting monotonicity and convexity, we show that the problem can be rewritten in an equivalent convex form.

To this end, assume a convex compact power set $\mathcal{P} \subseteq \mathbb{R}^K_{++}$. We rewrite (4.35) in an equivalent form

$$\min_{p \in \mathcal{P}} \sum_{l \in \mathcal{K}} p_l \quad \text{s.t.} \quad \gamma_k \mathcal{I}_k(\underline{p}) - p_k \leq 0, \text{ for all } k \in \mathcal{K}. \tag{4.38}$$

If $\mathcal{I}_k$ is convex, then (4.37) is a convex optimization problem. Property A4 ensures the existence of a nontrivial solution, provided that the targets γ_k are feasible.

Next, consider the case where $\mathcal{I}_k$ is strictly monotonic and *concave*. An example is the interference (4.5) resulting from beamforming, with either individual power constraints or a total power constraint. Then, problem (4.37) is nonconvex because inequality constraint functions are *concave*, but not convex.

This observation is in line with the literature on multiuser beamforming [BO01, WES06, HBO06], where it was observed that the corresponding problem is nonconvex. Fortunately, the beamforming problem could be solved by deriving equivalent convex reformulations, enabled by the specific interference structure resulting from beamforming receivers [BO01, WES06, HBO06].

An interesting question is: Does an equivalent convex reformulation also exist for the more general problem (4.37), which is only based on the axiomatic framework?

Indeed, if $\mathcal{I}_1, \ldots, \mathcal{I}_K$ are concave and strictly monotonic interference functions, then problem (4.37) is equivalent to

$$\max_{p \in \mathcal{P}} \sum_{l \in \mathcal{K}} p_l \quad \text{s.t.} \quad p_k - \gamma_k \mathcal{I}_k(\underline{p}) \leq 0, \quad \forall k \in \mathcal{K}. \tag{4.39}$$

Thanks to the monotonicity property of the interference functions, it can be shown (see [BS10]) that both problems (4.37) and (4.38) are solved by the same fixed-point $\boldsymbol{p}^*$. In this sense they are equivalent. Problem (4.38) is convex and $\boldsymbol{p}^*$ can be found efficiently using standard algorithms from convex optimization theory.

This result sheds some new light on the problem of multiuser beamforming, which is contained as a special case. It turns out that this problem has a generic convex reformulation (4.38). This insight possibly helps to better understand the convex reformulations observed in the beamforming literature [BO01, WES06].

It should be emphasized that the reformulation (4.38) holds for arbitrary concave standard interference functions, not just the beamforming case. This also includes other receive strategies that aim at optimizing the SINR. Examples are CDMA [KYH95, UY98] or base station assignment [Han95, YC95]. Also, it is easy to incorporate additional constraints on the receivers, like the *shaping constraints* studied in [HBO06].

4.3 JOINT INTERFERENCE MITIGATION AND RESOURCE ALLOCATION

It was shown in the previous section that the framework A1, A2, A3 offers some interesting analytical opportunities. In this section, we show how *concavity* can be exploited for the development of efficient iterative algorithms. To this end, consider an arbitrary interference function $\mathcal{I} : \mathbb{R}^L_{++} \mapsto \mathbb{R}_+$, defined by the axioms A1, A2, A3 in Section 4.2.1.

It was shown in [BS08b] that $\mathcal{I}$ is concave if and only if there exists a closed upward-comprehensive convex set $\mathcal{V} \subset \mathbb{R}^K_+$ such that

$$\mathcal{I}(\boldsymbol{r}) = \min_{\boldsymbol{v} \in \mathcal{V}} \sum_{l=1}^{L} v_l r_l, \quad \text{for all } \boldsymbol{r} > 0. \tag{4.40}$$

Similar results can be shown for *convex interference functions*. For more details, the reader is referred to [BS08b].

This result opens up new perspectives for a more general understanding of interference functions. For example, (4.39) has an interpretation as the optimum of a weighted cost minimization problem from some strategy set $\mathcal{V}$, with weighting factors r_k. The shape of the set $\mathcal{V}$ depends on the way the "users" or "players" of the system compete with each other. This notion of interference abstracts away from its original physical meaning. Related concepts are known from cooperative game theory, where one is often interested in utility tradeoffs that are modeled as comprehensive sets [Pet92].

4.3.1 Adaptive Transmit and Receive Strategies

In the remainder of this section, we will return to the aforementioned SINR balancing example. We show that the representation (4.39) is useful for developing an algorithmic solution for the power minimization problem (4.37). To this end, consider the extended power model $\underline{\boldsymbol{p}}$ with $\underline{p}_{K+1} = \sigma_n^2$. The K interference functions $\mathcal{I}_1, \ldots, \mathcal{I}_K$ are associated with coefficient sets $\mathcal{V}_1, \ldots, \mathcal{V}_K \subset \mathbb{R}^{K+1}_+$. From (4.39) we know that for any given $\underline{p}$, we have

$$\mathcal{I}_k(\underline{p}) = \min_{v_k \in \mathcal{V}_k} \left(\sum_{l \in \mathcal{K}} [v_k]_l \cdot p_l + [v_k]_{K+1} \cdot \sigma_n^2 \right). \tag{4.41}$$

The first K components of v_k determine the interference coupling between the users, while the last component determines the noise amplification. This is a typical structure for many interference scenarios involving adaptive receive strategies. In this case, the interference coupling v_k depends on the chosen *receive strategy*. An example is the interference (4.5).

Let z_k denote some receive strategy from a compact strategy set $\mathcal{Z}_k$, then the interference coupling vector is $v_k = v_k(z_k)$. The entire system coupling is described by the parameter-dependent $K \times (K+1)$ coupling matrix

$$\underline{V}(z) = \begin{bmatrix} v_1^T(z_1) \\ \vdots \\ v_K^T(z_K) \end{bmatrix}.$$

With this new parametrization, we can rewrite the interference function (4.40) as the minimum of all possible receive strategies.

$$\mathcal{I}_k(\underline{p}) = \min_{z_k \in \mathcal{Z}_k} \left(\sum_{l \in \mathcal{K}} \underline{V}_{k,l}(z) \cdot p_l + \underline{V}_{k,K+1}(z) \cdot \sigma_n^2 \right). \tag{4.42}$$

Note, that the kth row of $\underline{V}(z)$ only depends on the receive strategy z_k, which means that the interference functions are decoupled with respect to the receive strategies. A special case of (4.41) is the beamforming example (4.5).

Thus, every convex interference function has a representation (4.41), which is determined by the set $\mathcal{Z}_k$. In the context of power control, the set $\mathcal{Z}_k$ has an interpretation as a set of possible *receive strategies*, because minimizing interference is the typical task of a receiver.

Another way of minimizing interference is to employ a *transmit strategy*, which aims at avoiding interference by processing the signals prior to transmission. However, the choice of a particular transmit strategy can affect the interference experienced by other users. An example is transmit beamforming (see Section 4.1.3), where the interference power of one link depends on the beams of all transmitters. As a consequence, the users are coupled by transmit powers *and* transmit strategies. This prevents a direct application of the interference function framework.

Fortunately, we can exploit the duality principles from Section 4.1.3. By optimizing the "dual" system based on the transpose coupling matrix, we obtain an equivalent problem reformulation, where the transmit strategies behave like receive strategies. Hence, the functions (4.41) can also be used for optimizing transmit strategies. However, the joint optimization of transmit and receive strategies is still a difficult problem in general.

The following section shows how the structure (4.41) can be exploited for algorithmic solutions.

4.3.2 Optimization under Individual QoS Constraints

Consider again problem (4.37), where the goal is to achieve some feasible SINR targets $\boldsymbol{\gamma}$ with minimum total transmission power. It was already shown that concave interference functions enable an equivalent convex reformulation of the problem. Now we discuss how the structure of interference functions can be exploited for the development of algorithms.

One way to solve the power minimization problem is by using the fixed-point iteration (4.36). It was shown in [HY98, BS08d] that the fixed-point iteration has linear convergence. That is, there exists a constant $C > 0$ such that

$$\lim_{n\to\infty} \frac{\|\boldsymbol{p}^{(n+1)} - \boldsymbol{p}^*\|}{\|\boldsymbol{p}^{(n)} - \boldsymbol{p}^*\|} = C. \tag{4.43}$$

A better convergence can be achieved if there is additional knowledge about the structure of the problem at hand. From Section 4.2.3 we already know that an equivalent convex reformulation exists, since our interference functions are concave. Thus, standard convex optimization techniques can be applied to solve problem (4.37).

But instead of following a standard approach, we can do better by exploiting the specific form (4.41), as proposed in [BS08d]. To this end, we introduce the function

$$\boldsymbol{d}(\boldsymbol{p}) = \boldsymbol{p} - \boldsymbol{\Gamma}\boldsymbol{\mathcal{I}}(\underline{\boldsymbol{p}})$$

where $\boldsymbol{\Gamma} := \text{diag}\{\boldsymbol{\gamma}\}$ and $\boldsymbol{\mathcal{I}} = [\mathcal{I}_1, \ldots, \mathcal{I}_K]^T$. Here, the variable is $\boldsymbol{p}$. The last component $\underline{\boldsymbol{p}} = \sigma_n^2$ (the noise) is constant. To simplify the notation, we assume $\sigma_n^2 = 1$. With $\boldsymbol{d}(\boldsymbol{p})$, the constraints in (4.37) can be rewritten as

$$\boldsymbol{d}(\boldsymbol{p}) \geq 0.$$

From Section 4.2.3, it is clear that the unique optimum of (4.37) is the fixed-point for which $\boldsymbol{d}(\boldsymbol{p}) = 0$ (component-wise). In order to achieve this point, we can exploit that the function $\boldsymbol{d}(\boldsymbol{p})$ is convex, and the underlying interference functions have the form (4.41). Defining $\underline{\boldsymbol{V}}(z) = \left[\boldsymbol{V}(z) \mid \boldsymbol{n}(z)\right]$, where $\boldsymbol{V}(z) \in \mathbb{R}_+^{K\times K}$ determines the interference coupling and $\boldsymbol{n}(z) \in \mathbb{R}_+^K$ is the noise amplification factor, we can rewrite (4.41) as

$$\mathcal{I}_k(\underline{\boldsymbol{p}}) = \min_{z_k \in \mathcal{Z}_k} \left[\boldsymbol{V}(z)\boldsymbol{p} + \boldsymbol{n}(z)\right]_k. \tag{4.44}$$

Thus, the Jacobi matrix containing the partial derivatives can be written in terms of the coupling matrix $\boldsymbol{V}$, which depends on a receive strategy z. Assuming that a given power vector $\boldsymbol{p}$ corresponds exactly to one receive strategy $z(\boldsymbol{p})$, the Jacobi matrix can be written as

$$\nabla \boldsymbol{d}(\boldsymbol{p}) = \boldsymbol{I} - \boldsymbol{\Gamma}\boldsymbol{V}\big(z(\boldsymbol{p})\big). \tag{4.45}$$

In general, however, different optimum receivers can exist for the same $\boldsymbol{p}$. As a consequence, the function $\boldsymbol{d}(\boldsymbol{p})$ is generally not continuously differentiable. This prevents the direct application of Newton's method, for example.

Fortunately, the framework of interference functions have some nice properties, which allow to show that the following "Newton-type" iteration converges to the global optimum [BS08d]. We need an initialization $z^{(n)}$ such that the SINR targets $\boldsymbol{\Gamma}$ are feasible. One way of obtaining such an initialization will be discussed in the following Section 4.3.3.

$$\boldsymbol{p}^{(n+1)} = \left(\boldsymbol{\Gamma}^{-1} - \boldsymbol{V}(z^{(n)})\right)^{-1}\boldsymbol{n}\left(z^{(n)}\right) \tag{4.46}$$

$$\text{with} \quad z_k^{(n)} = \arg\min_{z_k \in \mathcal{Z}_k}\left[\boldsymbol{V}(z)\boldsymbol{p}^{(n)} + \boldsymbol{n}(z)\right]_k \quad \forall k \in \mathcal{K}. \tag{4.47}$$

This iteration is component-wise monotonic, and it converges to the unique global optimizer with super-linear convergence, i.e.,

$$\lim_{n\to\infty} \frac{\|\boldsymbol{p}^{(n+1)} - \boldsymbol{p}^*\|}{\|\boldsymbol{p}^{(n)} - \boldsymbol{p}^*\|} = 0. \tag{4.48}$$

Note, that this holds for any multiuser system where users are coupled via interference functions satisfying the axioms A1–A4 plus concavity. An example is the interference function resulting from optimum combining (4.5). In this case, problem (4.37) is known as the *multiuser beamforming* problem, which was studied and solved in [FN98, RLT98, BO01, SB04, SB06, HBO06, WES06]. The discussion here shows that this problem can be understood in a more general context of interference functions. It also brings out clearly the key properties that enable efficient algorithmic solutions.

Similar results can be shown for convex interference functions, which occur, for example in the context of robust optimization (e.g., [BSG04, PPIL07]).

We have assumed that the receive strategy is from a compact set $\mathcal{Z}_k$. For the beamforming case, this is typically the unit sphere. However, the above results show that $\mathcal{Z}_k$ can be arbitrary, so arbitrary constraints can be imposed on the beamformers. The only requirement is that $\mathcal{Z}_k$ is compact, to ensure that the minimum exists. Thus, additional constraints, like the shaping constraints [HBO06], can be added without affecting the convergence properties of the iteration. Of course, substep (4.46) can become more complicated. In the conventional beamforming case, this step is easily solved by an eigenvalue decomposition. A different approach might be necessary under additional constraints.

Note, that the algorithm (4.45) requires knowledge of the matrix $\boldsymbol{V}(z^{(n)})$, which consists of the coupling coefficients of all users. This knowledge requires cooperation between the users, which is not always available.

The fixed-point iteration (4.36) performs a user-wise update of the transmission powers, thus it is more amenable for decentralized implementations. However, this depends on how the interference functions are defined. For example, the interference function resulting from optimum combining (4.5) require knowledge of the other users' channels. Also, in this case the fixed-point iteration requires cooperation between users, which complicates a decentralized implementation.

4.3.3 Max-Min Balancing

Consider the min-max optimum $f(\boldsymbol{\gamma})$, defined by (4.28) in Section 4.2.2. The function $f(\boldsymbol{\gamma})$ is an indicator for feasibility. In (4.28) the function $f(\boldsymbol{\gamma})$ is defined in terms of the SINR. But since the infimum in (4.28) is taken over an unconstrained set, we can equivalently define $f(\boldsymbol{\gamma})$ by replacing SINR by the signal-to-interference ratio (SIR), leading to the SINR balancing problem [BS09]. SINR targets $\boldsymbol{\gamma}$ can be attained if and only if $f(\boldsymbol{\gamma}) < 1$. Then, there exists a power vector $\boldsymbol{p} > 0$ such that $\text{SINR}_k(\boldsymbol{p}) = \gamma_k$ for all $k \in \mathcal{K}$.

The indicator $f(\boldsymbol{\gamma})$ can be computed efficiently if the underlying interference functions are concave. Then we can exploit the form (4.43). We need to assume that all users are coupled by interference. More precisely, we require that the coupling matrix $\boldsymbol{V}(z)$ is irreducible for any z. This assumption is necessary because $f(\boldsymbol{\gamma})$ is defined without power constraints, so feasibility is only determined by the effects of interference coupling.

An algorithm for finding $f(\boldsymbol{\gamma})$ not only provides an indicator of feasibility, it also provides an initialization for the algorithm discussed in the previous section.

A special case is obtained for linear interference functions $\mathcal{I}_k(\boldsymbol{p}) = [\boldsymbol{V}\boldsymbol{p}]_k$, where $\boldsymbol{V} \in \mathbb{R}_+^{K \times K}$ is irreducible. Then $f(\boldsymbol{\gamma}) = \lambda_{\max}(\boldsymbol{\Gamma}\boldsymbol{V})$ (the maximal eigenvalue). In the context of non-negative irreducible matrices this is known as the *Perron root*.

If the underlying interference functions are concave, then we can use the representation (4.41). Then $\boldsymbol{V}$ depends on a receive strategy z, and $f(\boldsymbol{\gamma})$ is given as

$$f(\boldsymbol{\gamma}) = \min_{z_1 \in \mathcal{Z}_1, \ldots, z_K \in \mathcal{Z}_K} \lambda_{\max}(\boldsymbol{\Gamma}\boldsymbol{V}(z)). \tag{4.49}$$

The Perron root minimization problem (4.48) and the min-max balancing problem (4.28) offer two alternative ways of characterizing feasibility. The former approach is geared towards finding an optimal receive strategy z, while the latter approach focuses on finding the optimal power vector $\boldsymbol{p}^*$.

Note, that the matrix $\boldsymbol{V}$ is generally nonsymmetric. Such eigenvalue optimization problems are considered as complicated [HOY02]. But thanks to the special structure of the concave interference functions (4.43), the problem can be solved by a globally convergent algorithm involving alternating optimization of powers and receive strategies. This Principal Eigenvector (PEV) Iteration is as follows [BS09].

$$\boldsymbol{p}^{(n+1)} = \text{pev}\left(\boldsymbol{\Gamma}\boldsymbol{V}(z^{(n)})\right) \quad \text{(principal eigenvector)} \tag{4.50}$$

$$\text{where} \quad z_k^{(n)} = \arg\min_{z_k \in \mathcal{Z}_k} \sum_{l \in \mathcal{K}} V_{kl}(z_k) \cdot p_l^{(n)}, \quad \text{for all } k \in \mathcal{K}.$$

For an arbitrary initialization $\boldsymbol{p}^{(0)} > 0$, we have

$$\lambda_{\max}\big(\boldsymbol{\Gamma V}(z^{(n+1)})\big) \leq \lambda_{\max}\big(\boldsymbol{\Gamma V}(z^{(n)})\big) \quad \text{for all } n \in \mathbb{N}, \tag{4.51}$$

$$\lim_{n\to\infty} \lambda_{\max}\big(\boldsymbol{\Gamma V}(z^{(n)})\big) = f(\boldsymbol{\gamma}), \tag{4.52}$$

$$\lim_{n\to\infty} \boldsymbol{p}^{(n)} = \boldsymbol{p}^*. \tag{4.53}$$

That is, we have component-wise convergence to the unique (up to a scaling) max-min optimizer $\boldsymbol{p}^*$. Although the algorithm shows excellent convergence properties, analyzing the convergence speed is still an open problem.

Again, the beamforming interference function (4.5) is contained as a special case. This provides a more complete understanding of the max-min SIR balancing problem that was studied in the context of downlink beamforming [GP96, MS98, SB02].

4.4 IMPLEMENTATION ASPECTS

So far, this chapter has discussed multiuser strategies from a primarily theoretical viewpoint. By assuming simple, abstract models, we were able to handle multiuser problems in an analytical way. This approach is helpful for developing deeper understanding and intuition about problems which are otherwise difficult to handle. The optimal algorithms resulting from this approach are not only useful as a benchmark for the achievable system performance, they also provide a road-map for the development of practical techniques, which take into account the constraints imposed by real-world system architectures.

One of the main challenges in implementing these algorithms is the availability of channel knowledge. All the interference mitigation techniques discussed in this chapter rely on knowledge about the wireless channel. In some cases, like (4.7), instantaneous channel knowledge is required. Other solutions are based on statistical channel knowledge, as in (4.3). But all MIMO techniques for interference mitigation/avoidance have in common that they exploit the spatial structure of the multiuser channel. This is in contrast to most space-time coding techniques which only exploit the *diversity* offered by statistically independent propagation paths.

Obtaining channel knowledge is especially problematic for downlink channels, where spatial pre-processing is carried out at the base station, prior to transmission. There are two main ways to obtain channel knowledge. One is to estimate the channel at the receiver and to use a feedback channel to convey this information to the transmitter. The other way is to exploit the reciprocity between uplink and downlink channel, if available. That is, uplink channel estimates are used for transmission in the downlink.

The problem with the latter approach is that uplink and downlink are duplexed using either frequency division (FDD) or time division (TDD). Consequently, the validity of the assumed reciprocity very much depends on the variability of the channel parameters in the frequency or time domain, respectively. In many deployment scenarios reciprocity does not hold, at least not with regard to the small-scale channel characteristics caused by multipath propagation. The frequency gap between uplink and downlink is typically larger than the coherence bandwidth of the channel. Moreover, the employed radio front-ends must be reciprocal as well, therefore additional calibration is required.

The main problem with the feedback approach is the large amount of MIMO channel data, which grows with the number of antennas. Thus, feeding back full channel information is prohibitively costly in terms of bandwidth overhead. A practical, LTE-compliant strategy consists of feeding back only a precoding matrix index (PMI). This index selects a precoding matrix that is suitable for combining the antennas at the transmitter. The precoding matrix is chosen from a predefined *codebook*, which is known to both the transmitter and the receiver. This approach drastically reduces the feedback overhead, while still achieving some of the advantages of MIMO precoding. However, the disadvantage of the PMI approach is its lack of flexibility, especially when dealing with interference.

Interference will be the key performance limiting factor for many future deployment scenarios. But the development of dynamic techniques for interference mitigation/avoidance and subchannel allocation hinges on how efficiently channel state information can be exchanged in the network. This is especially true for distributed networks, where information exchange between nodes is very limited. Examples are macro cells with unplanned pico/femto cells [CA08], relays, or ad-hoc networks.

Besides the fundamental issue of channel knowledge, there are many other challenges that complicate the application of the theoretical results to existing systems. Depending on the system architecture, there can be additional constraints on powers, MIMO filters, and other parameters. For example, the duality approach discussed in Section 4.1.3 holds under a sum-power constraint, but not for individual power constraints.

In conclusion, there are many aspects of multiuser systems that still offer room for improvement. The theory of interference functions is a suitable framework for modeling mutual inter-dependencies caused by interference. However, much further research is required in order to take full advantage of the analytical structure of given interference scenarios. The ultimate goal is the development of decentralized networks, which deal with multiuser interference in a self-organized manner.

PART

Implementation

CHAPTER

5 Advanced Transmitter and Receiver Design

Harold H. Sneessens, Luc Vandendorpe, Cédric Herzet, Xavier Wautelet, Onur Oguz

5.1 INTRODUCTION

This chapter addresses the design of receivers for multiple-input multiple-output (MIMO) communication. It covers the decoding of error-correcting codes, the mitigation of intersymbol and cross-antenna interference, and the synchronization, that is, the estimation of channel parameters.

The techniques presented in this chapter are based on the turbo principle. The turbo principle originates from the turbo decoding algorithm, which has shown how to approximate an intractable optimal decoder with a simpler, iterative decoder. The receivers presented in this chapter use the same design idea, referred to as the turbo principle, to carry out interference mitigation and synchronization. This results in high-performance iterative receivers that address decoding globally by iterating between tasks, and that approximate closely the optimal but intractable decoders.

This introduction presents essential techniques and concepts used in the remaining of the chapter: convolutional codes and the BCJR decoding algorithm, turbo codes and the turbo decoding algorithm, and finally the turbo principle.

Next, Section 5.2 introduces turbo equalization, a technique to mitigate intersymbol interference. Section 5.3 extends turbo equalization to MIMO channels, where the interference comes both from adjacent symbols and from other antennas. Finally, Section 5.4 presents turbo synchronization, extended to MIMO channels in Section 5.5.

5.1.1 Convolutional Codes and BCJR Decoding

A convolutional code is an error-correcting code based on a finite-state machine. The finite-state machine takes as input a sequence of information bits, and produces as output a sequence of coded bits. The process is such that the coded bits are redundant, so as to provide protection against transmission errors.

MIMO. DOI: 10.1016/B978-0-12-382194-2.00005-8

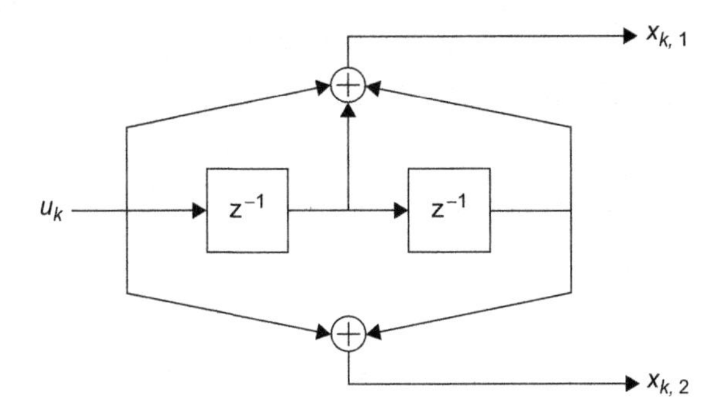

FIGURE 5.1

Example of nonrecursive convolutional encoder. The output bits $x_{k,1}$ and $x_{k,2}$ are a function of the input bit u_k and of the state $\epsilon_{k-1} = [u_{k-1}, u_{k-2}]$. ($z^{-1}$ is a delay operator)

The input/output relationship of a convolutional encoder is as follows. For so-called nonrecursive codes, the sequence of input bits u_k ($k = 1, \ldots, K$) results in P sequences of coded bits $x_{k,p}$ ($k = 1, \ldots, K$ and $p = 1, \ldots, P$) calculated as:

$$x_{k,p} = \sum_{m=0}^{M} f_m^{(p)} \; u_{k-m} \mod 2 \tag{5.1}$$

where the $f_m^{(p)}$ are binary coefficients. The parameter M is the memory of the encoder: the output bit $x_{k,p}$ depends on the input bit u_k and on the M previous inputs $u_{k-1}, \ldots, u_{k-M}$ (see Figure 5.1). These M previous inputs are referred to as the state of the encoder, written $\epsilon_{k-1} = [u_{k-1}, \ldots, u_{k-M}]$. The binary coefficients of the code are usually represented in a more compact octal notation: for each output sequence p, the series of binary coefficients $f_m^{(p)}$ ($m = 0, \ldots, M$) is written as an octal number. For instance, the code of Figure 5.1, with $\left[f_0^{(1)} f_1^{(1)} f_2^{(1)}\right] = [1\,1\,1]$ and $\left[f_0^{(2)} f_1^{(2)} f_2^{(2)}\right] = [1\,0\,1]$, has an octal representation $(7_8, 5_8)$. For so-called recursive codes, the state of the encoder is no longer made of input bits, but rather of a recursive function of the input bits (see Figure 5.2). Let $\epsilon_{k-1} = [a_{k-1}, \ldots, a_{k-M}]$ be the state of the encoder before time k. At each new input u_k, the encoder calculates:

$$a_k = u_k + \sum_{m=1}^{M} g_m \; a_{k-m} \mod 2, \tag{5.2}$$

for some binary coefficients g_m, $m = 1, \ldots, M$. The state variable a_k is thus a recursive function of the input bits and the encoder state. Next the output bits $x_{k,p}$ are calculated according to:

$$x_{k,p} = \sum_{m=0}^{M} f_m^{(p)} \; a_{k-m} \mod 2, \tag{5.3}$$

that is, they are a function of a_k and the previous state $\epsilon_{k-1} = [a_{k-1}, \ldots, a_{k-M}]$.

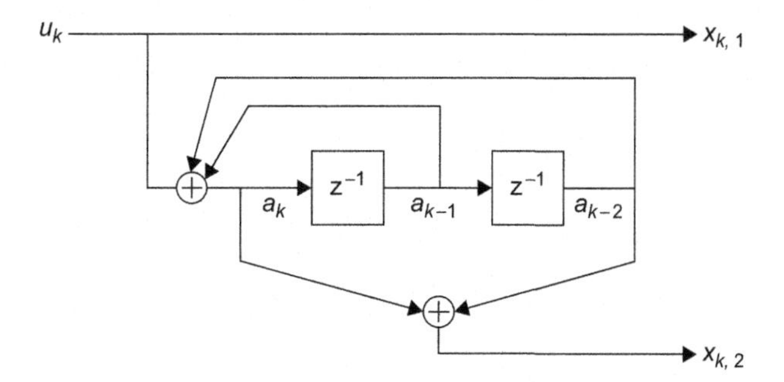

FIGURE 5.2

Example of a recursive convolutional encoder. The code is systematic, since the output bit $x_{k,1}$ is equal to the input bit u_k. The output bit $x_{k,2}$ is a function of a_k and of the encoder state $\epsilon_{k-1} = [a_{k-1}, \ldots, a_{k-M}]$. ($z^{-1}$ is a delay operator)

A convolutional code can have an output bit equal to the input bit u_k (as for example in Figure 5.2). In that case it is called *systematic*.

Several algorithms can decode convolutional codes, according to different criteria. A very common option is the Viterbi algorithm [Pro95], which finds the coded sequence with maximal a posteriori probability. In the context of turbo processing and for this chapter, the algorithm of interest is the BCJR algorithm [BCJR74], which calculates the a posteriori probability of each information bit.

The BCJR algorithm takes as inputs the channel observations, say a sequence $\mathbf{y}$, and possibly a priori probabilities on the bits u_k, say $P_a(u_k)$ for $k = 1, \ldots, K$. Equivalently, the a priori information can be expressed as log-ratios:

$$L_a(u_k) = \ln \frac{P_a(u_k = 1)}{P_a(u_k = 0)}. \tag{5.4}$$

As outputs, the BCJR algorithm produces a posteriori probabilities $P(u_k = 0|\mathbf{y})$ and $P(u_k = 1|\mathbf{y})$, or equivalently the log-ratios:

$$L_p(u_k) = \ln \frac{P(u_k = 1|\mathbf{y})}{P(u_k = 0|\mathbf{y})}. \tag{5.5}$$

From these log-ratios, the bits can be detected according to the rule:

$$\hat{u}_k = \begin{cases} 1 & \text{if} \quad L_p(u_k) \geq 0 \\ 0 & \text{if} \quad L_p(u_k) < 0. \end{cases} \tag{5.6}$$

The BCJR algorithm calculates the log-ratios $L_p(u_k)$ as follows. First, note that the initial state $\epsilon_{k-1} = \epsilon'$ and the final state $\epsilon_k = \epsilon$ of a transition in the encoder determine the input bit u_k. The a posteriori log-ratio (5.5) can thus be written:

$$L_p(u_k) = \ln \frac{\sum_{\mathcal{E}^+} P(\epsilon_{k-1} = \epsilon', \epsilon_k = \epsilon|\mathbf{y})}{\sum_{\mathcal{E}^-} P(\epsilon_{k-1} = \epsilon', \epsilon_k = \epsilon|\mathbf{y})} \tag{5.7}$$

where $\mathcal{E}^+$ (resp. $\mathcal{E}^-$) is the set of transitions $(\epsilon_{k-1} = \epsilon', \epsilon_k = \epsilon)$ corresponding to a bit $u_k = 1$ (resp. $u_k = 0$). Equivalently, the log-ratios can be written:

$$L_p(u_k) = \ln \frac{\sum_{\mathcal{E}^+} P(\epsilon_{k-1} = \epsilon', \epsilon_k = \epsilon, \mathbf{y})}{\sum_{\mathcal{E}^-} P(\epsilon_{k-1} = \epsilon', \epsilon_k = \epsilon, \mathbf{y})}. \tag{5.8}$$

The BCJR algorithm calculates the probabilities $P(\epsilon_{k-1} = \epsilon', \epsilon_k = \epsilon, \mathbf{y})$ by decomposing them in three factors [BCJR74]:

$$P(\epsilon_{k-1} = \epsilon', \epsilon_k = \epsilon, \mathbf{y}) = \alpha_{k-1}(\epsilon') \cdot \gamma_k(\epsilon', \epsilon) \cdot \beta_k(\epsilon). \tag{5.9}$$

The functions $\alpha_k(\epsilon)$, $\gamma_k(\epsilon', \epsilon)$ and $\beta_k(\epsilon)$ are evaluated as follows (for more details, see [BCJR74]).

1. The functions $\alpha_k(\epsilon)$ form a forward recursion expressed by:

$$\alpha_k(\epsilon) = \sum_{\epsilon' \in \mathcal{E}} \alpha_{k-1}(\epsilon') \cdot \gamma_k(\epsilon', \epsilon) \tag{5.10}$$

for $k = 1, \ldots, K-1$. The recursion is initiated by:

$$\alpha_0(\epsilon = \epsilon_0) = 1 \text{ and } \alpha_0(\epsilon \neq \epsilon_0) = 0, \tag{5.11}$$

where the encoder state is assumed to be ϵ_0 before beginning to encode.

2. The functions $\beta_k(\epsilon)$ form the following backward recursion:

$$\beta_{k-1}(\epsilon') = \sum_{\epsilon \in \mathcal{E}} \beta_k(\epsilon) \cdot \gamma_k(\epsilon', \epsilon) \tag{5.12}$$

for $k = 2, \ldots, K$. If the encoder uses trellis termination (i.e., the encoder adds M bits at the end of the sequence to bring its state back to some predetermined value ϵ_K), the recursion is initiated by:

$$\beta_K(\epsilon = \epsilon_K) = 1 \text{ and } \beta_K(\epsilon \neq \epsilon_K) = 0. \tag{5.13}$$

Otherwise, the initial condition is:

$$\beta_K(\epsilon) = \frac{1}{2M} \quad \forall \epsilon \tag{5.14}$$

which means that all 2^M values of the encoder state are equiprobable.

3. The function $\gamma_k(\epsilon', \epsilon)$ describes the probability of the transition between states $\epsilon_{k-1} = \epsilon'$ and $\epsilon_k = \epsilon$:

$$\gamma_k(\epsilon', \epsilon) = P(\mathbf{y_k} | \epsilon_{k-1} = \epsilon', \epsilon_k = \epsilon) \cdot P(\epsilon_k = \epsilon | \epsilon_{k-1} = \epsilon'), \tag{5.15}$$

where $\mathbf{y_k}$ are the channel observations corresponding to the kth information bit (i.e., the channel observations of all coded bits produced by the encoder at time k). Equivalently, this expression can be written:

$$\gamma_k(\epsilon', \epsilon) = P(\mathbf{y_k} | u_k, \epsilon_{k-1} = \epsilon') \cdot P(u_k), \tag{5.16}$$

since the initial and final states of the transition determine u_k. The first factor depends on the received symbols and the channel model, while the second is the available a priori information, i.e., $P(u_k) = P_a(u_k)$.

To sum up, the BCJR algorithm first evaluates the transition metrics $\gamma_k(\epsilon', \epsilon)$, then the forward recursion $\alpha_k(\epsilon)$ and the backward recursion $\beta_k(\epsilon)$, and finally the a posteriori log-ratios:

$$L_p(u_k) = \ln \frac{\sum_{\mathcal{E}^+} \alpha_{k-1}(\epsilon') \cdot \gamma_k(\epsilon', \epsilon) \cdot \beta_k(\epsilon)}{\sum_{\mathcal{E}^-} \alpha_{k-1}(\epsilon') \cdot \gamma_k(\epsilon', \epsilon) \cdot \beta_k(\epsilon)}. \tag{5.17}$$

A fundamental property of the BCJR algorithm is that the a posteriori log-ratios $L_p(u_k)$ can be decomposed in two terms: the first term is a function of the inputs relative to the bit u_k only (i.e., the a priori information on u_k and the channel observation of u_k), and the second term combines information on other bits by exploiting the code constraints. This second term is referred to as *extrinsic information*. Formally, this can be written

$$L_p(u_k) = f(L_a(u_k), y_k) + L_e(u_k), \tag{5.18}$$

for some function $f(\cdot)$, and where y_k denotes the channel observation of u_k which exists only for systematic codes.

Logarithmic Expression of the BCJR Algorithm

When implemented as above, the BCJR algorithm suffers from numerical problems due to the finite precision of the representation of numbers. As shown in [RHV99, RHV97], these problems can be avoided by working in the logarithmic domain, i.e., by using the logarithm of the functions above. With the definitions:

$$\bar{\alpha}_k(\epsilon) = \ln(\alpha_k(\epsilon)) \tag{5.19}$$

$$\bar{\beta}_k(\epsilon) = \ln(\beta_k(\epsilon)) \tag{5.20}$$

$$\bar{\gamma}_k(\epsilon', \epsilon) = \ln(\gamma_k(\epsilon', \epsilon)), \tag{5.21}$$

the a posteriori log-ratio (5.17) becomes:

$$\begin{aligned} L_p(u_k) = \ln &\left(\sum_{\mathcal{E}^+} \exp\left(\bar{\alpha}_{k-1}(\epsilon') + \bar{\gamma}_k(\epsilon', \epsilon) + \bar{\beta}_k(\epsilon)\right) \right) \\ - \ln &\left(\sum_{\mathcal{E}^-} \exp\left(\bar{\alpha}_{k-1}(\epsilon') + \bar{\gamma}_k(\epsilon', \epsilon) + \bar{\beta}_k(\epsilon)\right) \right). \end{aligned} \tag{5.22}$$

The BCJR algorithm in the log-domain has a more concise expression when using the following function:

$$\begin{aligned} \overset{*}{\max}(x, y) &\triangleq \ln(\exp(x) + \exp(y)) \\ &= \max(x, y) + \ln(1 + \exp(-|x - y|)), \end{aligned} \tag{5.23}$$

extended recursively for more than two arguments:

$$\overset{*}{\max}(x,y,z) \triangleq \ln(\exp(x)+\exp(y)+\exp z) \qquad (5.24)$$
$$= \overset{*}{\max}(\overset{*}{\max}(x,y),z).$$

Then (5.22) becomes:

$$L_p(u_k) = \overset{*}{\max_{\mathcal{E}^+}}\left(\bar{\alpha}_{k-1}(\epsilon')+\bar{\gamma}_k(\epsilon',\epsilon)+\bar{\beta}_k(\epsilon)\right)$$
$$-\overset{*}{\max_{\mathcal{E}^-}}\left(\bar{\alpha}_{k-1}(\epsilon')+\bar{\gamma}_k(\epsilon',\epsilon)+\bar{\beta}_k(\epsilon)\right), \qquad (5.25)$$

and the functions $\bar{\alpha}_k(\epsilon)$, $\bar{\beta}_k(\epsilon)$ and $\bar{\gamma}_k(\epsilon',\epsilon)$ can be calculated as:

$$\bar{\alpha}_k(\epsilon) = \overset{*}{\max_{\epsilon'\in\mathcal{E}}}(\bar{\alpha}_{k-1}(\epsilon')+\bar{\gamma}_k(\epsilon',\epsilon)) \qquad (5.26)$$

$$\bar{\beta}_{k-1}(\epsilon') = \overset{*}{\max_{\epsilon\in\mathcal{E}}}(\bar{\beta}_k(\epsilon)+\bar{\gamma}_k(\epsilon',\epsilon)) \qquad (5.27)$$

$$\bar{\gamma}_k(\epsilon',\epsilon) = \ln P(\mathbf{y}_k|u_k,\epsilon_{k-1}=\epsilon')+\ln P(u_k). \qquad (5.28)$$

The initial conditions of the recursions are of course modified accordingly.

5.1.2 Turbo Codes and Turbo Decoding

A turbo code is an error-correcting code formed by the concatenation of two constituent convolutional codes, separated by an interleaver [BG96]. The interleaver induces a complex redundancy structure, that protects efficiently against transmission errors.

The two constituent codes can be concatenated in parallel, or in series (see [BDMP98, BM96] for a detailed presentation of the latter option). Figure 5.3 depicts an example of turbo code with parallel concatenation. In this encoder, the sequence of information bits u_k ($k=1,\ldots,K$) is encoded into two sequences of coded bits $x_{k,1}$

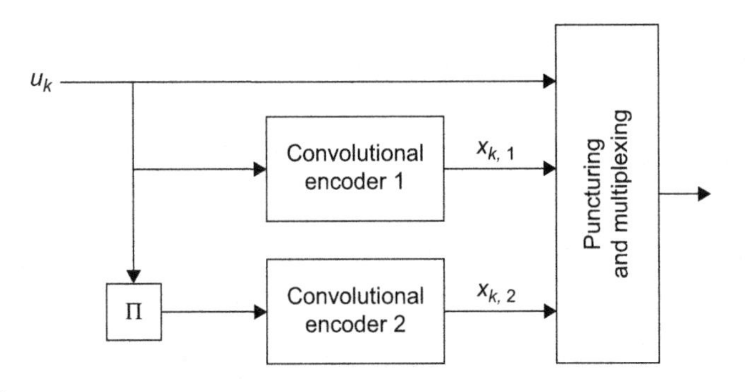

FIGURE 5.3

Turbo encoder. In this diagram, $\mathbf{\Pi}$ denotes the interleaver.

and $x_{k,2}$. These sequences of information bits and coded bits are then multiplexed, and possibly punctured, before transmission.

Optimal decoding of a turbo code is intractable, because of the complexity of the code structure. But a suboptimal iterative algorithm, termed the turbo decoding algorithm, offers close-to-optimal decoding performance at a reasonable complexity.

The turbo decoding algorithm uses the decoders of the constituent codes iteratively. In turn, each decoder produces so-called extrinsic information and provides it as a priori information to the other decoder; the repeated exchange of information progressively refines the knowledge of the transmitted sequence. The turbo decoding algorithm was at first a heuristic approach to decoding [BG96], but it was later justified in the factor graphs and sum-product algorithm framework [MM98, FM97]. According to that framework, the algorithm produces an approximation of the a posteriori probabilities of the signal.

For example, Figure 5.4 depicts the turbo decoder corresponding to the encoder of Figure 5.3. The channel observations y_k^0, y_k^1, and y_k^2 correspond to the bits u_k, x_k^1, and x_k^2 ($k = 1,\ldots,K$). Each convolutional decoder $i = 1,2$ calculates extrinsic information L_e^i from the observations y_k^0 and y_k^i, as well as from the a priori information L_a^i. The a priori information used by a constituent decoder is the extrinsic information produced by the other decoder, interleaved or de-interleaved as required. After several iterations, the second convolutional decoder outputs an approximation of the a posteriori probabilities $P(u_k|\mathbf{y})$.

The interleaver plays a crucial role to ensure that the approximation stays close to the actual a posteriori probabilities. If the decoders were sources of independent information for each other, the turbo decoding algorithm would produce exact a posteriori probabilities. In practice the information exchanged is not independent, but stays fairly uncorrelated along iterations thanks to the interleaver. The interleaver, indeed, scrambles the sequence and so ensures that the constraints induced by each constituent code are quite uncorrelated locally. Therefore the information exchanged between the constituent decoders is fairly uncorrelated as well.

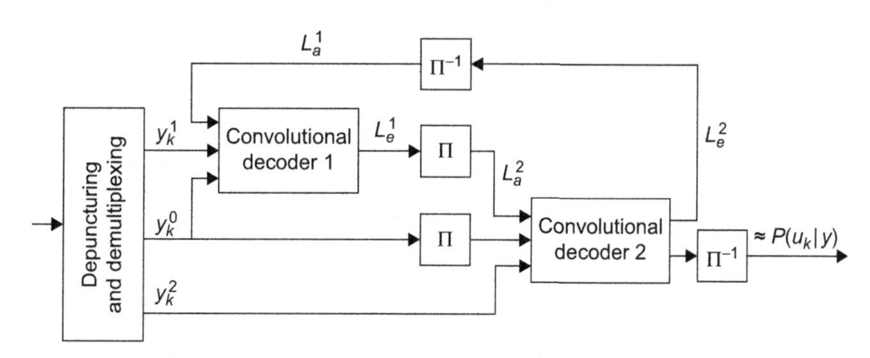

FIGURE 5.4

Block diagram of a turbo decoder.

Exchanging extrinsic information instead of a posteriori information also enables to keep the exchanges as uncorrelated as possible, since the extrinsic information produced for a bit does not depend on the a priori information received for that same bit.

5.1.3 The Turbo Principle

The turbo decoding algorithm uses the decoders of the constituent codes iteratively, to provide a simple approximate solution to an otherwise intractable decoding problem. This idea of approximating a complex optimal decoder by an iterative combination of simpler decoding functions has been used successfully for other decoding problems; it is now referred to as the *turbo principle*.

The turbo principle tells that a complex decoder can be approximated by iterating between simpler soft-input soft-output (SI-SO) decoding stages. For example, the intractable optimal decoder for turbo codes is approximated by an iterative decoder based on simple BCJR constituent decoders.

The constituent decoders exchange so-called extrinsic information: the extrinsic information produced by a decoder becomes the a priori information for the other decoder, which in turn produces extrinsic information, and so on. As seen above, the extrinsic information on a variable (e.g. on a bit) is the difference between the a posteriori information and the a priori information on this variable. It is the additional amount of information calculated by the constituent decoder on a variable, independent of the a priori information on this variable. Only that additional information is exchanged between constituent decoders.

The turbo principle works well if the coding scheme contains an interleaver placed in such a way that, in the receiver, the constituent decoders are separated by an interleaver. From the receiver perspective, this ensures that the a priori information is locally uncorrelated, since interleaving breaks the correlation in the sequence of extrinsic information. In a turbo code, the interleaver before the second convolutional encoder induces the use of an interleaver in the receiver, between the two constituent decoders. As a consequence, the sequence of a priori information provided to a constituent decoder looks uncorrelated (locally), as assumed by the decoder.

The turbo principle has been successfully applied to many decoding problems. In the sequel, it is applied to turbo equalization and to turbo synchronization.

5.2 TURBO EQUALIZATION

Signals transmitted on frequency-selective channels suffer from inter-symbol interference (ISI). To decode the data, the receiver must both suppress the intersymbol interference and decode the error-correcting code.

The optimal approach would be to address these two problems jointly. The decoder would have to evaluate the a posteriori probability (APP) of each information bit, taking into account the effects of the error-correcting code and of the interference altogether. This approach is in general much too complex, because a received symbol contains contributions from many information bits through a complex relationship dependent on both the error-correcting code and the channel interference.

A conventional sub-optimal approach is to separate the decoding problem in two tasks: First equalization, that is, remove the interference from the received symbols and obtain APPs on the coded bits; and then decoding the error-correcting code.

A much better suboptimal approach though, is to perform the equalization and decoding tasks iteratively. This approach, called *turbo equalization* because it is based on the turbo principle, provides drastic performance improvements over the conventional disjoint approach [DJB95, BKH97, LGL01, TKS02, DV02]. This section presents turbo equalization for a Bit-Interleaved Coded Modulation (BICM) transmission.

5.2.1 Transmission Scheme

The transmitter sends K information bits u_k $(k = 1,\ldots,K)$, encoded and mapped into N symbols s_n $(n = 1,\ldots,N)$ (see Figure 5.5). The encoding and mapping scheme is BICM [Zeh92, CTB98, LR99]. From the K information bits, a nonrecursive systematic convolutional encoder produces $K \cdot P$ coded bits $x_{k,p}$ $(p = 1,\ldots,P)$. These coded bits are interleaved and written $x_{n,q}$ $(n = 1,\ldots,N$ and $q = 1,\ldots,Q$ with $N \cdot Q = K \cdot P)$. The interleaved bits $x_{n,q}$ are then mapped onto symbols s_n taken in an alphabet $\mathcal{S}$ of 2^Q symbols. The information sequence, the coded sequence and the symbol sequence are written $\boldsymbol{u}, \boldsymbol{x}$ and $\boldsymbol{s}$ respectively.

The last step before transmission is to shape the sequence of symbols s_n into a continuous-time signal $s(t)$. In baseband representation, this is done as follows:

$$s(t) = \sum_{n=1}^{N} s_n p(t - nT) \tag{5.29}$$

where $p(t)$ is the shaping filter.

This signal is transmitted on a frequency-selective channel characterized by its impulse response $c(t)$. This impulse response is such that the signal undergoes intersymbol interference. At the reception, the signal is also corrupted by additive white Gaussian noise, with power spectral density $N_0/2$.

The receiver filters the signal with a matched filter, samples it at symbol rate and filters the discrete sequence with a noise-whitening filter. The outputs y_n of this

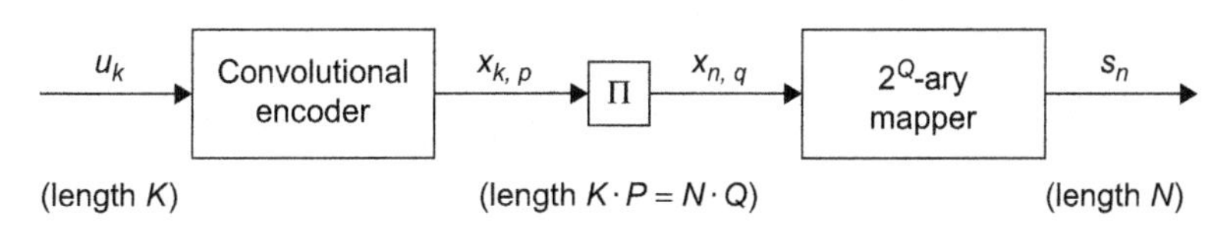

FIGURE 5.5

Transmission scheme with BICM.

process can be written as:

$$y_n = \sum_{l=0}^{L} g_l \, s_{n-l} + z_n, \tag{5.30}$$

where the first term contains the symbols s_n filtered by a causal filter with coefficients g_l $(l = 0,\ldots,L)$ and the second term consists of additive white Gaussian noise samples z_n of variance σ^2. The equivalent discrete-time filter coefficients g_l are assumed real and perfectly known. The sequence of observations y_n is denoted $\boldsymbol{y}$.

5.2.2 Separation of Equalization and Decoding

The receiver has to estimate the information sequence $\boldsymbol{u}$, given the received sequence $\boldsymbol{y}$. The optimal receiver is intractable, but the turbo principle suggests a receiver that decodes the code and equalizes the channel iteratively. This receiver is referred to as a turbo equalizer.

Note that the coding scheme is well suited to the turbo principle: the discrete equivalent model for the channel is a finite state machine (see Figure 5.6), as is the convolutional code, and they are separated by an interleaver. The resulting iterative algorithm resembles that of a serial-concatenated turbo code: the encoder and the channel play the roles of the outer and inner codes, respectively.

The turbo equalizer (Figure 5.7) comprises a SI-SO equalizer/demapper and a SI-SO decoder, separated by the appropriate interleaver and deinterleaver. The extrinsic information produced by each SI-SO stage becomes the a priori information for the other stage. As in a turbo decoder, these iterative exchanges of extrinsic information progressively refine the estimation of the information bits.

The next sections detail the SI-SO equalizer/demapper and the SI-SO decoder.

5.2.3 SI-SO Equalization and Demapping

The role of the SI-SO equalizer/demapper is to produce, from the sequence of channel observations $\boldsymbol{y}$ and from a priori information, a sequence of information on the coded bits that the SI-SO decoder can exploit (see overview in Figure 5.8). This involves

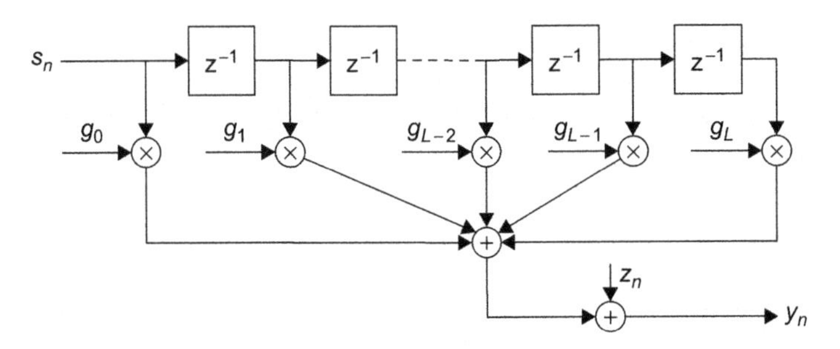

FIGURE 5.6

Channel model as a finite state machine.

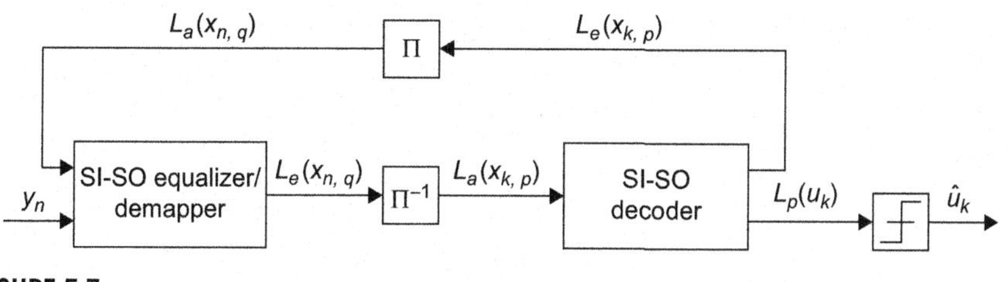

FIGURE 5.7

Turbo equalizer.

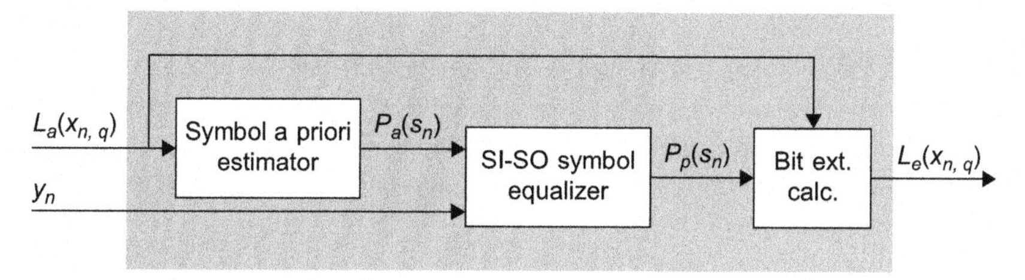

FIGURE 5.8

SI-SO equalizer/demapper.

mitigating the intersymbol interference, i.e., equalizing the signal, as well as converting information on symbols into information on bits, i.e., demapping the signal. These two operations can be split, assuming local independence of the coded bits [MMDdC00]. This assumption holds thanks to the presence of the bit interleaver at the transmitter.

This section details the calculations involved in these operations.

Inputs and Outputs of the SI-SO Equalizer/Demapper

As shown in Figure 5.8, the SI-SO equalizer/demapper takes as inputs, first, the received samples y_n ($n = 1,\ldots,N$) and second, the a priori probabilities $P_a(x_{n,q})$ or equivalently the a priori log-ratios $L_a(x_{n,q})$ for the coded bits $x_{n,q}$:

$$L_a(x_{n,q}) = \ln\left(\frac{P_a(x_{n,q}=1)}{P_a(x_{n,q}=0)}\right). \tag{5.31}$$

This a priori log-ratios sequence is obtained by interleaving the extrinsic information sequence calculated by the SI-SO decoder at the previous iteration. At the first iteration, the symbols $x_{n,q}$ are assumed equiprobable, so the a priori log-ratios are all zero.

As outputs, the SI-SO equalizer produces extrinsic log–ratios $L_e(x_{n,q})$ on the coded bits $x_{n,q}$. As it is the case for turbo decoding, the extrinsic information is obtained by subtracting the a priori information from the a posteriori information:

$$L_e(x_{n,q}) = L_p(x_{n,q}) - L_a(x_{n,q}), \tag{5.32}$$

where

$$L_p(x_{n,q}) = \ln\left(\frac{P(x_{n,q}=1|\mathbf{y})}{P(x_{n,q}=0|\mathbf{y})}\right). \tag{5.33}$$

The derivation below shows that the extrinsic information depends on the received samples y_n ($n = 1,\ldots,N$) and on the a priori information available for all bits but $x_{n,q}$, i.e., $L_a(x_{n',q'})$ for $n' = 1,\ldots,N$ and $q' = 1,\ldots,Q$ with $(n',q') \neq (n,q)$.

Calculation of a Priori Information on Symbols

From the a priori probabilities on bits, the symbol a priori estimator produces a priori probabilities on symbols (see Figure 5.8). Assuming the independence of the coded bits $x_{n,q}$, the a priori probability of a symbol is the product of the a priori probabilities of the corresponding bits:

$$P_a(s_n) = \prod_{q=1}^{Q} P_a(x_{n,q}) \tag{5.34}$$

where the bits $x_{n,q}$, $q = 1,\ldots,Q$ take the values corresponding to the value of the symbol s_n.

SI-SO Symbol Equalization

The SI-SO symbol equalizer uses the BCJR algorithm to calculate a posteriori probabilities $P_p(s_n)$ on the symbols, based on the a priori probabilities $P_a(s_n)$ and on the channel observations y_n. The BCJR algorithm solves the general problem of estimating the APP of the states and transitions of a Markov process observed through memoryless noise. In this case, the Markov process is the following; at each time n, the state ϵ_n is a function of the L last transmitted symbols:

$$\epsilon_n = (s_n,\ldots,s_{n-L+1}). \tag{5.35}$$

At time n, the input s_n causes a transition from state ϵ_{n-1} to state ϵ_n. This transition produces as output the received sample y_n.

The transmission of the whole sequence can be associated with a path in the trellis representing the channel, exactly as with a convolutional code.

Given these definitions, we can use the BCJR algorithm as defined in 5.1.1. The a posteriori probabilities at the output of the SI-SO equalizer are:

$$P_p(s_n = S) = \sum_{\mathcal{E}_S} \alpha_{n-1}(\epsilon')\cdot\gamma_n(\epsilon',\epsilon)\cdot\beta_n(\epsilon), \tag{5.36}$$

where $\mathcal{E}_S$ is the set of transitions $(\epsilon_{n-1} = \epsilon', \epsilon_n = \epsilon)$ caused by a symbol value $s_n = S$.

The functions $\alpha_n(\epsilon)$, $\beta_n(\epsilon)$, and $\gamma_n(\epsilon)$, are evaluated according to (5.10), (5.12), and (5.15). The transition metric can be developed as:

$$\begin{aligned}\gamma_n(\epsilon',\epsilon) &= P(y_n|\epsilon_{n-1}=\epsilon',\epsilon_n=\epsilon)\cdot P(\epsilon_n=\epsilon|\epsilon_{n-1}=\epsilon') \\ &= P(y_n|s_n,\epsilon_{n-1}=\epsilon')\cdot P_a(s_n) \\ &= P(y_n|s_n,s_{n-1},\ldots,s_{n-L})\cdot P_a(s_n),\end{aligned} \tag{5.37}$$

to show that the transition metric contains the channel metric $P(y_n|s_n,s_{n-1},\ldots,s_{n-L})$ and the a priori probability $P_a(s_n)$.

Calculation of the Extrinsic Information on Coded Bits

The last step of the SI-SO equalizer/demapper is to produce extrinsic information $L_e(x_{n,q})$ on the coded bits, from the a posteriori probabilities $P_p(s_n)$ and a priori log-ratios $L_a(x_{n,q})$ (see fig. Figure 5.8).

The probabilities $P_p(s_n)$ on the symbols give probabilities $P_p(x_{n,q})$ on the coded bits as follows:

$$P_p(x_{n,q}) = \sum_{s_n:x_{n,q}} P_p(s_n) \tag{5.38}$$

where the notation $s_n : x_{n,q}$ indicates that the sum is over all symbol values s_n corresponding to a bit value $x_{n,q}$.

From the a posteriori probabilities $P_p(x_{n,q})$, the extrinsic probabilities $P_e(x_{n,q})$ are defined as

$$P_e(x_{n,q}) = \frac{P_p(x_{n,q})}{P_a(x_{n,q})}, \tag{5.39}$$

or equivalently in terms of log-ratios:

$$L_e(x_{n,q}) = L_p(x_{n,q}) - L_a(x_{n,q}). \tag{5.40}$$

The extrinsic probability $P_e(x_{n,q})$ (and the corresponding log-ratio) does not depend on the a priori probability on $x_{n,q}$, as required by the turbo principle; indeed $P_e(x_{n,q})$ can be developed as follows:

$$\begin{aligned}P_e(x_{n,q}) &= \frac{P_p(x_{n,q})}{P_a(x_{n,q})} \\ &= \frac{\sum_{s_n:x_{n,q}} P_p(s_n)}{P_a(x_{n,q})} \\ &= \frac{\sum_{s_n:x_{n,q}} \sum_{\mathcal{E}_S} \alpha_{n-1}(\epsilon')\cdot\gamma_n(\epsilon',\epsilon)\cdot\beta_n(\epsilon)}{P_a(x_{n,q})} \\ &= \frac{\sum_{s_n:x_{n,q}} \sum_{\mathcal{E}_S} \alpha_{n-1}(\epsilon')\cdot P(y_n|s_n,s_{n-1},\ldots,s_{n-L})\cdot P_a(s_n)\cdot\beta_n(\epsilon)}{P_a(x_{n,q})}\end{aligned}$$

$$= \sum_{s_n:x_{n,q}} \sum_{\mathcal{E}_S} \alpha_{n-1}(\epsilon') \cdot P(y_n|s_n, s_{n-1}, \ldots, s_{n-L}) \cdot \beta_n(\epsilon) \frac{P_a(s_n)}{P_a(x_{n,q})}$$

$$= \sum_{s_n:x_{n,q}} \sum_{\mathcal{E}_S} \alpha_{n-1}(\epsilon') \cdot P(y_n|s_n, s_{n-1}, \ldots, s_{n-L}) \cdot \beta_n(\epsilon)$$

$$\times \prod_{\substack{q'=1,\ldots,Q \\ q' \neq q}} P_a(x_{n,q'}),$$

using the relationships of this section. The last equality confirms that the extrinsic probability $P_e(x_{n,q})$ is not dependent on $P_a(x_{n,q})$.

Extrinsic Probability of the Symbols

As a final remark on SI-SO equalization, we define an extrinsic probability $P_e(s_n)$ of the symbols and relate it to the extrinsic probability $P_e(x_{n,q})$ of the coded bits. This extrinsic probability $P_e(s_n)$ is useful below in the design of an approximate SI-SO equalizer. We define the extrinsic probability of the symbols as:

$$P_e(s_n) \triangleq \frac{P_p(s_n)}{P_a(s_n)}. \tag{5.41}$$

Using this definition, the extrinsic probability of the bits can be written:

$$\begin{aligned} P_e(x_{n,q}) &= \frac{P_p(x_{n,q})}{P_a(x_{n,q})} \\ &= \frac{\sum_{s_n:x_{n,q}} P_p(s_n)}{P_a(x_{n,q})} \\ &= \frac{\sum_{s_n:x_{n,q}} P_e(s_n) \cdot P_a(s_n)}{P_a(x_{n,q})} \\ &= \sum_{s_n:x_{n,q}} P_e(s_n) \left(\prod_{\substack{q'=1,\ldots,Q \\ q' \neq q}} P_a\left(x_{n,q'}\right) \right), \end{aligned}$$

which describes the relationship between the extrinsic probabilities $P_e(x_{n,q})$ and $P_e(s_n)$.

5.2.4 SI-SO Decoding

Inputs and Outputs of the SI-SO Decoder

The SI-SO decoder uses as input the sequence of a priori log-ratios $L_a(x_{k,p})$ on the coded bits $x_{k,p}$:

$$L_a(x_{k,p}) = \ln\left(\frac{P_a(x_{k,p}=1)}{P_a(x_{k,p}=0)}\right). \tag{5.42}$$

This a priori information is the extrinsic information produced by the equalizer, deinterleaved to ensure time coherence.

From these a priori log-ratios, the SI-SO decoder produces as outputs a sequence of a posteriori log-ratios on the information bits:

$$L_p(u_k) = \ln\left(\frac{P_p(u_k = 1)}{P_p(u_k = 0)}\right), \tag{5.43}$$

and a sequence of extrinsic log-ratios on the coded bits:

$$L_e(x_{k,p}) = \ln\left(\frac{P_e(x_{k,p} = 1)}{P_e(x_{k,p} = 0)}\right). \tag{5.44}$$

After the last iteration, the a posteriori log-ratios on the information bits yield hard decisions as follows:

$$u_k = \begin{cases} 1 & \text{if} \quad L_p(u_k) \geq 0 \\ 0 & \text{if} \quad L_p(u_k) < 0 \end{cases} \tag{5.45}$$

BCJR Algorithm for Decoding

Again, the convolutional encoder is a Markov process with a state dependent on previous input bits. In this case the state at time k is:

$$\epsilon_k = (u_k, \ldots, u_{k-M+1}). \tag{5.46}$$

Applied to this Markov process, the BCJR algorithm can calculate the a posteriori log-ratios $L_p(u_k)$ and $L_p(x_{k,p})$. We have

$$\begin{aligned} L_p(u_k) = & \overset{*}{\max_{\mathcal{E}^+}} \left(\bar{\alpha}_{k-1}(\epsilon') + \bar{\gamma}_k(\epsilon',\epsilon) + \bar{\beta}_k(\epsilon)\right) \\ & - \overset{*}{\max_{\mathcal{E}^-}} \left(\bar{\alpha}_{k-1}(\epsilon') + \bar{\gamma}_k(\epsilon',\epsilon) + \bar{\beta}_k(\epsilon)\right) \end{aligned} \tag{5.47}$$

and

$$\begin{aligned} L_p(x_{k,p}) = & \overset{*}{\max_{\mathcal{E}_p^+}} \left(\bar{\alpha}_{k-1}(\epsilon') + \bar{\gamma}_k(\epsilon',\epsilon) + \bar{\beta}_k(\epsilon)\right) \\ & - \overset{*}{\max_{\mathcal{E}_p^-}} \left(\bar{\alpha}_{k-1}(\epsilon') + \bar{\gamma}_k(\epsilon',\epsilon) + \bar{\beta}_k(\epsilon)\right). \end{aligned} \tag{5.48}$$

The difference between 5.47 and 5.48 is the subset of transitions considered. In 5.47, $\mathcal{E}^+$ (resp. $\mathcal{E}^-$) is the subset of transitions $(\epsilon_{k-1} = \epsilon', \epsilon_k = \epsilon)$ caused by a symbol $u_k = 1$ (resp $u_k = 0$). In 5.48, $\mathcal{E}_p^+$ (resp. $\mathcal{E}_p^-$) is the subset of transitions $(\epsilon_{k-1} = \epsilon', \epsilon_k = \epsilon)$ caused by a symbol $x_{k,p} = 1$ (resp $x_{k,p} = 0$).

Detail of the Transition Metric

The α and β recursions are calculated as in Section 5.1.1. The metric $\bar{\gamma}_k$ for the transition from state $\epsilon_{k-1} = \epsilon'$ to state $\epsilon_k = \epsilon$ depends on the a priori information received from the equalizer, not directly on the channel observations (which are only available to the equalizer). The transition metric is thus:

$$\bar{\gamma}_k(\epsilon',\epsilon) = \ln\left(P(\epsilon_k = \epsilon | \epsilon_{k-1} = \epsilon')\right). \tag{5.49}$$

The transition parametrized by the pair $(\epsilon_k, \epsilon_{k-1})$ can equivalently be parametrized by the coded bits output during the transition, that is, the coded bits $x_{k,p}$ $(p = 1,\ldots,P)$. Under the assumption that the coded bits are independent, the above metric becomes:

$$\bar{\gamma}_k(\epsilon',\epsilon) = \sum_{p=1}^{P} \ln\left(P_a(x_{k,p})\right). \tag{5.50}$$

The probabilities $P_a(x_{k,p})$ appearing above can be evaluated from the available a priori log-ratios $L_a(x_{k,p})$ $(p = 1,\ldots,P)$, since:

$$L_a(x_{k,p}) = \ln\left(\frac{P_a(x_{k,p} = 1)}{P_a(x_{k,p} = 0)}\right). \tag{5.51}$$

and thus:

$$P_a(x_{k,p}) = \begin{cases} \dfrac{\exp(L_a(x_{k,p}))}{1+\exp(L_a(x_{k,p}))} & \text{if } x_{k,p} = 1 \\ \dfrac{1}{1+\exp(L_a(x_{k,p}))} & \text{if } x_{k,p} = 0. \end{cases} \tag{5.52}$$

Substituting into (5.50) and suppressing the terms common to all hypotheses, the transition metric becomes:

$$\bar{\gamma}_k(\epsilon',\epsilon) = \sum_{p=1}^{P} L_a(x_{k,p})\, x_{k,p}. \tag{5.53}$$

This transition metric along with the recursions $\bar{\alpha}_k(\cdot)$ and $\bar{\beta}_k(\cdot)$ enables to calculate the a posteriori log-ratios $L_p(u_k)$ and $L_p(x_{k,p})$.

Calculation of the Extrinsic Information

We now rewrite the log-ratios $L_p(x_{k,p})$ to reveal the extrinsic information. Defining:

$$\gamma^e_{k,p'}(\epsilon',\epsilon) = \sum_{p=1;p\neq p'}^{P} L_a(x_{k,p})\, x_{k,p}, \tag{5.54}$$

the transition metric can be written:

$$\gamma_k(\epsilon',\epsilon) = L_a(x_{k,p'})x_{k,p'} + \gamma^e_{k,p'}(\epsilon',\epsilon). \tag{5.55}$$

Since the set $\mathcal{E}_{p'}^{+}$ (respectively $\mathcal{E}_{p'}^{-}$) is the subset of transitions causing $x_{k,p'} = 1$ (respectively $x_{k,p'} = 0$), the a posteriori log-ratio can be expressed as:

$$L_p(x_{k,p'}) = \overset{*}{\max_{\mathcal{E}_{p'}^{+}}} (\bar{\alpha}_{k-1}(\epsilon') + \bar{\gamma}_{k,p'}^{e}(\epsilon',\epsilon) + \bar{\beta}_k(\epsilon) + 1 \cdot L_a(x_{k,p'}))$$
$$- \overset{*}{\max_{\mathcal{E}_{p'}^{-}}} (\bar{\alpha}_{k-1}(\epsilon') + \bar{\gamma}_{k,p'}^{e}(\epsilon',\epsilon) + \bar{\beta}_k(\epsilon) + 0 \cdot L_a(x_{k,p'})). \quad (5.56)$$

According to the definition of the modified maximum operation $\overset{*}{\max}$, this last operation can be rewritten:

$$L_p(x_{k,p'}) = L_a(x_{k,p'}) + \overset{*}{\max_{\mathcal{E}_{p'}^{+}}} (\bar{\alpha}_{k-1}(\epsilon') + \bar{\gamma}_{k,p'}^{e}(\epsilon',\epsilon) + \bar{\beta}_k(\epsilon))$$
$$- \overset{*}{\max_{\mathcal{E}_{p'}^{-}}} (\bar{\alpha}_{k-1}(\epsilon') + \bar{\gamma}_{k,p'}^{e}(\epsilon',\epsilon) + \bar{\beta}_k(\epsilon)). \quad (5.57)$$

Defining the last two terms as the extrinsic information $L_e(x_{k,p'})$, we have

$$L_e(x_{k,p'}) = L_p(x_{k,p'}) - L_a(x_{k,p'}). \quad (5.58)$$

5.2.5 Reduced Complexity SI-SO Equalization

The above implementation of the SI-SO equalizer leads to a complexity per transmitted bit and per iteration that evolves as $\mathcal{O}(2^{Q \cdot L})$, where 2^Q is the mapping alphabet size and L the discrete-time equivalent channel length. The equalization becomes thus rapidly intractable for usual mappings and channels. This section presents instead a lower complexity approximation of the SI-SO equalizer above, based on minimum mean square error (MMSE) estimation [WP99, TSK02, DV02].

The MMSE equalizer takes as inputs the channel observations and a priori information on the symbols. It produces a linear MMSE estimation of the received symbols, and from these, estimates the extrinsic probabilities on the symbols (Figure 5.9).

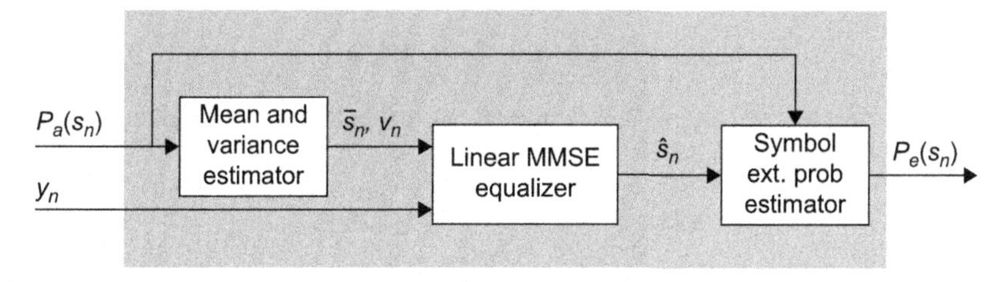

FIGURE 5.9

MMSE equalizer.

Sliding-Window Model

To simplify MMSE equalization, the estimation of a symbol considers interference from its closest neighboring symbols only. For the estimation of symbol s_n, the equalizer takes into account the N_1 channel observations received before and the N_2 observations received after s_n. This yields an equalizer of length $N = 1 + N_1 + N_2$. Accordingly, the derivation of the MMSE equalizer is based on the following sliding-window model of the transmission:

$$\boldsymbol{y_n} = \boldsymbol{G}\boldsymbol{s_n} + \boldsymbol{z_n} \tag{5.59}$$

with:

$$\boldsymbol{y_n} \triangleq \left[y_{n-N_1} \cdots y_n \cdots y_{n+N_2}\right]^T_{N\times 1} \tag{5.60}$$

$$\boldsymbol{s_n} \triangleq \left[s_{n-N_1-L} \cdots s_n \cdots s_{n+N_2}\right]^T_{(N+L)\times 1} \tag{5.61}$$

$$\boldsymbol{z_n} \triangleq \left[z_{n-N_1} \cdots z_n \cdots z_{n+N_2}\right]^T_{N\times 1} \tag{5.62}$$

and the channel matrix $\boldsymbol{G}$ defined as:

$$\boldsymbol{G} = \begin{pmatrix} g_L & \cdots & g_0 & 0 & \cdots & \cdots & 0 \\ 0 & g_L & \cdots & g_0 & 0 & \cdots & 0 \\ \vdots & \ddots & \ddots & \ddots & \ddots & \ddots & \vdots \\ 0 & \cdots & \cdots & 0 & g_L & \cdots & g_0 \end{pmatrix}_{N\times(N+L)} . \tag{5.63}$$

MMSE Symbol Estimation

The MMSE estimator of a given symbol s_n, given the observation $\boldsymbol{y_n}$, can be expressed as follows:

$$\hat{s}_n = \phi(\boldsymbol{y_n}) \tag{5.64}$$

where the function ϕ minimizes the mean square error:

$$J = \mathrm{E}\left\{[s_n - \hat{s}_n][s_n - \hat{s}_n]^*\right\}, \tag{5.65}$$

with x^* designating the conjugate of x. The best linear solution to this problem is, as shown in [Men95],

$$\hat{s}_n = \mathrm{E}\{s_n\} + \boldsymbol{w_n}^H[\boldsymbol{y_n} - \mathrm{E}\{\boldsymbol{y_n}\}], \tag{5.66}$$

where $\boldsymbol{w_n}^H$ denotes the hermitian (conjugate transpose) of the equalization time-varying filter $\boldsymbol{w_n}$ defined as:

$$\boldsymbol{w_n} = \mathrm{cov}\{\boldsymbol{y_n}, \boldsymbol{y_n}\}^{-1}\mathrm{cov}\{\boldsymbol{y_n}, s_n\}. \tag{5.67}$$

The expectations and covariances appearing in (5.66) and (5.67) are estimated using the a priori probabilities $P_a(s_{n'})$ $(n' = n - N_1 - L, \ldots, n-1, n+1, \ldots, n+N_2)$ available at the input of the equalizer. Note that since the goal is to calculate the extrinsic probability for the symbol s_n, the a priori probability $P_a(s_n)$ is not taken into account, according to the turbo principle. Consequently, we have:

$$\mathrm{E}\{s_n\} = 0 \tag{5.68}$$

whereas for all other symbols $s_{n'}$ $(n' = n - N_1 - L, \ldots, n-1, n+1, \ldots, n+N_2)$, we have:

$$\bar{s}_{n'} \triangleq \mathrm{E}\{s_{n'}\} = \sum_{s_{n'} \in \mathcal{S}} s_{n'} P_a(s_{n'}). \tag{5.69}$$

Note that at the first iteration, the a priori probabilities $P_a(s_{n'})$ are equiprobable (no a priori information). The average value of the channel observation is:

$$\mathrm{E}\{\boldsymbol{y_n}\} = \mathrm{E}\{\boldsymbol{G}\boldsymbol{s_n} + \boldsymbol{z_n}\} = \boldsymbol{G}\bar{\boldsymbol{s}}_{\boldsymbol{n}} \tag{5.70}$$

where the noise term disappears because it has a zero mean, and where:

$$\begin{aligned}\bar{\boldsymbol{s}}_{\boldsymbol{n}} &\triangleq \mathrm{E}\{\boldsymbol{s_n}\} \\ &= [\bar{s}_{n-N_1-L} \quad \cdots \quad \bar{s}_{n-1} \quad 0 \quad \bar{s}_{n+1} \quad \cdots \quad \bar{s}_{n+N_2}]^T.\end{aligned} \tag{5.71}$$

As for the covariances, we have:

$$\mathrm{cov}\{\boldsymbol{y_n}, \boldsymbol{y_n}\} = \boldsymbol{G}\,\mathrm{cov}\{\boldsymbol{s_n}, \boldsymbol{s_n}\}\boldsymbol{G}^H + \sigma^2 \boldsymbol{I} \tag{5.72}$$

where the covariance $\mathrm{cov}\{\boldsymbol{s_n}, \boldsymbol{s_n}\}$ is again estimated using the a priori information available for all symbols but s_n, thus:

$$\begin{aligned}\mathrm{cov}\{\boldsymbol{s_n}, \boldsymbol{s_n}\} &= \mathrm{E}\left\{[\boldsymbol{s_n} - \bar{\boldsymbol{s}}_{\boldsymbol{n}}][\boldsymbol{s_n} - \bar{\boldsymbol{s}}_{\boldsymbol{n}}]^H\right\} \\ &= \mathrm{diag}\big[v_{n-N_1-L} \quad \cdots \quad v_{n-1} \\ &\qquad \sigma_s^2 \quad v_{n+1} \quad \cdots \quad v_{n+N_2}\big] \\ &\triangleq \boldsymbol{R_{ss,n}}\end{aligned} \tag{5.73}$$

where σ_s^2 is the variance of the symbols alphabet $\mathcal{S}$; i.e., the variance when no a priori information is available, and where:

$$\begin{aligned}v_{n'} &\triangleq \mathrm{var}\{s_{n'}\} \\ &= \mathrm{E}\left\{|s_{n'}|^2\right\} - \mathrm{E}\{|s_{n'}|\}^2 \\ &= \sum_{s_{n'} \in \mathcal{S}} |s_{n'}|^2 P_a(s_{n'}) - |\bar{s}_{n'}|^2.\end{aligned} \tag{5.74}$$

Finally, we have:

$$\begin{aligned}\operatorname{cov}\{\boldsymbol{y_n}, s_n\} &= \boldsymbol{G}\operatorname{cov}\{\boldsymbol{s_n}, s_n\}\\ &= \boldsymbol{G}\,\mathrm{E}\left\{[\boldsymbol{s_n} - \bar{\boldsymbol{s}}_{\boldsymbol{n}}]\, s_n^*]\right\}\\ &= \boldsymbol{G}\,\mathrm{E}\left\{\boldsymbol{s_n} s_n^*\right\}\\ &= \boldsymbol{g}\,\sigma_s^2,\end{aligned} \tag{5.75}$$

where $\boldsymbol{g}$ is the $N_1 + L + 1^{\text{th}}$ column of the matrix $\boldsymbol{G}$.

Substituting these values in (5.66)–(5.67), the MMSE filter becomes:

$$\boldsymbol{w_n} = [\boldsymbol{G}\boldsymbol{R}_{\boldsymbol{ss},\boldsymbol{n}}\boldsymbol{G}^H + \sigma^2\boldsymbol{I}]^{-1}\boldsymbol{g}\sigma_s^2, \tag{5.76}$$

and the MMSE estimate becomes:

$$\hat{s}_n = \boldsymbol{w_n}^H[\boldsymbol{y_n} - \boldsymbol{G}\bar{\boldsymbol{s}}_{\boldsymbol{n}}] \tag{5.77}$$

$$= \sigma_s^2\, \boldsymbol{g}^H[\boldsymbol{G}\boldsymbol{R}_{\boldsymbol{ss},\boldsymbol{n}}\boldsymbol{G}^H + \sigma^2\boldsymbol{I}]^{-1}\,[\boldsymbol{y_n} - \boldsymbol{G}\bar{\boldsymbol{s}}_{\boldsymbol{n}}]. \tag{5.78}$$

Estimation of the Extrinsic Information from the MMSE Estimates

From the MMSE estimates $\hat{s}_n$, the equalizer produces extrinsic probabilities using the approximation $P_e(s_n) \approx P(\hat{s}_n)$. To simplify the latter distribution, we model the symbol estimates $\hat{s}_n$ as the outputs of an equivalent Additive White Gaussian Noise (AWGN) channel having symbols s_n at its input:

$$\hat{s}_n = \mu_n\, s_n + \eta_n \tag{5.79}$$

with $\eta_n \approx \mathcal{N}_c(0, \nu_n^2)$. The equivalent noise η_n accounts for the channel noise and the residual intersymbol interference.

The parameters μ_n and ν_n^2 of this model can be calculated for each symbol n as a function of the filter $\boldsymbol{w_n}$. The mean μ_n is calculated by evaluating:

$$\mathrm{E}\left\{\hat{s}_n s_n^*\right\} = \mu_n\,\mathrm{E}\left\{|s_n|^2\right\} = \mu_n\,\sigma_s^2 \tag{5.80}$$

and requiring that it is equal to:

$$\begin{aligned}\mathrm{E}\left\{\hat{s}_n s_n^*\right\} &= \boldsymbol{w_n}^H\,\mathrm{E}\left\{[\boldsymbol{y_n} - \boldsymbol{G}\bar{\boldsymbol{s}}_{\boldsymbol{n}}]\, s_n^*\right\}\\ &= \boldsymbol{w_n}^H\,\boldsymbol{G}\,\mathrm{E}\left\{[\boldsymbol{s_n} - \bar{\boldsymbol{s}}_{\boldsymbol{n}}]\, s_n^*\right\}\\ &= \boldsymbol{w_n}^H\,\boldsymbol{g}\,\sigma_s^2.\end{aligned} \tag{5.81}$$

Combining these two expressions leads to:

$$\mu_n = \boldsymbol{w_n}^H\,\boldsymbol{g}. \tag{5.82}$$

As for the variance ν_n^2, we have:

$$\nu_n^2 = \mathrm{E}\left\{|\eta_n|^2\right\} = \mathrm{E}\left\{|\hat{s}_n - \mu_n s_n|^2\right\} = \mathrm{E}\left\{|\hat{s}_n|^2\right\} - \mu_n^2\sigma_s^2. \tag{5.83}$$

Substituting (5.77), the expression above becomes:

$$\begin{aligned}\nu_n^2 &= \boldsymbol{w_n}^H \mathrm{E}\{[\boldsymbol{y_n} - \boldsymbol{G}\bar{\boldsymbol{s}}_{\boldsymbol{n}}][\boldsymbol{y_n} - \boldsymbol{G}\bar{\boldsymbol{s}}_{\boldsymbol{n}}]^H\}\boldsymbol{w_n} - \mu_n^2\sigma_s^2 \\ &= \boldsymbol{w_n}^H \operatorname{cov}\{\boldsymbol{y_n}, \boldsymbol{y_n}\}\boldsymbol{w_n} - \mu_n^2\sigma_s^2. \end{aligned} \tag{5.84}$$

Given the expression (5.67) of the filter $\boldsymbol{w_n}$, this becomes:

$$\begin{aligned}\nu_n^2 &= \boldsymbol{w_n}^H \operatorname{cov}\{\boldsymbol{y_n}, s_n\} - \mu_n^2\sigma_s^2 \\ &= \boldsymbol{w_n}^H \boldsymbol{h}\sigma_s^2 - \mu_n^2\sigma_s^2. \end{aligned} \tag{5.85}$$

Finally, this last expression is equivalent to the following:

$$\nu_n^2 = \mu_n\sigma_s^2 - \mu_n^2\sigma_s^2. \tag{5.86}$$

Given the equivalent Gaussian channel assumption in (5.79), the extrinsic probability is calculated from the estimates $\hat{s}_n$ as:

$$P_e(s_n) = \frac{1}{\nu_n^2\pi}\exp\left(-\frac{|\hat{s}_n - \mu_n s_n|^2}{\nu_n^2}\right). \tag{5.87}$$

Note that these extrinsic probabilities are obtained without using the a priori information $P_a(s_n)$ on the symbol s_n, in accordance with the turbo principle.

An Approximate Implementation of the MMSE Equalizer

The MMSE estimate $\hat{s}_n$ defined by (5.77)–(5.78) uses the time-varying filter $\boldsymbol{w_n}$ defined by (5.76). To avoid computing this filter for each symbol, a close approximate solution can be obtained by using its average value over the whole frame. This amounts to replace $\boldsymbol{R_{ss,n}}$ by its empirical average $\bar{\boldsymbol{R}}_{\boldsymbol{ss}}$. If N is the total number of symbols in the frame, the average a priori variance of the symbols sent in the frame is:

$$\bar{\nu} = \frac{1}{N}\sum_{n'=1}^{N}\nu_{n'}, \tag{5.88}$$

where $\nu_{n'}$ is defined by (5.74). The empirical average of the covariance matrix $\boldsymbol{R_{ss,n}}$ is then:

$$\bar{\boldsymbol{R}}_{\boldsymbol{ss}} = \operatorname{diag}\left[\bar{\nu}\cdots\bar{\nu} \quad \sigma_s^2 \quad \bar{\nu}\cdots\bar{\nu}\right], \tag{5.89}$$

and the MMSE filter becomes:

$$\boldsymbol{w} = [\boldsymbol{G}\bar{\boldsymbol{R}}_{\boldsymbol{ss}}\boldsymbol{G}^H + \sigma^2\boldsymbol{I}]^{-1}\,\boldsymbol{g}\,\sigma_s^2. \tag{5.90}$$

It is constant, that is, it does not depend on n.

5.2.6 Simulations

This section illustrates the considered turbo equalization scheme with simulation results. As seen in Figure 5.10, the turbo equalizer manages to almost suppress the effects of intersymbol interference, even for moderate signal-to-noise ratios. The simulation uses the following parameters. The channel coefficients are equal to $\left[\sqrt{0.45}\,\sqrt{0.25}\,\sqrt{0.15}\,\sqrt{0.10}\,\sqrt{0.05}\right]$, they are real and normalized to $\sum_{l=0}^{L} g_l^2 = 1$. The channel memory L is thus 4, and the associated finite state machine has $2^L = 16$ states. The transmitter sends frames of $K = 2048$ information bits encoded with a rate-1/2, memory-4 convolutional code with octal representation $(23_8, 35_8)$. The mapper uses binary phase shift keying (BPSK). The receiver uses the SI-SO equalizer of Section 5.2.3, implemented in the logarithmic domain.

Figure 5.10 presents the bit error rate as a function of the E_b/N_0 ratio, where E_b is the energy per transmitted information bit and $N_0/2$ the double-sided power spectral density of the additive white Gaussian noise. The figure comprises the following curves:

- The bit error rate achieved by the turbo equalization scheme, at each of the first five iterations.
- The bit error rate achieved with the same coding scheme on an AWGN channel, without intersymbol interference. This curve is labeled *No ISI*.
- The bit error rate achieved with the same coding scheme on the same channel, with a noniterative receiver. The receiver comprises an optimal equalizer and an optimal decoder, exchanging hard information. This curve is labeled *Hard decisions*.
- The bit error rate without error-correcting code. The receiver consists of an optimal equalizer. This curve is referenced *No channel coding* in the figure.

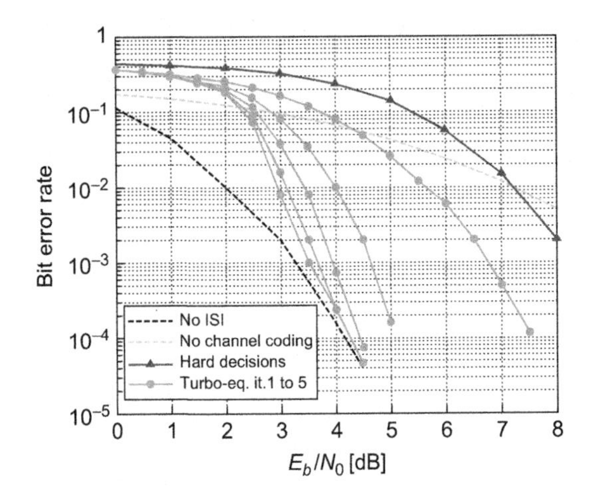

FIGURE 5.10

The turbo equalizer achieves much lower bit error rates than a classic non-iterative equalizer producing hard decisions. For E_b/N_0 larger than 4.5 dB, the turbo equalizer manages to suppress the interference in 4 iterations.

For E_b/N_0 ratios above 4.5 dB, the turbo equalizer is able to completely suppress the ISI in only 4 iterations: the bit error rate is then equal to the bit error rate on an AWGN channel. Compared to the classic non-iterative receiver with an equalizer providing hard decisions to a decoder, the turbo equalizer gains more than 4 dB for bit error rates below 10^{-2}. Comparing the first iteration of the turbo equalizer with the non-iterative hard receiver shows that exchanging soft decisions instead of hard decisions already provides a gain of 1.5 dB. Finally, the comparison with the bit error rate in the absence of error-correcting code shows how important is the gain of using a relatively simple error-correcting code and the turbo principle at the receiver.

The MMSE equalizer of Section 5.2.5 behaves almost as well as the optimal SI-SO equalizer based on the BCJR algorithm, as shown in Figure 5.11. The

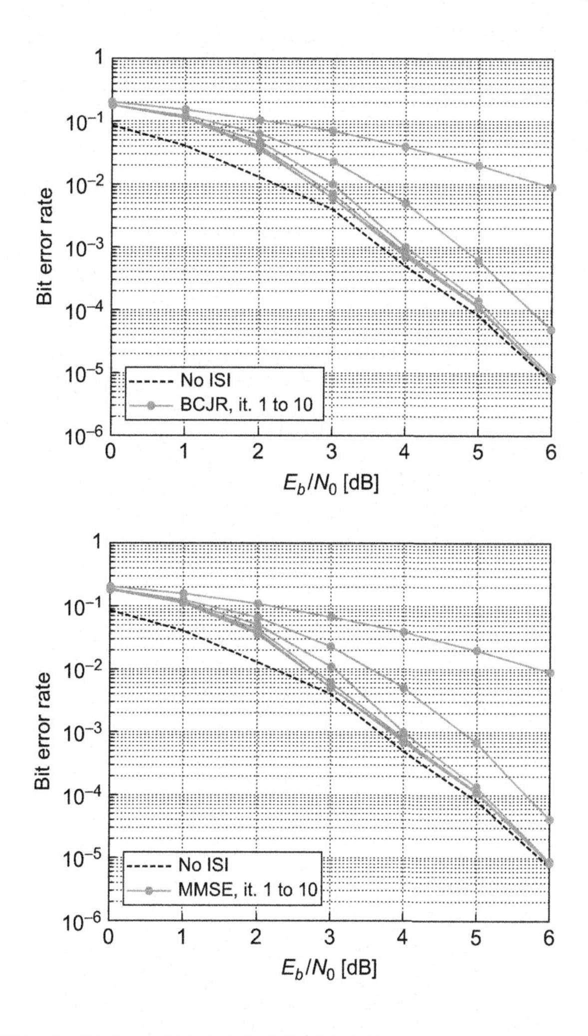

FIGURE 5.11

The equalizer based on the BCJR algorithm and the MMSE equalizer achieve about the same performance. The complexity of the MMSE equalizer is lower though.

simulation uses a channel of length $L = 5$ with complex coefficients $[2 - 0.4\iota, 1.5 + 1.8\iota, 1, 1.2 - 1.3\iota, 0.8 + 1.6\iota]$. The code is a nonrecursive systematic convolutional code of rate 1/2, with memory $M = 2$ and octal representation $(7_8, 5_8)$. The frame length is $K = 1800$ and the mapper uses Quadrature Phase-Shift Keying (QPSK). For the case of the MMSE equalizer, the sliding window has parameters $N_1 = 2$ and $N_2 = 5$.

The bit error rate achieved with the BCJR equalizer (upper figure) is only marginally better than the bit error rate of the simpler MMSE equalizer (lower figure). The first iterations provide the largest improvements: after about 4 iterations the bit error rate decreases only slightly. Moreover, from $E_b/N_0 = 3$ dB the intersymbol interference is almost suppressed, as the curves get close to the bit error rate curve of an AWGN channel (labeled *no ISI*).

A comparison of the MMSE equalizer and the approximate MMSE equalizer shows that the latter performs almost as well as the former (Figure 5.12). In this simulation, the channel of length $L = 11$ has coefficients $[0.04, -0.05, 0.07, -0.21, -0.5, 0.72, 0.36, 0, 0.21, 0.03, 0.07]$. The code is again the nonrecursive systematic convolutional code of rate 1/2, memory $M = 2$ and octal representation $(7_8, 5_8)$. The frame length is $K = 1998$ and the mapping is 8-PSK with set partitioning. The equalizers use a sliding windows with parameters $N_1 = 0$ and $N_2 = 10$ and the turbo equalizer performs 10 iterations (bit error rate given for iterations $1 - 4$ and 10).

The MMSE equalizer performs slightly better than the approximate MMSE equalizer, but the latter has a significantly lower complexity. For E_b/N_0 larger than 4 dB, both equalizers manage to suppress the ISI in 10 iterations.

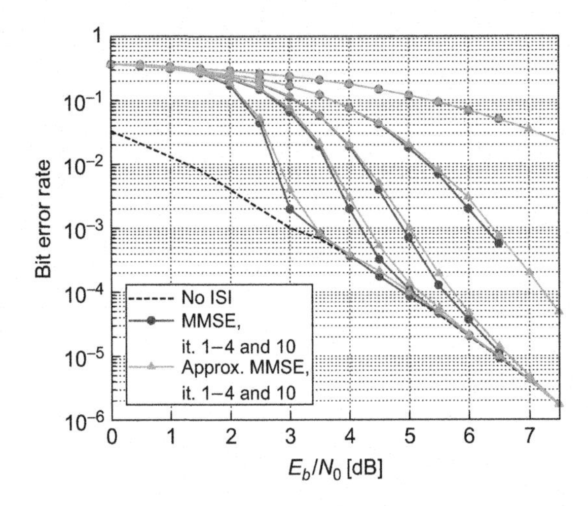

FIGURE 5.12

The approximate implementation of the MMSE equalizer yields higher bit error rates, but only marginally so. Both equalizers suppress the intersymbol interference for E_b/N_0 larger than 4 dB.

5.3 TURBO EQUALIZATION ON FREQUENCY-SELECTIVE MIMO CHANNELS

On frequency-selective MIMO channels, the signal is corrupted by intersymbol interference and by cross-antenna interference. This section extends the turbo equalization method above to such channels [W04, Ton00b, Ton00a, LW00, SH01]. The coding scheme is here space-time BICM (ST-BICM), in which BICM encoded symbols are multiplexed to multiple transmit antennas. ST-BICM is a pragmatic coding approach that enables to easily trade rate versus diversity.

5.3.1 Transmission Scheme

In this section we focus on a frequency-selective channel with multiple inputs and multiple outputs (MIMO), i.e., the transmitter and the receiver have multiple antennas. To exploit diversity on such channels, one can use space-time block coding, or space-time trellis coding. Another powerful approach is to use space-time BICM: the data is encoded with a convolutional code, then interleaved, distributed to the transmitter antennas, modulated into complex symbols and transmitted (Figure 5.13). This scheme is very flexible because the encoder and the mapping can be chosen independently. For instance, a more powerful but more complex system can use turbo codes instead of convolutional codes.

The receiver can use the turbo principle to decode the signal, thanks to the presence of the interleaver. The transmission scheme is indeed very close to the transmission scheme of the preceding section. The main difference is the channel model, as shown below.

The transmission scheme is as follows. As above, the information bits u_k ($k = 1,\ldots,N$) are encoded into coded bits $x_{k,p}$ ($k = 1,\ldots,K$ and $p = 1\ldots,P$ for a convolutional code of rate $r = 1/P$). These coded bits are interleaved and split into n_T sub-blocks, where n_T is the number of antennas at the transmitter. Within each block $i = 1,\ldots,n_T$, the coded bits, written $x_{n,q}^{(i)}$ (with $n = 1,\ldots,N$ and $q = 1,\ldots,Q$), are mapped onto N symbols $s_n^{(i)}$. The symbol alphabet $\mathcal{S}$ contains 2^Q symbol values.

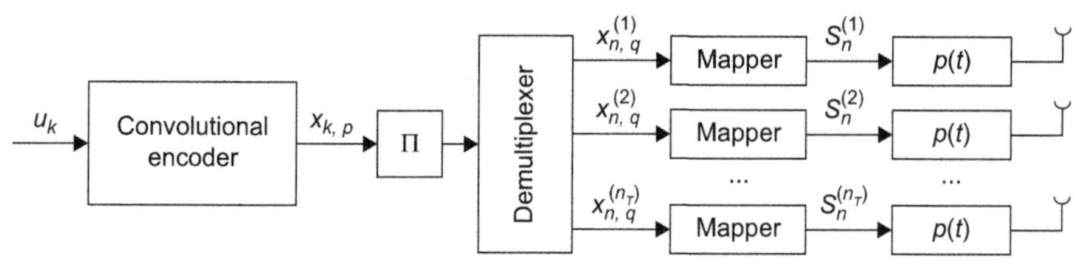

FIGURE 5.13

Space-time BICM transmitter.

The sequence of symbols $s_n^{(i)}$ is sent from antenna (i) after filtering by a pulse-shaping filter $p(t)$. The signal sent is, in baseband representation,

$$s^{(i)}(t) = \sum_{n=1}^{N} s_n^{(i)} p(t-nT). \tag{5.91}$$

The receiver observes the channel with n_R reception antennas. The signal $s^{(i)}(t)$ is sent from transmit antenna i to reception antenna $j = 1,\ldots,n_R$ through a frequency-selective channel with impulse response $c^{(i,j)}(t)$. We assume that this channel produces intersymbol interference. At the receiver, the signal is also corrupted by additive white Gaussian noise. This noise is white in time, and in space (i.e., independent across antennas).

The transmission can be represented with the whitened matched-filter discrete-time model. At each reception antenna, the received signal is filtered by a matched filter, sampled at symbol frequency, and filtered by a noise-whitening filter. The outputs of this process contain cross-antenna interference, intersymbol interference, and additive white Gaussian noise. The output signal can be written, at time $n = 1,\ldots,N$:

$$\begin{pmatrix} y_n^{(1)} \\ \vdots \\ y_n^{(n_R)} \end{pmatrix} = \sum_{l=0}^{L} \boldsymbol{G}_l \begin{pmatrix} s_{n-l}^{(1)} \\ \vdots \\ s_{n-l}^{(n_T)} \end{pmatrix} + \begin{pmatrix} z_n^{(1)} \\ \vdots \\ z_n^{(n_R)} \end{pmatrix} \tag{5.92}$$

where $z_n^{(1)},\ldots,z_n^{(n_R)}$ are additive white Gaussian noise samples, and where the discrete-time equivalent impulse response of the transmission and processing at the reception is a L-length filter $\boldsymbol{G}_l$, for $l = 0,\ldots,L$, with:

$$\boldsymbol{G}_l = \begin{pmatrix} g_l^{(1,1)} & \cdots & g_l^{(1,n_T)} \\ \vdots & \ddots & \vdots \\ g_l^{(n_R,1)} & \cdots & g_l^{(n_R,n_T)} \end{pmatrix}_{(n_R \times n_T)} . \tag{5.93}$$

5.3.2 Application of the Turbo Equalization Framework

The turbo equalization framework of Section 5.2 can handle MIMO signals as the one in (5.92), if properly adapted. The overall iterative decoder is depicted in Figure 5.14. The only differences are the presence of the demultiplexer, and a slight modification of the MMSE equalizer. The difference in the MMSE equalizer simply stems from a different expression for the sliding-window signal model, as detailed below.

For the MIMO case the sliding-window model for the received sample is as follows:

$$\boldsymbol{y}_n = \boldsymbol{G}\boldsymbol{s}_n + \boldsymbol{z}_n \tag{5.94}$$

where the signal vectors $\boldsymbol{y}_n$, $\boldsymbol{s}_n$ and $\boldsymbol{z}_n$ stack the samples received on the n_R antennas, and received during the whole equalizer window. For an equalizer considering the N_1

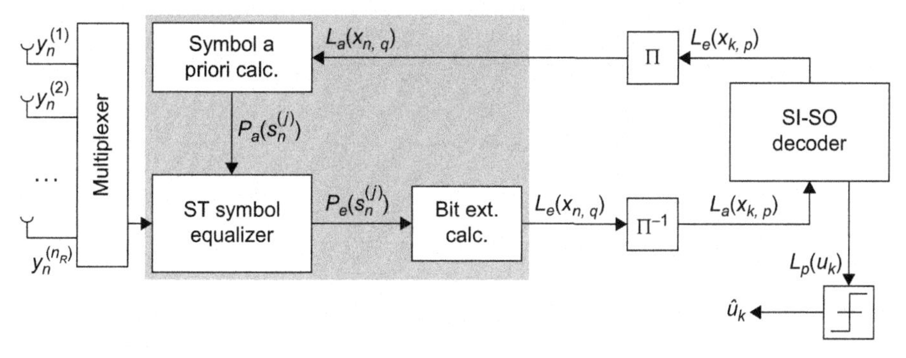

FIGURE 5.14

Turbo equalizer for ST-BICM on frequency-selective MIMO channels.

previous and N_2 next received samples, the signal vectors above are defined as:

$$\boldsymbol{y}_n \triangleq \left[y_{n-N_1}^{(1)}, \ldots, y_{n-N_1}^{(n_R)}, \ldots, y_{n+N_2}^{(1)}, \ldots, y_{n+N_2}^{(n_R)}\right] \tag{5.95}$$

$$\boldsymbol{s}_n \triangleq \left[s_{n-N_1-L}^{(1)}, \ldots, s_{n-N_1-L}^{(n_R)}, \ldots, s_{n+N_2}^{(1)}, \ldots, s_{n+N_2}^{(n_R)}\right] \tag{5.96}$$

$$\boldsymbol{z}_n \triangleq \left[z_{n-N_1}^{(1)}, \ldots, z_{n-N_1}^{(n_R)}, \ldots, z_{n+N_2}^{(1)}, \ldots, z_{n+N_2}^{(n_R)}\right]. \tag{5.97}$$

The channel matrix is then defined as:

$$\boldsymbol{G} = \begin{pmatrix} \boldsymbol{G}_L & \cdots & \boldsymbol{G}_0 & 0 & \cdots & \cdots & 0 \\ 0 & \boldsymbol{G}_L & \cdots & \boldsymbol{G}_0 & 0 & \cdots & 0 \\ \vdots & \ddots & \ddots & \ddots & \ddots & \ddots & \vdots \\ 0 & \cdots & \cdots & 0 & \boldsymbol{G}_L & \cdots & \boldsymbol{G}_0 \end{pmatrix}_{n_R N \times n_T (N+L)}. \tag{5.98}$$

with $N = 1 + N_1 + N_2$.

With this model definition, the problem is formally identical to the single-input single-output problem of the previous section. The only difference stems from the different definition of the channel matrix $\boldsymbol{G}$. The MMSE equalization operation is given by (5.76)–(5.78).

5.3.3 Simulations

This section illustrates the performance of turbo equalization on MIMO channels through simulations. These simulations are carried out for a frequency-selective slow Rayleigh-fading channel. The model we use is referred to as *GSM typical urban (TU)* and has been specified for Enhanced Data rates for GSM Evolution (EDGE) and

Table 5.1 GSM Typical Urban (TU) Channel Model

Path delay (μs)	0.0	0.2	0.5	1.6	2.3	5.0
Path power (dB)	−3.0	0.0	−2.0	−6.0	−8.0	−10.0

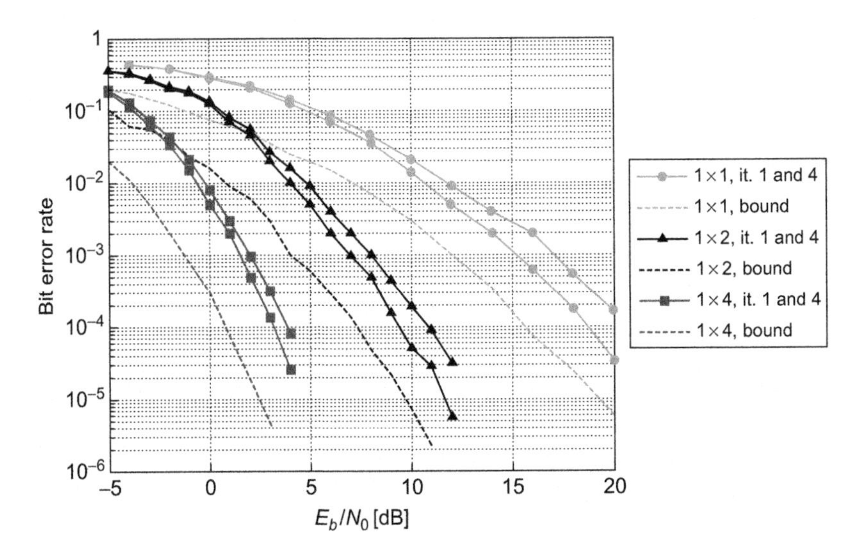

FIGURE 5.15

Turbo equalization on the GSM TU channel. The turbo equalizer is able to benefit from diversity and array gain, as shown by the performance improvement when the number of receive antennas increases.

Global System for Mobile Communications (GSM). The symbol period T is equal to 3.7 μs. Table 5.1 shows the path power and delay profiles of this channel. Each tap is modeled as an independent zero-mean complex Gaussian random variable, with power specified by the profile. Every channel between a transmit antenna and a receive antenna follows this model.

The transmitter sends frames of 400 information bits, encoded with the rate-1/2 nonrecursive systematic convolutional encoder with memory $M = 4$ and octal representation $(23_8, 35_8)$. The coded bits are mapped to 8-PSK symbols with Gray mapping. At the reception, the turbo equalizer uses the approximate implementation of the MMSE receiver with $N_1 = 4$ and $N_2 = 6$. In the figures, the bit error rates are compared to a lower bound based on Fano's inequality, calculated as in [W00].

The performance improves when the number n_R of receive antennas increases, as shown in Figure 5.15 for $n_R = 1, 2, 4$ and $n_T = 1$. The improvement is due to the

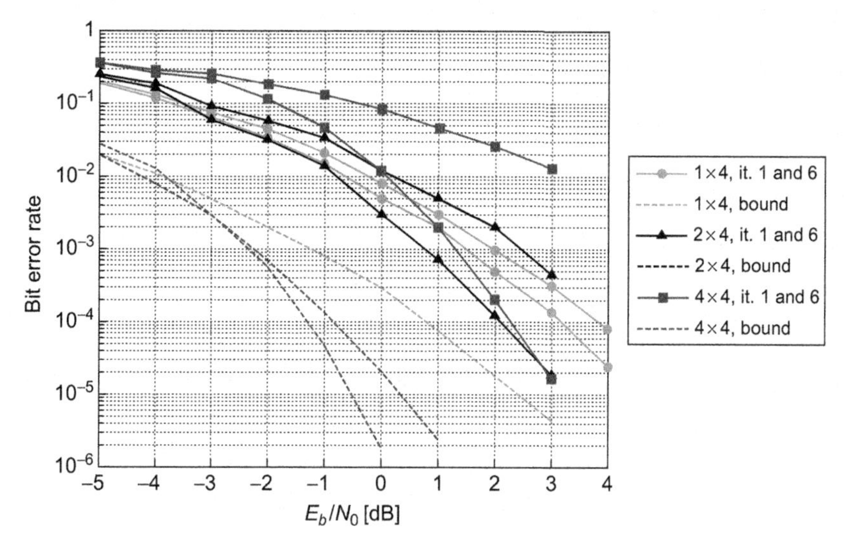

FIGURE 5.16

Turbo equalization on the GSM TU channel. The turbo equalizer is able to benefit from transmit diversity. However, since more transmit antennas means more cross-antenna interference, the turbo equalizer needs several iterations to benefit from transmit diversity. At the first iteration, the bit error rate is worse for transmitters with more antennas.

array gain, i.e., more receive antennas receive more power, and to diversity. At a bit error rate of 10^{-4}, the receiver with 2 antennas has a 9.5 dB gain with respect to the receiver with one antenna. The receiver with 4 antennas provides an additional 6.5 dB gain.

The turbo equalizer can also benefit from transmit diversity and suppress cross-antenna interference, as Figure 5.16 shows for $n_T = 1, 2$ and 4. At the first iteration, increasing the number of transmit antennas degrades the performance, because the turbo equalizer has not canceled out the cross-antenna interference yet. But after 6 iterations, the turbo equalizer manages to suppress a large part of the interference, and benefits from the diversity. Note that transmission with 4 antennas gives a higher bit error rate than transmission with 2 antennas for E_b/N_0 lower than 3 dB. This is because a higher number of transmit antennas also means more cross-antenna interference; the turbo equalizer need a higher E_b/N_0 to be able to suppress that interference.

Finally, Figure 5.17 shows the performance achieved with the same number of antennas at the transmitter and at the receiver ($n_T = n_R = 1, 2, 4$). Again, this figure shows that the turbo equalizer can benefit from transmit diversity, receive diversity and array gain when the number of antennas increases.

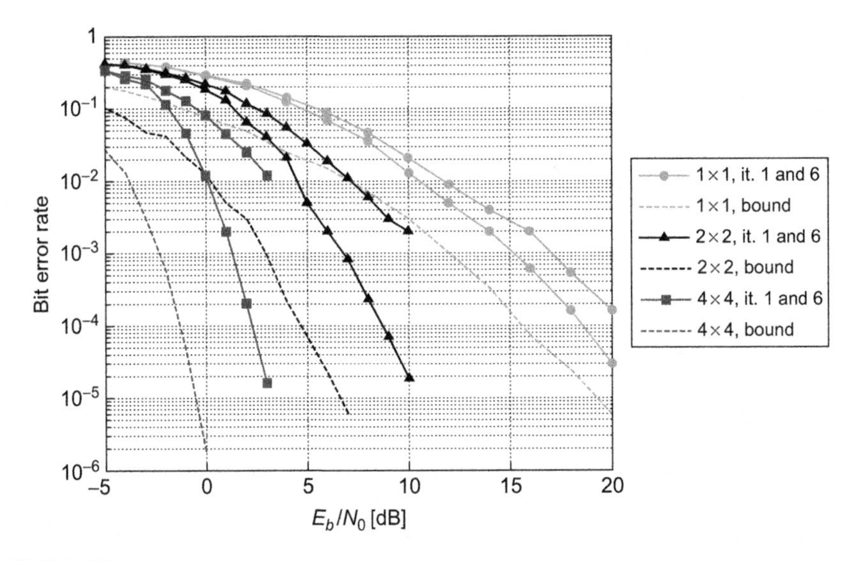

FIGURE 5.17

Turbo equalization on the GSM TU channel. The turbo equalizer benefits from transmit diversity, receive diversity and array gain.

5.4 TURBO SYNCHRONIZATION

The soft information delivered by soft decoders can also be used to revisit classical synchronization devices. If the classical approach consists in estimating the channel parameters, and then decoding the signal, the turbo principle suggests to iterate between these two stages. This section identifies the synchronization problem as a stage of the overall decoding process, and describes a family of iterative algorithm that achieve synchronization and decoding altogether. These algorithms are based on the Expectation-Maximization (EM) algorithm and on the turbo principle.

5.4.1 Transmission Scheme

The sampled signal model of Section 5.2.1 assumes perfect demodulation and sampling of the signal. In particular, the channel impulse response, the channel delay, the carrier frequency and phase are all considered as perfectly known. In general however, the channel impulse response and delay are unknown a priori, and the oscillators used at the transmitter and receivers can have frequency and phase offsets. These parameters, referred to as nuisance parameters because they add uncertainty to the information transmitted, need to be estimated at the receiver; this task is referred to as synchronization.

This section develops a more general signal model that accounts for these nuisance parameters, and enables the development of synchronization algorithms.

As previously, the transmitter encodes and maps information bits u_k, $k = 1,\ldots,K$ into coded symbols s_n, $n = 1,\ldots,N$. These symbols are shaped with a pulse $p(t)$ and modulated onto a carrier at frequency f. The passband continuous-time signal is:

$$s(t) = \Re\left\{\sum_{n=1}^{N} s_n p(t-nT)\, e^{j(2\pi ft)}\right\}. \tag{5.99}$$

This signal is transmitted on a channel with impulse response $c(t)$. For the sake of simplicity, we consider the simple channel model:

$$c(t) = c\,\delta(t-\tau), \tag{5.100}$$

which accounts for a modification of signal amplitude by a factor c, and a delay of duration τ. It does not account, however, for frequency selectivity.

At the receiver, the signal is demodulated with an oscillator at frequency $f' = f + \nu$ and phase θ'. The complex baseband signal can be modeled as:

$$r(t) = c\sum_{n=1}^{N} s_n p(t-nT-\tau)\, e^{j(2\pi\nu t+\theta)} + z(t) \tag{5.101}$$

where $z(t)$ is an additive white Gaussian noise with two-sided power spectral density $N_0/2$. This model accounts for the above-mentioned nuisance parameters: the carrier frequency offset $\nu = f' - f$, the carrier phase offset $\theta = -\theta' - f\tau$, the channel amplitude c and delay τ.

This baseband signal is then sampled at a frequency $1/T_s$ satisfying the Nyquist criterion. This yields a discrete-time signal

$$r(lT_s) = c\sum_{n=1}^{N} s_n p(lT_s-nT-\tau)\, e^{j(2\pi\nu lT_s+\theta)} + z(lT_s) \tag{5.102}$$

where the noise samples are Gaussian with variance $\sigma_z^2 = \frac{2N_0}{T_s}$. Since the Nyquist criterion is satisfied, these samples $r(lT_s)$ contain the same information as the continuous-time signal $r(t)$. The remaining operations for decoding the information bits u_k can thus be carried out digitally on the samples $r(lT_s)$. Hereafter the vector of received samples is denoted by $\mathbf{r}$.

From the samples $r(lT_s)$ it observes, the receiver can calculate the matched-filter outputs, written:

$$y_n(\nu,\tau) \triangleq \int_{-\infty}^{+\infty} r(t)\, e^{-j2\pi\nu t}\, p(t-nT-\tau)\,dt, \tag{5.103}$$

$$= T_s \sum_{l=-\infty}^{+\infty} r(lT_s)\, e^{-j2\pi\nu lT_s}\, p(lT_s-nT-\tau). \tag{5.104}$$

These matched-filter outputs are a sufficient statistic for synchronization and decoding; i.e., the matched-filter outputs $y_n(\nu,\tau)$ or the sampled signal $\boldsymbol{r}$ enable equally good estimation of the synchronization parameters and of the signal. Note that in Sections 5.2 and 5.3, the nuisance parameters ν and τ where considered perfectly known, so that the receiver could carry out matched-filtering before sampling (at symbol rate). Here however, the receiver cannot carry out matched-filtering because the nuisance parameters are unknown. Instead, it must sample the signal at a rate $1/T_s$, higher than the symbol rate, to obtain the signal $\boldsymbol{r}$. The matched-filter outputs $y_n(\nu,\tau)$ can then be calculated for every (ν,τ) with (5.104).

5.4.2 Separation of Synchronization and Decoding

From the samples $\boldsymbol{r}$ of the received signal, the digital receiver produces estimates $\hat{u}_k$ of the information bits u_k. To minimize the probability of error, that is, the probability that $\hat{u}_k \neq u_k$, the optimal decision rule is to choose the bit $\hat{u}_k$ that maximizes the a posteriori probability $P(u_k|\boldsymbol{r})$:

$$\hat{u}_k = \arg\max_{u_k} P(u_k|\boldsymbol{r}). \tag{5.105}$$

However, this optimization problem is intractable. The relationship between u_k and $\boldsymbol{r}$ is complex: it depends on the code, on the bit-to-symbol mapping, on the additive channel noise, and on the nuisance parameters.

To approximate the optimal solution, the maximization is carried out in two steps: first, in the synchronization step, the receiver estimates the nuisance parameters; second, in the decoding step, it uses these estimates to produce bit estimates $\hat{u}_k$. Formally, this separation in two tasks stems from the following approximation to the optimal maximization problem. The maximization problem can be rewritten to make the nuisance parameters $\boldsymbol{b} \triangleq [\theta,\nu,\tau,c]$ apparent:

$$\hat{u}_k = \arg\max_{u_k} \int_{\boldsymbol{b}} P(u_k,\boldsymbol{b}|\boldsymbol{r})d\boldsymbol{b} \tag{5.106}$$

$$= \arg\max_{u_k} \int_{\boldsymbol{b}} P(u_k|\boldsymbol{b},\boldsymbol{r})\, P(\boldsymbol{b}|\boldsymbol{r})\, d\boldsymbol{b}. \tag{5.107}$$

Then, the probability density function $P(\boldsymbol{b}|\boldsymbol{r})$ is approximated by a single-point mass distribution:

$$P(\boldsymbol{b}|\boldsymbol{r}) \simeq \delta(\boldsymbol{b}-\hat{\boldsymbol{b}}), \tag{5.108}$$

where $\hat{\boldsymbol{b}}$ is a well-chosen value for $\boldsymbol{b}$. The separation in two tasks results from this approximation. The synchronization task consists in choosing a suitable $\hat{\boldsymbol{b}}$; if the maximum a posteriori criterion is used, this amounts to solving:

$$\hat{\boldsymbol{b}} = \arg\max_{\boldsymbol{b}} P(\boldsymbol{b}|\boldsymbol{r}). \tag{5.109}$$

Then, with the approximation (5.108), the decoding task becomes:

$$\hat{u}_k \simeq \arg\max_{u_k} P(u_k|\boldsymbol{r},\boldsymbol{b}=\hat{\boldsymbol{b}}). \tag{5.110}$$

Note that the a posteriori probability above is what the BCJR algorithm calculates for convolutional codes, or what the turbo decoder approximates for turbo codes.

5.4.3 Approaches to Synchronization

This section develops three approaches to solving the synchronization problem (5.109). These three approaches differ in the assumptions or approximations made to the objective function $P(\boldsymbol{b}|\boldsymbol{r})$ in order to make the estimation tractable.

The a posteriori probability $P(\boldsymbol{b}|\boldsymbol{r})$ can be calculated as follows:

$$P(\boldsymbol{b}|\boldsymbol{r}) = \sum_{\boldsymbol{s}} \frac{P(\boldsymbol{r}|\boldsymbol{s},\boldsymbol{b})\,P(\boldsymbol{s})\,P(\boldsymbol{b})}{P(\boldsymbol{r})}. \tag{5.111}$$

Given our signal model, the probability $P(\boldsymbol{r}|\boldsymbol{s},\boldsymbol{b})$ is the Gaussian distribution describing the additive channel noise. The probability $P(\boldsymbol{b})$ can be considered as uniform, in the absence of a priori information on the nuisance parameters. Both $P(\boldsymbol{b})$ and $P(\boldsymbol{r})$ can be omitted for the maximization because they are constant whatever the value of the argument $\boldsymbol{b}$. Consequently, the synchronization problem (5.109) can equivalently be written:

$$\hat{\boldsymbol{b}} = \arg\max_{\boldsymbol{b}} \sum_{\boldsymbol{s}} P(\boldsymbol{r}|\boldsymbol{s},\boldsymbol{b})\,P(\boldsymbol{s}). \tag{5.112}$$

This problem has in general no explicit solution. To make the problem tractable, one usually resorts to one of three approximations of $P(\boldsymbol{s})$, each yielding a different synchronization mode. These synchronization modes are termed Data-Aided (DA), Non-Data-Aided (NDA), and Code-Aided (CA) modes.

Data-Aided Mode

In the DA mode, the transmitter sends, among the data symbols, a set of pilot symbols, i.e., symbols known to the receiver. The estimation of nuisance parameters is based solely on these pilot symbols. If the transmitter sends a sequence $\boldsymbol{s_p}$ of $N_p < N$ pilot symbols with value $\boldsymbol{s}_p^\star$, the probability $P(\boldsymbol{s_p})$ becomes for these:

$$P(\boldsymbol{s_p}) = \delta(\boldsymbol{s_p} - \boldsymbol{s}_p^\star) \tag{5.113}$$

i.e., $P(\boldsymbol{s_p})$ cancels out for all sequences but for $\boldsymbol{s_p} = \boldsymbol{s}_p^\star$. The synchronization problem becomes:

$$\hat{\boldsymbol{b}} = \arg\max_{\boldsymbol{b}} P(\boldsymbol{r}|\boldsymbol{s}_p^\star,\boldsymbol{b}). \tag{5.114}$$

In the maximization above, the probability density $P(\boldsymbol{r}|\boldsymbol{s}^{\star},\boldsymbol{b})$ is a Gaussian distribution. The resolution yields the following explicit solution:

$$[\hat{\nu}_{DA},\hat{\tau}_{DA}] = \arg\max_{\nu,\tau}\left\{\left|\sum_{n=1}^{N_p} s_n^* y_n(\nu,\tau)\right|\right\} \tag{5.115}$$

$$\hat{\theta}_{DA} = \arg\left\{\sum_{n=1}^{N_p} s_n^* y_n(\hat{\nu}_{DA},\hat{\tau}_{DA})\right\} \tag{5.116}$$

$$\hat{c}_{DA} = \frac{\left|\sum_{n=1}^{N_p} s_n^* y_n(\hat{\nu}_{DA},\hat{\tau}_{DA})\right|}{\sum_{n=1}^{N_p}|s_n|^2} \tag{5.117}$$

where $y_n(\nu,\tau)$ are the matched-filter outputs computed at time $nT+\tau$, with carrier frequency offset ν.

The advantage of the DA mode is that it yields an explicit solution. However, it requires the transmission of pilot symbols, that waste power and spectral efficiency without carrying information; and it does not use other symbols than the pilots to refine the synchronization estimates.

Non-Data-Aided Mode

In the NDA mode, the synchronizer considers all sequences $\boldsymbol{s}$ as equiprobable, that is:

$$P(\boldsymbol{s}) = \frac{1}{|\mathcal{S}|^N} \qquad \forall \boldsymbol{s} \in \mathcal{S}^N. \tag{5.118}$$

This approximation drops the coding information, since coding makes only some sequences probable, all others having a probability zero. The degradation due to this approximation may be important, particularly at low SNR. Nevertheless it usually leads to low-complexity synchronization methods, and is therefore well-suited to practical implementation. Two important examples are the Oerder and Meyr timing synchronizer [H88], and the Viterbi and Viterbi carrier phase synchronizer [H83].

Code-Aided Mode

The CA mode considers the exact probability $P(\boldsymbol{s})$, that is,

$$P(\boldsymbol{s}) = \frac{1}{|\mathcal{C}|} \qquad \forall \boldsymbol{s} \in \mathcal{C} \subset \mathcal{S}^N, \tag{5.119}$$

where $\mathcal{C}$ is the codebook. CA synchronizers are more complex, but usually yield better performance, especially at low SNR. The next section presents such CA synchronizers, based on the Expectation-Maximization (EM) algorithm. The EM algorithm is an iterative algorithm that can solve the problem above. Depending on the code, the solution can be exact, or approximate.

5.4.4 Code-Aided Synchronization with the EM Algorithm

Unfortunately, no simple analytical solution exists for the CA synchronization problem, because of the difficulty of limiting the sum to codewords only, in (5.112).

However, an iterative calculation of the nuisance parameters is possible with the Expectation-Maximization (EM) algorithm. The EM algorithm, first defined by Dempster, Laird and Rubin in [DLR77], is an iterative method for solving maximum likelihood (ML) problems:

$$\hat{\boldsymbol{b}}_{ML} = \arg\max_{\boldsymbol{b}} P(\boldsymbol{r}|\boldsymbol{b}). \tag{5.120}$$

It is well suited to ML estimation problems that would be straightforward to solve if some additional data was known.

The EM algorithm calculates a sequence of estimates $\{\hat{\boldsymbol{b}}^{(\lambda)}\}_{\lambda=1}^{\infty}$ by repeating the following two steps iteratively:

$$\text{Expectation step:} \quad \mathcal{Q}(\boldsymbol{b},\hat{\boldsymbol{b}}^{(\lambda)}) = \int_{\mathcal{A}(\boldsymbol{r})} P\left(\boldsymbol{a}|\boldsymbol{r},\hat{\boldsymbol{b}}^{(\lambda)}\right) \log P(\boldsymbol{a}|\boldsymbol{b})\, d\boldsymbol{a} \tag{5.121}$$

$$\text{Maximization step:} \quad \hat{\boldsymbol{b}}^{(\lambda+1)} = \arg\max_{\boldsymbol{b}} \left\{\mathcal{Q}\left(\boldsymbol{b},\hat{\boldsymbol{b}}^{(\lambda)}\right)\right\}. \tag{5.122}$$

The set $\mathcal{A}(\boldsymbol{r})$ of elements $\boldsymbol{a}$ represents the so-called *complete-data set*, and is defined by:

$$\mathcal{A}(\boldsymbol{r}) = \{\boldsymbol{a} : \boldsymbol{r} = f(\boldsymbol{a})\} \tag{5.123}$$

where $f(\cdot)$ is a many-to-one mapping. The complete-data set is chosen so that the ML problem is easy to solve when the variables of the complete-data set are known. The simplicity of the complete-data problem is then exploited in the maximization step. The sequence $\{\hat{\boldsymbol{b}}^{(\lambda)}\}_{\lambda=1}^{\infty}$ converges to the ML estimate $\hat{\boldsymbol{b}}_{ML}$ with a linear order of convergence, under some general conditions [Wu83].

For the code-aided synchronization problem, the complete-data set is chosen equal to $\mathcal{A}(\boldsymbol{r}) = \{\boldsymbol{a} : \boldsymbol{a} = (\boldsymbol{r},\boldsymbol{s})\}$. This choice is such that the synchronization problem is easy to solve when the variables $\boldsymbol{a} = (\boldsymbol{r},\boldsymbol{s})$ are known: indeed the problem becomes a simple data-aided synchronization problem, which has the explicit solution (5.115)–(5.117). With this definition of the complete-data set, we have:

$$\begin{aligned} P(\boldsymbol{a}|\boldsymbol{b}) &= P(\boldsymbol{r},\boldsymbol{s}|\boldsymbol{b}) && (5.124)\\ &= P(\boldsymbol{r}|\boldsymbol{s},\boldsymbol{b})\, P(\boldsymbol{s}|\boldsymbol{b}) && (5.125)\\ &= P(\boldsymbol{r}|\boldsymbol{s},\boldsymbol{b})\, P(\boldsymbol{s}), && (5.126) \end{aligned}$$

using the Bayes rule and taking into account the independence between the symbol sequence $\boldsymbol{s}$ and the nuisance parameters $\boldsymbol{b}$. The expectation step becomes:

$$\mathcal{Q}\left(\boldsymbol{b},\hat{\boldsymbol{b}}^{(\lambda)}\right) = \sum_{\boldsymbol{s}} P\left(\boldsymbol{s}|\boldsymbol{r},\hat{\boldsymbol{b}}^{(\lambda)}\right) \log P(\boldsymbol{r}|\boldsymbol{s},\boldsymbol{b}) + \sum_{\boldsymbol{s}} P\left(\boldsymbol{s}|\boldsymbol{r},\hat{\boldsymbol{b}}^{(\lambda)}\right) \log P(\boldsymbol{s}). \tag{5.127}$$

The second term above does not depend on $\boldsymbol{b}$, and therefore has no impact on the maximization step. It can be dropped, leaving:

$$\mathcal{Q}\left(\boldsymbol{b},\hat{\boldsymbol{b}}^{(\lambda)}\right) = \sum_{\boldsymbol{s}} P\left(\boldsymbol{s}|\boldsymbol{r},\hat{\boldsymbol{b}}^{(\lambda)}\right) \log P(\boldsymbol{r}|\boldsymbol{s},\boldsymbol{b}). \tag{5.128}$$

Given our transmission model, the second factor above is, after dropping the terms independent of $\boldsymbol{b}$:

$$\log P(\boldsymbol{r}|\boldsymbol{s},\boldsymbol{b}) \propto \frac{c}{N_0} \Re\left\{\sum_{n=1}^{N} s_n^* y_n(\nu,\tau) e^{-j\theta}\right\} - \frac{c^2}{2N_0} \sum_{n=1}^{N} |s_n|^2. \tag{5.129}$$

Substituting these values in (5.128) finally yields:

$$\mathcal{Q}(\boldsymbol{b},\hat{\boldsymbol{b}}^{(\lambda)}) \propto \frac{c}{N_0} \Re\left\{\sum_{n=1}^{N} \eta_n^*(\boldsymbol{r},\hat{\boldsymbol{b}}^{(\lambda)}) y_n(\nu,\tau) e^{-j\theta}\right\} - \frac{c^2}{2N_0} \sum_{n=1}^{N} \rho_n\left(\boldsymbol{r},\hat{\boldsymbol{b}}^{(\lambda)}\right) \tag{5.130}$$

where:

$$\eta_n(\boldsymbol{r},\hat{\boldsymbol{b}}^{(\lambda)}) \triangleq \sum_{\boldsymbol{s}\in\mathcal{S}^N} s_n P(\boldsymbol{s}|\boldsymbol{r},\hat{\boldsymbol{b}}^{(\lambda)}) \tag{5.131}$$

$$= \sum_{s_n\in\mathcal{S}} s_n P(s_n|\boldsymbol{r},\hat{\boldsymbol{b}}^{(\lambda)}) \tag{5.132}$$

and:

$$\rho_n(\boldsymbol{r},\hat{\boldsymbol{b}}^{(\lambda)}) \triangleq \sum_{\boldsymbol{s}\in\mathcal{S}^N} |s_n|^2 P(\boldsymbol{s}|\boldsymbol{r},\hat{\boldsymbol{b}}^{(\lambda)}) \tag{5.133}$$

$$= \sum_{s_n\in\mathcal{S}} |s_n|^2 P(s_n|\boldsymbol{r},\hat{\boldsymbol{b}}^{(\lambda)}). \tag{5.134}$$

During the maximization step, the EM algorithm calculates the argument $\hat{\boldsymbol{b}}^{(\lambda)}$ that maximizes the function $Q(\boldsymbol{b},\hat{\boldsymbol{b}}^{(\lambda)})$. The solution is:

$$[\hat{\nu}^{(\lambda+1)},\hat{\tau}^{(\lambda+1)}] = \arg\max_{\nu,\tau}\left\{\left|\sum_{n=1}^{N}\eta_n^*(\boldsymbol{r},\hat{\boldsymbol{b}}^{(\lambda)})\,y_n(\nu,\tau)\right|\right\} \tag{5.135}$$

$$\hat{\theta}^{(\lambda+1)} = \arg\left\{\sum_{n=1}^{N}\eta_n^*(\boldsymbol{r},\hat{\boldsymbol{b}}^{(\lambda)})\,y_n(\hat{\nu}^{(\lambda+1)},\hat{\tau}^{(\lambda+1)})\right\} \tag{5.136}$$

$$\hat{c}^{(\lambda+1)} = \frac{\left|\sum_{n=1}^{N}\eta_n^*(\boldsymbol{r},\hat{\boldsymbol{b}}^{(\lambda)})\,y_n(\hat{\nu}^{(\lambda+1)},\hat{\tau}^{(\lambda+1)})\right|}{\sum_{n=1}^{N}\rho_n(\boldsymbol{r},\hat{\boldsymbol{b}}^{(\lambda)})}. \tag{5.137}$$

Note the similarity between the solution of the maximization step of the EM algorithm, given by (5.135)–(5.137), with the solution (5.115)–(5.117) of the DA synchronization problem. The EM maximization step corresponds to the DA solution where the actual symbols s_n have been replaced by their a posteriori mean $\eta_n^*(\boldsymbol{r},\hat{\boldsymbol{b}}^{(\lambda)})$ or variance $\rho_n^*(\boldsymbol{r},\hat{\boldsymbol{b}}^{(\lambda)})$.

As for the EM expectation step, the calculation of the probabilities $P(s_n|\boldsymbol{r},\hat{\boldsymbol{b}}^{(\lambda)})$ is what SI-SO maximum a posteriori decoders calculate. For a convolutional code, the BCJR calculates these a posteriori probabilities exactly; for a turbo code or LDPC code, they are calculated approximatively.

5.4.5 Turbo Synchronization and Decoding

The CA synchronizer based on the EM algorithm iterates between the expectation and maximization steps. But the expectation step requires the calculation of the a posteriori probabilities $P(s_n|\boldsymbol{r},\hat{\boldsymbol{b}}^{(\lambda)})$, which can also involve an iterative algorithm, for example for turbo codes. The receiver can thus be doubly iterative.

Noels et al., have proposed a simplification of the EM synchronizer in the case of turbo codes or other iteratively decoded codes [NHD+03]. Their algorithm performs only one turbo decoding iteration in each EM iteration; and the turbo decoder initializes the a priori information with the last extrinsic information it has calculated. Albeit the resulting algorithm is no longer strictly speaking an EM algorithm, this approach has demonstrated its performance in a number of contributions [HRVM03, RHVM04, GWM04, GSWM05].

5.5 TURBO SYNCHRONIZATION ON FREQUENCY-SELECTIVE MIMO CHANNELS

This section extends the turbo synchronization framework of Section 5.4 to MIMO transmissions over frequency-selective channels [ZJP99, KBC01, BUV01, LGH01].

5.5.1 Transmission Scheme

The transmission scheme is the same as in Section 5.3. The signal is encoded and transmitted through n_T antennas. The information bits u_k, $k=1,\ldots,K$ are encoded and mapped to n_T symbol streams $s_n^{(i)}$ of length N, with $n=1,\ldots,N$, $i=1,\ldots,n_T$. The baseband continuous-time signal transmitted by antenna $i=1,\ldots,n_T$ is:

$$s^{(i)}(t)=\sum_{n=1}^{N}s_n^{(i)}p(t-nT) \tag{5.138}$$

where $p(t)$ is the pulse-shaping filter and $1/T$ the symbol rate.

This signal propagates to the n_R reception antennas along multiple paths. The following channel model describes the propagation from antenna (i) to antenna (j) along $P^{(i,j)}$ paths; each path $l=1,\ldots,P^{(i,j)}$ has an attenuation factor $c_l^{(i,j)}$ and a delay $\tau_l^{(i,j)}$:

$$c^{(i,j)}(t)=\sum_{l=1}^{P^{(i,j)}}c_l^{(i,j)}\,\delta\left(t-\tau_l^{(i,j)}\right) \tag{5.139}$$

with $i=1,\ldots,n_T$ and $j=1,\ldots,n_R$. The channel is assumed quasi-static, i.e., it remains static during the transmission of a frame.

Each reception antenna receives the signal propagated along these multiple paths, and corrupted by additive noise. We assume additive white complex Gaussian noise with power spectral density $2N_0$ in baseband. The signal received at antenna (j) can be modeled as:

$$r^{(j)}(t)=\sum_{i=1}^{n_T}(s^{(i)}\otimes c^{(i,j)})(t)+z^{(j)}(t) \tag{5.140}$$

$$=\sum_{i=1}^{n_T}\sum_{n=1}^{N}s_n^{(i)}\left(p\otimes c^{(i,j)}\right)(t-nT)+z^{(j)}(t) \tag{5.141}$$

$$=\sum_{i=1}^{n_T}\sum_{n=1}^{N}s_n^{(i)}h^{(i,j)}(t-nT)+z^{(j)}(t) \tag{5.142}$$

where $z^{(j)}$ is the additive Gaussian noise and where $h^{(i,j)}(t)=\left(p\otimes c^{(i,j)}\right)(t)$. Note that our baseband model assumes that the passband signal is demodulated using an oscillator without frequency offset.

We consider a fractionally–spaced receiver. Such a receiver is less expensive to implement than the combination of matched filtering and noise whitening, and it is much more convenient here since the channel parameters are a priori unknown to the receiver. The fractionally-spaced receiver samples the signal $r^{(j)}(t)$ of every antenna at a rate M_s/T, where the oversampling factor M_s is chosen so as to satisfy the Nyquist criterion. To avoid spectral aliasing, the signal goes first through a low-pass filter with cutoff frequency $M_s/2T$. We represent the signal sampled at rate M_s/T

by its M_s polyphase components at rate $1/T$, defined as follows:

$$\begin{aligned} r^{(j)}_{n',m} &\triangleq r^{(j)}(n'T + mT/M_s) \\ &= \sum_{i=1}^{n_T}\sum_{n=1}^{N} s^{(i)}_n h^{(i,j)}(n'T + mT/M_s - nT) + z^{(j)}(n'T + mT/M_s). \end{aligned} \tag{5.143}$$

for $m = 0, \ldots, M_s - 1$.

To obtain a more compact expression, we define the channel taps:

$$h^{(i,j)}_{n',m} \triangleq h^{(i,j)}(n'T + mT/M_s) \tag{5.144}$$

and the noise samples:

$$z^{(j)}_{n',m} \triangleq z^{(j)}(n'T + mT/M_s) \tag{5.145}$$

so as to write the mth polyphase component of the signal received at antenna (j) as follows:

$$r^{(j)}_{n',m} = \sum_{i=1}^{n_T}\sum_{n=1}^{N} s^{(i)}_n h^{(i,j)}_{n'-n,m} + z^{(j)}_{n',m}. \tag{5.146}$$

The observation $r^{(j)}_{n',m}$ adds up terms with each of the $n_T \cdot N$ symbols $s^{(i)}_n$, but part of these terms are negligible. The discrete impulse response $h^{(i,j)}_{n',m}$ can indeed be truncated, because both the pulse-shaping filter $p(t)$ and the channel impulse response $c^{(i,j)}(t)$ have a finite length or can be truncated. We choose to truncate $h^{(i,j)}_{n',m}$ to $L^{(i,j)}_m + 1$ samples: without loss of generality, in (5.146) we take $h^{(i,j)}_{n'-n,m} = 0$ for $n' - n < 0$ and for $n' - n > L^{(i,j)}_m$. Changing indices to $l = n' - n$, the observation model becomes:

$$r^{(j)}_{n',m} = \sum_{i=1}^{n_T}\sum_{l=0}^{L^{(i,j)}_m} s^{(i)}_{n'-l} h^{(i,j)}_{l,m} + z^{(j)}_{n',m} \tag{5.147}$$

Note that each polyphase component $m = 0, \ldots, M_s - 1$ uses a different channel impulse response $h^{(i,j)}_{l,m}$.

The objective is to estimate the M_s channel impulse responses $h^{(i,j)}_{l,m}$ for the channel between the antenna pair (i, j), and to estimate the variance of the noise $z^{(j)}_{n',m}$ at reception antenna (j).

For the sake of estimation, we now represent the observation model (5.147) in a more convenient, matrix form. We stack the observations in a vector, but discard the first $L^{(i,j)}_m$ ones that are corrupted by symbols from the previous frame:

$$\mathbf{r}^{(j)}_{\mathbf{m}} = [r^{(j)}_{L+1,m}, \ldots, r^{(j)}_{N,m}]^T. \tag{5.148}$$

Note that we have dropped the indices of $L_m^{(i,j)}$ to write simply L. Then the observation model can be written:

$$\boldsymbol{r}_m^{(j)} = \sum_{i=1}^{n_T} \boldsymbol{S}^{(i)} \boldsymbol{h}_m^{(i,j)} + \boldsymbol{z}_m^{(j)} \tag{5.149}$$

with the following definitions of the Toeplitz matrix of symbols sent from antenna i:

$$\boldsymbol{S}^{(i)} = \begin{pmatrix} s_{L+1}^{(i)} & s_L^{(i)} & \cdots & s_1^{(i)} \\ s_{L+2}^{(i)} & s_{L+1}^{(i)} & \cdots & s_2^{(i)} \\ \ddots & \ddots & \ddots & \ddots \\ s_N^{(i)} & s_{N-1}^{(i)} & \cdots & s_{N-L}^{(i)} \end{pmatrix}, \tag{5.150}$$

the channel impulse response:

$$\boldsymbol{h}_m^{(i,j)} = [h_{0,m}^{(i,j)}, \ldots, h_{L,m}^{(i,j)}]^T, \tag{5.151}$$

and the noise vector:

$$\boldsymbol{z}_m^{(j)} = [z_{L+1,m}^{(j)}, \ldots, z_{N,m}^{(j)}]^T. \tag{5.152}$$

Finally, stacking the matrices $\boldsymbol{S}^{(i)}$ and the impulse responses $\boldsymbol{h}_m^{(i,j)}$ for $i = 1, \ldots, n_T$ as follows:

$$\boldsymbol{S} = [\boldsymbol{S}^{(1)}, \ldots, \boldsymbol{S}^{(n_T)}], \tag{5.153}$$

$$\boldsymbol{h}_m^{(j)} = [\boldsymbol{h}_m^{(1,j)}, \ldots, \boldsymbol{h}_m^{(n_T,j)}]^T, \tag{5.154}$$

we get the following observation model, that is convenient for estimation:

$$\boldsymbol{r}_m^{(j)} = \boldsymbol{S} \boldsymbol{h}_m^{(j)} + \boldsymbol{z}_m^{(j)}. \tag{5.155}$$

5.5.2 Data-Aided Synchronization

In the DA mode, the transmitter sends N_p pilot symbols from each of the n_T antennas. The estimation of the channels impulse responses and of the noise variance are based on these pilots only. We write the model (5.155) restricted to the pilot symbols as follows:

$$\boldsymbol{r}_{p,m}^{(j)} = \boldsymbol{S}_p \boldsymbol{h}_m^{(j)} + \boldsymbol{z}_{p,m}^{(j)}. \tag{5.156}$$

The channel observations vector $\boldsymbol{r}_{p,m}^{(j)}$ has length $N_r = N_p - L$, because the L samples corrupted by interference from unknown data symbols are discarded.

The ML estimation of the M_s discrete channel impulse responses can be carried out separately for each $\boldsymbol{h}_m^{(j)}$, because the noise is white in time and space. Moreover, the joint ML estimation of the discrete channel impulse response and of the noise variance yields a channel estimate that does not depend on the noise estimate. The estimation of the channel impulse response can be formulated as:

$$\hat{\boldsymbol{h}}^{(j)}_{m,DA} = \arg\max_{\boldsymbol{h}^{(j)}_m} P(\boldsymbol{r}^{(j)}_{p,m} | \boldsymbol{S}_p, \boldsymbol{h}^{(j)}_m), \tag{5.157}$$

and its solution is:

$$\hat{\boldsymbol{h}}^{(j)}_{m,DA} = \left(\boldsymbol{S}_p{}^H \boldsymbol{S}_p\right)^{-1} \boldsymbol{S}_p{}^H \boldsymbol{r}^{(j)}_{p,m} \tag{5.158}$$

when the pilot-sequence length satisfies $N_p \geq (L+1)n_T + L$. This estimate is equivalent to the least square estimate in this case. It is an efficient estimator of channels taps, i.e., it is unbiased and its variance is minimal [Men95].

The quality of the estimate depends on the pilot matrix $\boldsymbol{S}_p$. For more details about the generation of this matrix, see [CM01, NLM98]. The pilot symbols that are optimal for estimation purposes are such that the matrix $\left(\boldsymbol{S}_p{}^H \boldsymbol{S}_p\right)$ is diagonal.

As for the noise variance, its ML estimate is:

$$\hat{\sigma}^2_{DA} = \frac{1}{n_R M_s N_r} \sum_{j=1}^{n_R} \sum_{m=0}^{M_s-1} \left(\boldsymbol{r}^{(j)}_{p,m} - \boldsymbol{S}_p \hat{\boldsymbol{h}}^{(j)}_{m,DA}\right)^H \left(\boldsymbol{r}^{(j)}_{p,m} - \boldsymbol{S}_p \hat{\boldsymbol{h}}^{(j)}_{m,DA}\right). \tag{5.159}$$

This estimate is biased, because of the joint estimation with estimate $\hat{\boldsymbol{h}}^{(j)}_{m,DA}$. This is similar to the bias on the estimation of variance of a random variable, when the mean and variance are estimated jointly. The bias can be removed as in the latter case; it is easy to show that the following estimate is unbiased:

$$\hat{\sigma}^2_{UDA} = \frac{N_r}{N_r - n_T(L+1)} \hat{\sigma}^2_{DA} \tag{5.160}$$

The biased estimator $\hat{\sigma}^2_{DA}$ and the unbiased estimator $\hat{\sigma}^2_{UDA}$ are equal, asymptotically in the number of pilots. Note that one needs $N_r > n_T(L+1)$, which translates into a constraint $N_p \geq (n_T+1)(L+1)$ on the number of pilot symbols.

The drawback of the DA estimation mode is that it requires a large number of pilots to get satisfactory estimates. This uses power and spectral efficiency without carrying information. Moreover, the data symbols are not used to refine the synchronization estimate.

5.5.3 Code-Aided Synchronization with the EM Algorithm

As opposed to DA estimation, CA estimation has in general no explicit solution. This section presents iterative CA estimation algorithms, based on the EM algorithm, to compute ML estimates or approximations thereof.

For CA synchronization, the relevant model is (5.155):

$$\boldsymbol{r}^{(j)}_m = \boldsymbol{S}\boldsymbol{h}^{(j)}_m + \boldsymbol{z}^{(j)}_m,$$

where the observation vector $\boldsymbol{r}^{(j)}_m$ has length $N_r = N - L$. The EM framework is presented in Section 5.4.4. Here, the complete data set is $\mathcal{A}\left(\boldsymbol{r}^{(j)}_m\right) = \{\boldsymbol{a} : \boldsymbol{a} = (\boldsymbol{r}^{(j)}_m, \boldsymbol{S})\}$, and the nuisance parameters are $\boldsymbol{b} = (\boldsymbol{h}_m^{(j)}, \sigma^2)$, i.e., the channel coefficients and the noise variance σ^2. In this case as well, the Expectation step (5.121) is equivalent to (5.128); with the notations of this section, (5.128) becomes:

$$\mathcal{Q}\left(\boldsymbol{b}, \hat{\boldsymbol{b}}^{(\lambda)}\right) = \sum_{S} P(S|\boldsymbol{r}_m^{(j)}, \hat{\boldsymbol{b}}^{(\lambda)}) \log P\left(\boldsymbol{r}_m^{(j)}|S, \boldsymbol{b}\right). \tag{5.161}$$

Given the observation model, and after dropping the terms independent of $\boldsymbol{b}$, the second factor particularizes to:

$$\log P(\boldsymbol{r}_m^{(j)}|S, \boldsymbol{b}) \propto -n_R M_s N_r \log\left(\pi\sigma^2\right) - \frac{1}{\sigma^2} \sum_{j=1}^{n_R} \sum_{m=0}^{M_s-1} \left(\boldsymbol{r}_m^{(j)} - S\boldsymbol{h}_m^{(j)}\right)^H \left(\boldsymbol{r}_m^{(j)} - S\boldsymbol{h}_m^{(j)}\right). \tag{5.162}$$

The Maximization step (5.122) at iteration $\lambda + 1$ gives:

$$\hat{\boldsymbol{h}}_{m,EM}^{(j)(\lambda+1)} = \left(\bar{\boldsymbol{R}}_{ss}(\boldsymbol{r}_m^{(j)}, \hat{\boldsymbol{b}}^{(\lambda)})\right)^{-1} \bar{\boldsymbol{S}}(\boldsymbol{r}_m^{(j)}, \hat{\boldsymbol{b}}^{(\lambda)})^H \boldsymbol{r}_m^{(j)} \tag{5.163}$$

and:

$$\hat{\sigma}_{EM}^{2(\lambda+1)} = \frac{1}{n_R M_s N_r} \sum_{j=1}^{n_R} \sum_{m=0}^{M_s-1} \Big(\boldsymbol{r}_m^{(j)H} \boldsymbol{r}_m^{(j)} + \hat{\boldsymbol{h}}_{m,EM}^{(j)(\lambda)H} \bar{\boldsymbol{R}}_{ss}\left(\boldsymbol{r}_m^{(j)}, \hat{\boldsymbol{b}}^{(\lambda)}\right) \hat{\boldsymbol{h}}_{m,EM}^{(j)(\lambda)} - 2\Re\left\{\boldsymbol{r}_m^{(j)H} \bar{\boldsymbol{S}}\left(\boldsymbol{r}_m^{(j)}, \hat{\boldsymbol{b}}^{(\lambda)}\right) \hat{\boldsymbol{h}}_{m,EM}^{(j)(\lambda)}\right\}\Big) \tag{5.164}$$

with the definitions:

$$\bar{\boldsymbol{S}}(\boldsymbol{r}_m^{(j)}, \hat{\boldsymbol{b}}^{(\lambda)}) = \sum_{S} S\, P\left(S|\boldsymbol{r}_m^{(j)}, \hat{\boldsymbol{b}}^{(\lambda)}\right) \tag{5.165}$$

and:

$$\bar{\boldsymbol{R}}_{ss}(\boldsymbol{r}_m^{(j)}, \hat{\boldsymbol{b}}^{(\lambda)}) = \sum_{S} S^H S\, P(S|\boldsymbol{r}_m^{(j)}, \hat{\boldsymbol{b}}^{(\lambda)}). \tag{5.166}$$

Again, the solution of the estimation problem is partly decoupled: the estimation of the channel taps does not depend on the estimation of the noise variance. The channel taps estimate can be calculated first, and then the noise variance estimate.

In (5.165)–(5.166), the average is with respect to the a posteriori distribution $P(S|\boldsymbol{r}_m^{(j)}, \hat{\boldsymbol{b}}^{(\lambda)})$. The calculation of this probability distribution is the role of the decoder, and is addressed in Section 5.5.5 below.

The overall EM process is iterative; at each iteration, the synchronizer carries out the expectation step (5.165)–(5.166) and the maximization step (5.163)–(5.164). During the expectation step, the decoder calculates an updated version of the a posteriori distribution $P(S|\boldsymbol{r}_m^{(j)}, \hat{\boldsymbol{b}}^{(\lambda)})$. To initialize the iterations of the EM algorithm, the synchronizer can use DA estimates calculated with pilot symbols.

Unfortunately the EM estimates of the channel taps and of the noise variance are biased, as suggested in [KBC01]. The bias is due to the joint estimation of the channel taps, the noise variance, and the symbols, as seen below.

5.5.4 Modifications of the EM Algorithm to Reduce the Bias

Calculating the bias of the EM estimates is in general not tractable, except in very simple cases. In our case, where the sequence of symbols is coded, the difficulty is to calculate the average of $P(\boldsymbol{S}|\boldsymbol{r}_m^{(j)},\hat{\boldsymbol{b}}^{(\lambda)})$ over the possible noise realizations.

To approximate the calculation of the bias, we replace $P(\boldsymbol{S}|\boldsymbol{r}_m^{(j)},\hat{\boldsymbol{b}}^{(\lambda)})$ by an a priori distribution on symbols $\mathcal{P}(\boldsymbol{S})$, independent from the channel observations, and define:

$$\bar{S}(\mathcal{P}) = \sum_{S} \boldsymbol{S}\,\mathcal{P}(\boldsymbol{S}) \tag{5.167}$$

$$\bar{\boldsymbol{R}}_{ss}(\mathcal{P}) = \sum_{S} \boldsymbol{S}^H \boldsymbol{S}\,\mathcal{P}(\boldsymbol{S}) \tag{5.168}$$

With this approximation, averaging the estimates (5.163)–(5.164) over the noise distribution and the symbols distribution becomes tractable.

Unbiased Channel Taps Estimation

The average value of the estimator $\hat{\boldsymbol{h}}_{m,EM}^{(j)(\lambda+1)}$, with respect to the noise and the symbols (assumed to be distributed as $\mathcal{P}(\boldsymbol{S})$), is:

$$\begin{aligned}\mathrm{E}\left\{\hat{\boldsymbol{h}}_{m,EM}^{(j)(\lambda+1)}\right\} &= \left(\bar{\boldsymbol{R}}_{ss}(\mathcal{P})\right)^{-1}\bar{S}(\mathcal{P})^H\,\mathrm{E}\left\{\boldsymbol{r}_m^{(j)}\right\}\\ &= \left(\bar{\boldsymbol{R}}_{ss}(\mathcal{P})\right)^{-1}\bar{S}(\mathcal{P})^H\,\bar{S}(\mathcal{P})\,\boldsymbol{h}_m^{(j)},\end{aligned} \tag{5.169}$$

assuming that the noise has zero mean. The average value of the estimator is thus not the actual channel value; the bias factor is $\left(\bar{\boldsymbol{R}}_{ss}(\mathcal{P})\right)^{-1}\bar{S}(\mathcal{P})^H\,\bar{S}(\mathcal{P})$. This bias appears because the symbols are unknown; if they were known, the bias factor would be equal to 1. Multiplying the EM estimator by the bias inverse produces an unbiased estimator; the resulting estimator is:

$$\hat{\boldsymbol{h}}_{m,UEM}^{(j)} = \left(\bar{S}(\mathcal{P})^H\,\bar{S}(\mathcal{P})\right)^{-1}\bar{S}(\mathcal{P})^H\,\boldsymbol{r}_m^{(j)}. \tag{5.170}$$

Note that this estimator can rigorously be considered unbiased only when $\mathcal{P}(\boldsymbol{S})$ replaces $P(\boldsymbol{S}|\boldsymbol{r}_m^{(j)},\hat{\boldsymbol{b}}^{(\lambda)})$. However, it suggests a way to reduce the bias when $P(\boldsymbol{S}|\boldsymbol{r}_m^{(j)},\hat{\boldsymbol{b}}^{(\lambda)})$ is used, by analogy.

Noise Variance Estimation with Reduced Bias

The EM estimation of the noise variance is biased as well. In this case, the bias is due to the joint estimation with the channel taps (as in the DA case) and to the joint estimation with the symbols. To put aside the part of bias due to the joint estimation with the channel taps, we consider the channel taps $\boldsymbol{h}_{\boldsymbol{m}}^{(j)}$ as perfectly known in this section.

The average value of the noise variance estimator, with respect to the noise and symbols distributions, is as follows:

$$\mathrm{E}\left\{\hat{\sigma}_{EM}^2\right\} = \sigma^2 + \frac{2}{n_R M_s N_r} \sum_{j=1}^{n_R} \sum_{m=0}^{M_s-1} \boldsymbol{h}_{\boldsymbol{m}}^{(j)} \left(\bar{\boldsymbol{R}}_{ss}(\mathcal{P}) - \bar{S}(\mathcal{P})^H \bar{S}(\mathcal{P})\right) \boldsymbol{h}_{\boldsymbol{m}}^{(j)} \tag{5.171}$$

The second term is the bias; again it is clearly due to the uncertainty on the symbols since the bias term would be equal to zero if the symbols were perfectly known. An unbiased estimate could be calculated as follows:

$$\hat{\sigma}_{UEM}^2 = \frac{1}{n_R M_s N_r} \sum_{j=1}^{n_R} \sum_{m=0}^{M_s-1} \boldsymbol{r}_{\boldsymbol{m}}^{(j)H} \boldsymbol{r}_{\boldsymbol{m}}^{(j)} - \boldsymbol{h}_{\boldsymbol{m}}^{(j)H} \bar{\boldsymbol{R}}_{ss}(\mathcal{P})\, \boldsymbol{h}_{\boldsymbol{m}}^{(j)} \tag{5.172}$$

but this estimate can be negative, and is therefore useless. Another option is to use the following estimator, which has half the bias of the EM estimator in (5.171); this estimator is referred to as half biased EM (HEM) estimator:

$$\hat{\sigma}_{HEM}^2 = \frac{1}{n_R M_s N_r} \sum_{j=1}^{n_R} \sum_{m=0}^{M_s-1} \left(\boldsymbol{r}_{\boldsymbol{m}}^{(j)} - \bar{S}(\mathcal{P})\, \boldsymbol{h}_{\boldsymbol{m}}^{(j)}\right)^H \left(\boldsymbol{r}_{\boldsymbol{m}}^{(j)} - \bar{S}(\mathcal{P})\, \boldsymbol{h}_{\boldsymbol{m}}^{(j)}\right).$$

Again, the bias reduction is obtained when using $\mathcal{P}(\boldsymbol{S})$ instead of $P(\boldsymbol{S}|\boldsymbol{r}_{\boldsymbol{m}}^{(j)}, \hat{\boldsymbol{b}}^{(\lambda)})$; it is not proven that the HEM estimator has a smaller bias than the EM estimator when using $P(\boldsymbol{S}|\boldsymbol{r}_{\boldsymbol{m}}^{(j)}, \hat{\boldsymbol{b}}^{(\lambda)})$. The reasoning is but a justification for using the HEM noise variance estimator.

5.5.5 Turbo Synchronization and Decoding

In this section we apply the EM code-aided synchronizer to the ST-BICM coding scheme of Section 5.3. Figure 5.18 depicts the overall receiver. To reduce the complexity, only one iteration of turbo equalization is performed per EM iteration, as in Section 5.4.5. The iterations are initialized with a data-aided estimate produced thanks to a sequence of pilot symbols.

The synchronizer of Section 5.5.3 has assumed that the joint a posteriori probability $P(\boldsymbol{S}|\boldsymbol{r}_{\boldsymbol{m}}^{(j)}, \hat{\boldsymbol{b}}^{(\lambda)})$ of the matrix of symbols $\boldsymbol{S}$ could be calculated by the decoder. However what the turbo equalizer calculates is different for two reasons.

First, the turbo equalizer is designed to calculate the a posteriori probabilities $P(s_n^{(i)}|\boldsymbol{r}_{\boldsymbol{m}}^{(j)}, \hat{\boldsymbol{b}}^{(\lambda)})$ for individual symbols $s_n^{(i)}$, and not the joint probability of all symbols in $\boldsymbol{S}$. While the probabilities $P(s_n^{(i)}|\boldsymbol{r}_{\boldsymbol{m}}^{(j)}, \hat{\boldsymbol{b}}^{(\lambda)})$ suffice to calculate the averages in (5.165), the joint probability $P(s_n^{(i)}, s_{n'}^{(i')}|\boldsymbol{r}_{\boldsymbol{m}}^{(j)}, \hat{\boldsymbol{b}}^{(\lambda)})$ of pairs of symbols is required to calculate the covariances in (5.166). To overcome this problem, the joint probability is approximated by the product of marginal probabilities $P(s_n^{(i)}|\boldsymbol{r}_{\boldsymbol{m}}^{(j)}, \hat{\boldsymbol{b}}^{(\lambda)}) \cdot P(s_{n'}^{(i')}|\boldsymbol{r}_{\boldsymbol{m}}^{(j)}, \hat{\boldsymbol{b}}^{(\lambda)})$ ($\forall n' \neq n$ and $\forall i' \neq i$).

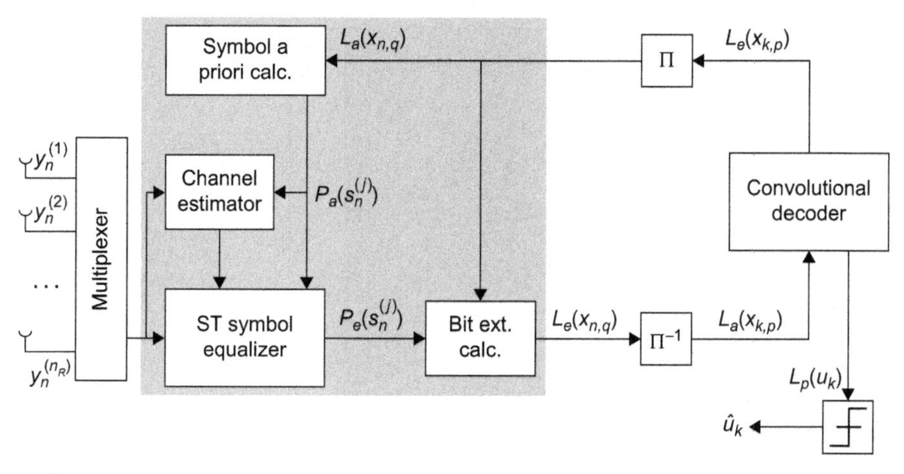

FIGURE 5.18

Turbo synchronization and decoding for ST-BICM on frequency-selective MIMO channels.

Second, the turbo equalizer only computes approximations of these a posteriori probabilities—though these approximations are widely believed to be close to the actual probabilities.

Because of these approximations, the conditions that ensure the convergence of the EM algorithm are no longer fulfilled [Wu83]. Therefore the convergence of the turbo synchronization estimate to the ML estimate cannot be ensured.

5.5.6 Simulations

This section illustrates the synchronization algorithms above for an ST-BICM coding scheme on MIMO channels. The simulations address both the estimation of the channel impulse response and of the noise variance. For all simulations, the channel is subject to slow fading, and the results presented here are averaged over a large number of channel realizations.

Figure 5.19 compares several estimation methods for the estimation of the channel taps. The channel is a 4-input 4-output flat Rayleigh-fading channel. The receiver uses matched-filtering and symbol-rate sampling, instead of the fractional receiver above, because the channel is not frequency-selective. The transmitter sends frames of 2000 information bits, encoded by a rate-$1/2$ convolutional encoder with octal representation $(23_8, 35_8)$. The coded bits are interleaved by a random interleaver and mapped to 8-PSK symbols, with Gray mapping. We assume that each frame starts with 5 pilots on each of the 4 transmit antennas, for simulating an acquisition mode. The training sequence is designed according to [CM01].

Figure 5.19 shows the normalized Mean Square Error (MSE) obtained after 10 iterations of the turbo synchronizer, as a function of the average E_b/N_0 per receive antenna. The normalized MSE is the average value of the square distance between

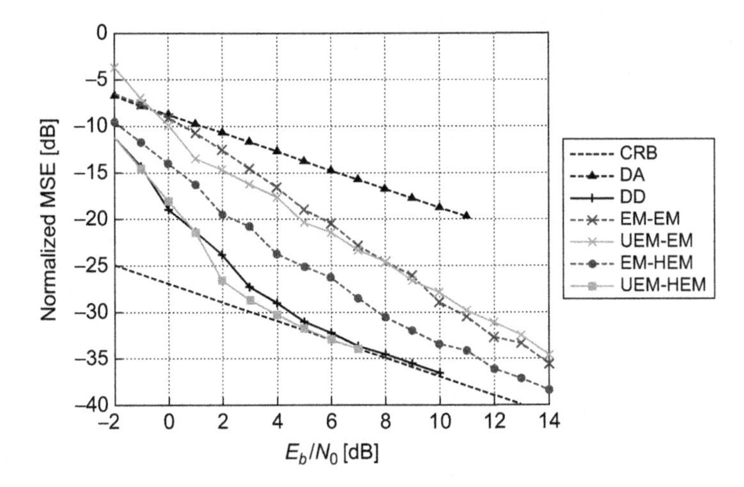

FIGURE 5.19

Normalized mean square error on the channel taps, for various estimation algorithms. Flat MIMO channel with $n_T = n_R = 4$ antennas.

the channel-tap estimate and its true value, divided by the average total power of the channel impulse response. The normalized MSE is the average over many frames of the Normalized Square Error (NSE) given by

$$\mathrm{NSE} = \frac{\sum_{i=1}^{n_T} \sum_{j=1}^{n_R} \sum_{m=0}^{M_s-1} \sum_{l=0}^{L} |\hat{h}_{l,m}^{(i,j)} - h_{l,m}^{(i,j)}|^2}{n_R n_T M_s L \sum_{i=1}^{n_T} \sum_{j=1}^{n_R} \sum_{m=0}^{M_s-1} \sum_{l=0}^{L} \mathrm{E}\left\{|h_{l,m}^{(i,j)}|^2\right\}}. \tag{5.173}$$

The curve labeled *CRB* is the Cramer Rao bound (CRB) when all the symbols of the frame are known, that is, the DA CRB. The MSE of the UEM-HEM estimator (UEM estimator for the channel taps and HEM estimator for the noise variance) achieves the CRB bound for E_b/N_0 larger than 6 dB. Another estimator that performs close to the CRB is the Decision-Directed (DD) estimator of [MMF98], an estimator that uses hard decisions instead of soft information on the symbols. The MSEs of the other three EM-based estimators are between those of the DD and DA estimators; the worse performance compared to the UEM-HEM solution shows the need for unbiasing the estimates of the EM algorithm. Finally, all the iterative algorithms manage to perform better than the DA estimator, which is not iterative.

For the same scenario, Figure 5.20 shows the bit error rate after 10 iterations. As above, the UEM-HEM and DD algorithm perform best, close to the bit error rate achieved when the nuisance parameters are perfectly known (curve labeled *Perf. sync.*). As for the EM-EM algorithm, it yields a higher bit error rate than the DA method, in apparent contradiction with its better performance in terms of MSE. Without getting into the details here, this difference is due to the fact that the EM algorithm

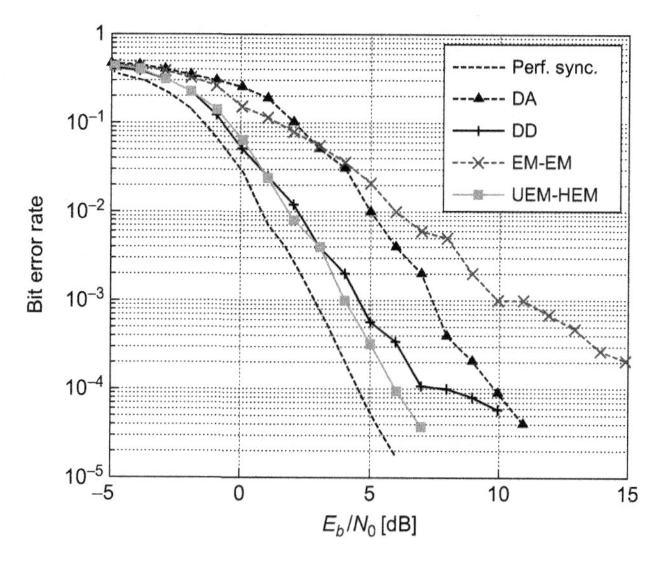

FIGURE 5.20

Bit error rates, for various estimation algorithms. Flat MIMO channel with $n_T = n_R = 4$ antennas.

does not always converge in this setup. When it does not converge, it produces frames with high bit error rate, thereby increasing the average bit error rate.

Figure 5.21 shows the performance of the same scheme on the 4-input 4-output frequency-selective GSM TU channel model (see Section 5.3.3). The number of pilot symbols is 55 for each of the 4 transmit antennas. The use of pilot symbols induces a penalty of 1.15 dB in terms of E_b/N_0, that needs to be considered in the figure. The oversampling factor is $M_s = 2$ and the sliding window of the equalizer spans 11 symbol periods. The bit error rate shown corresponds to the 6th iteration. All the considered iterative algorithms perform better than the DA ML method, and have very similar performance.

Note that we have chosen the smallest number of pilots to enable channel taps and noise variance estimation. This number is large however, since there must be at least $4 \cdot 55$ pilot symbols. As an alternative to using pilots, one can use as initial estimates the estimates from the previous frame, if the channel does not vary substantially from frame to frame. Figures 5.22 and 5.23 represent such a scenario: the algorithms are initiated by the true channel tap values plus Gaussian noise, so as to get a normalized MSE equal to -25 dB. The channel is a 2-input 2-output *Hiperlan/2 B* Rayleigh-fading channel, with the power and delay profile given in Table 5.2. The symbol period T is equal to $50\,ns$. The transmitter uses the nonrecursive systematic convolutional encoder with octal representation $(7_8, 5_8)$, and 16-QAM with Gray mapping. The sliding window of the space-time equalizer spans 18 symbol periods.

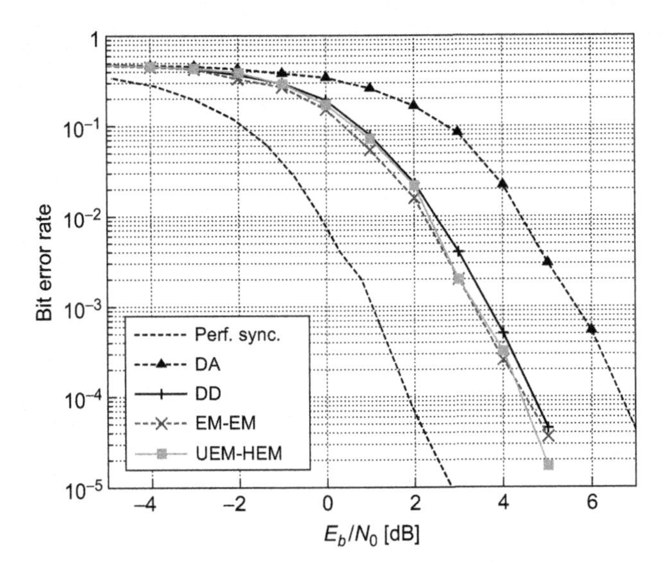

FIGURE 5.21

Bit error rates, for various estimation algorithms. GSM TU MIMO channel with $n_T = n_R = 4$ antennas.

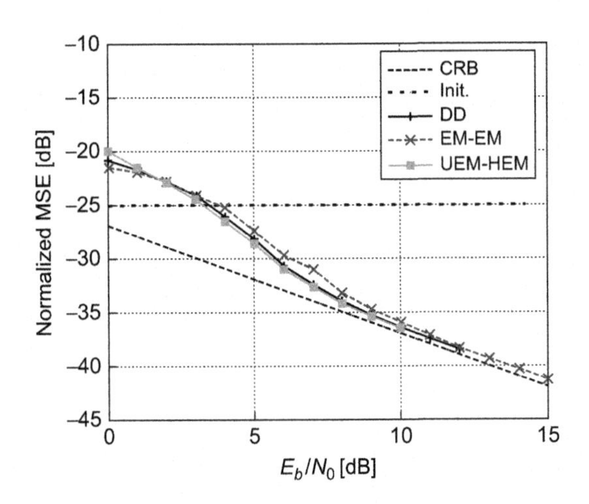

FIGURE 5.22

Normalized MSE on the channel taps, for various estimation algorithms. Hiperlan/2 B MIMO channel with $n_T = n_R = 2$ antennas.

Figure 5.22 shows that after 6 iterations, all the iterative estimation algorithms result in a normalized MSE close to the CRB when E_b/N_0 is larger than 8 dB. In terms of bit error rate, all algorithms achieve lower error rates than with the initial estimate,

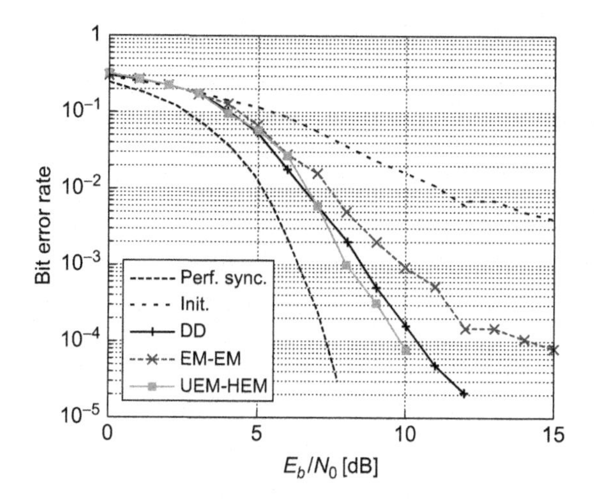

FIGURE 5.23

Bit error rates, for various estimation algorithms. Hiperlan/2 B MIMO channel with $n_T = n_R = 2$ antennas.

Table 5.2 Hiperlan/2 B Channel Model

Path delay (*ns*)	0	10	20	30	50	80
Path power (dB)	−2.6	−3.0	−3.5	−3.9	0.0	−1.3
Path delay (*ns*)	110	140	180	230	280	330
Path power (dB)	−2.6	−3.9	−3.4	−5.6	−7.7	−9.9
Path delay (*ns*)	380	430	490	560	640	730
Path power (dB)	−12.1	−14.3	−15.4	−18.4	−20.7	−24.6

and the best performance is achieved by the UEM-HEM and the DD algorithms (see Figure 5.23). The poorer performance of the EM-EM algorithm is due to a small number of instances in which an inaccurate estimation of the channel corrupts a whole frame; this has a negligible impact on the MSE but a more significant one on the bit error rate.

CHAPTER

Implementing Scalable List Detectors for MIMO-SDM in LTE

Claude Desset, Eduardo Lopez Estraviz, Min Li, Robert Fasthuber

6.1 INTRODUCTION

Since the beginning of the 21st century, wireless communication systems have been booming, with a continuous growth in rate, performance, and mobility. Next to increasing the bandwidth allocation and the spectral efficiency of modulation and coding schemes, the use of MIMO has been at the source of an additional dimension exploited to strongly increase the data rates.

Wireless systems with 2 to 4 antennas are currently used for local area networks (802.11n) or developed for cellular systems such as 3GPP-LTE and LTE advanced. Multiple antennas can bring diversity in order to enable more robust communication under tough link conditions, but more often they are used in order to increase the number of data streams within a given band, thanks to Space Division Multiplexing (MIMO-SDM).

In order to use space-division multiplexing, the transmitter side is relatively simple in terms of signal processing complexity. At most a precoder is implemented, based on channel feedback in order to ensure more orthogonality between the various streams and less interference. In other cases no preprocessing at all is performed and the different streams are simply transmitted in parallel on the different antennas.

The detection of such signals poses much more challenges. In a MIMO-SDM system with N_t transmit and N_r receive antennas, the MIMO detector recovers the multiple transmitted data streams. Its task is to recover the symbol vector $\mathbf{s}$ that was sent by the transmitter based on the received vector. The transmitted symbol vector consists of N_t complex symbols when working at full load, each being one of the M possible points when using M-QAM constellation. The received signal $\mathbf{y}$ is a vector of length $N_r \geq N_t$, based on the basic system equation where $\mathbf{H}$ is the $N_t \times N_r$ channel matrix and $\mathbf{n}$ the noise vector of length N_r, all variables having complex values:

$$\mathbf{y} = \mathbf{Hs} + \mathbf{n} \tag{6.1}$$

Unless a perfect channel diagonalization (SVD) is performed prior to the transmission, which is generally not practical to implement (amount of feedback

MIMO. DOI: 10.1016/B978-0-12-382194-2.00006-X

information, stability of the channel, estimation accuracy...), some inter-stream interference will be present and it may lead to strong degradations. This is especially the case when some streams are strongly attenuated by the channel or when multiple streams are not separable due to a quasi-singular channel matrix. Under such conditions, traditional simple detectors (hard-output ZF and MMSE) lead to a strong degradation. They are based on the following equations[1]:

$$\hat{\mathbf{s}}_{ZF} = \mathbf{H}^{-1}\mathbf{y} \tag{6.2}$$

$$\hat{\mathbf{s}}_{MMSE} = \left(\mathbf{H}^H\mathbf{H} + \sigma_n^2\mathbf{I}\right)^{-1}\mathbf{H}^H\mathbf{y}, \tag{6.3}$$

where $(\ldots)^H$ denotes the conjugated transpose, $\mathbf{I}$ is the identity matrix of dimensions $N_t \times N_t$, and σ_n^2 is the noise variance. Alternatively, the MMSE detector may be improved by an additional scaling step according to (6.5). This makes sure that the detected constellation is scaled back to the initial constellation and provides a small performance improvement (0.2 to 0.3 dB for 16-QAM). $\mathbf{S}$ is the scaling matrix, out of which only the diagonal elements S_{ii} are needed.

$$\mathbf{S} = \left(\mathbf{H}^H\mathbf{H} + \sigma_n^2\mathbf{I}\right)^{-1}\mathbf{H}^H\mathbf{H} \tag{6.4}$$

$$\hat{\mathbf{s}}_{i,Scaled} = \frac{\hat{\mathbf{s}}_{i,MMSE}}{S_{ii}} \tag{6.5}$$

A strong improvement comes when we use maximum likelihood detection, situated at the other extreme in terms of performance and complexity. By considering all possible transmitted sets of symbols and computing a metric for each of them at the output of the channel, we can determine the most likely value for each transmitted symbol or bit. The gap in performance between linear and maximum-likelihood detection can actually be pretty large (close to 10 dB at BER 10^{-3}, as shown on Figure 6.1). This comes from the fact that the ML detector exploits the full diversity of the channel, while the linear detectors are limited to a diversity one (in case $N_t = N_r$).

For ML detection, the MIMO detector is designed to solve:

$$\hat{\mathbf{s}} = \arg\min_{\mathbf{s}\in\Omega^{N_t}} \|\mathbf{y} - \mathbf{H}\mathbf{s}\|^2 \tag{6.6}$$

where Ω^{N_t} is the set containing all possible vector signals $\mathbf{s}$ of length N_t. Solving (6.6) corresponds to an exhaustive search. For near-ML detection, not all, but only a limited number of vector signals $\mathbf{s}$ are considered in the search.

Indeed, ML detection can be prohibitively complex. As an example, receiving four 64-QAM symbols on a 4×4 MIMO system requires to test some 16 million points in order to guarantee ML performance, which is totally not compatible with any real-time implementation. This is the main motivation to provide intermediate

[1]The ZF equation (6.2) only holds when $\mathbf{H}$ is square, i.e., $N_r = N_t$. If $N_r > N_t$ the pseudo-inverse $\left(\mathbf{H}^H\mathbf{H}\right)^{-1}\mathbf{H}^H$ should be used, and practically for almost the same complexity the MMSE detector will be used in such cases.

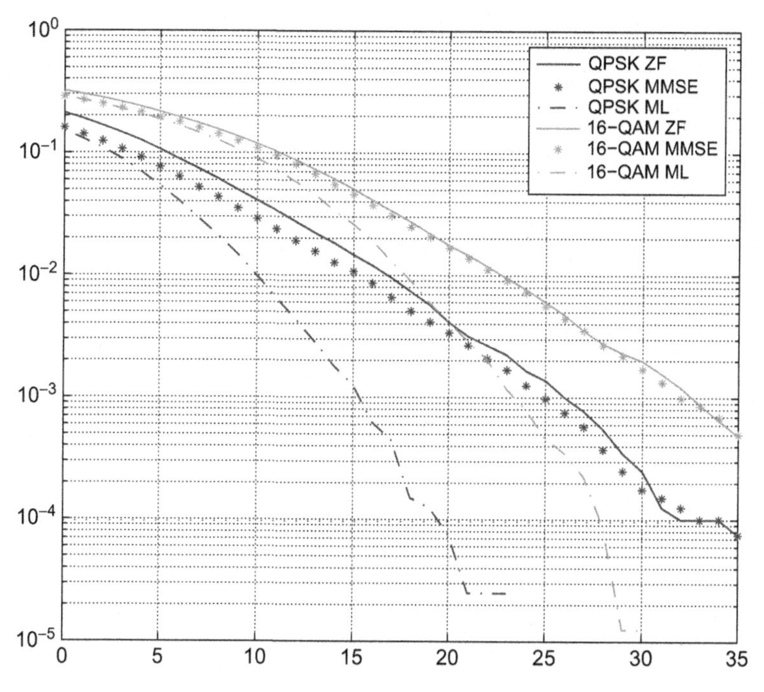

FIGURE 6.1

Comparison of BER performance as function of SNR for 2 × 2 MIMO-SDM on Rayleigh fading channel with independent components for QPSK and 16-QAM with ZF, MMSE, and ML detection; 10000 couples of symbols are simulated.

solutions, leading to better performance than linear detectors, but at a more reasonable complexity than the ML solution. For the implementation of this detector, a wide range of different detection algorithms are available [BGPvdV06, PGNB04]. Some are closer to the linear detectors, such as the Successive Interference Cancellation (SIC). Others are closer to the ML approach, so-called near-ML solutions being the most popular ones, leading to a performance close to ML while strongly reducing its complexity. Most are based on generating a list of most likely candidate symbol vectors, this list being a subset we can choose out of the exhaustive list of the ML detector; hence, the name *scalable list detectors* we consider in this chapter.

Like many other standards such as WLAN (802.11) and WiMAX (802.16), the 3GPP-LTE standard is based on OFDM. This means that the band is split into a number of carriers. The system is designed in such a way that the fading on a single carrier can be assumed to be flat, while different channel matrices are observed across different carriers, more or less correlated depending on the carrier separation in frequency with respect to the coherence bandwidth of the channel. Thanks to the orthogonality property, each carrier can be processed independently.

Section 6.2 presents a first solution for near-ML detection. Based on a single parameter called the search radius, it automatically adapts the search depth to the channel state on all the carriers. Section 6.3 describes the mapping of this solution, starting from the initial algorithm and proposing several modifications in order to make it better fit the selected processor architecture with minimal performance degradation. Finally, Section 6.4 presents an alternative approach, based on co-design between algorithm and architecture. It also considers the extension to soft-output. Indeed, convolutional, turbo, and LDPC channel codes provide a better performance when their decoders can exploit soft information, and those codes are used in all the modern wireless systems. Soft-output MIMO detectors not only provide the most likely symbol vector **s** as hard-output detectors would do, but also give the Log-Likelihood-Ratio (LLR) of each bit, which gives the probability that a bit is a logical 0 or 1. The presented implementations target hand-held devices or mobile phones; this implies a strong concern for limiting the complexity and power consumption of the device.

6.2 RADIUS-BASED DETECTOR ALGORITHM

In this section, a scalable list-based detector is proposed. It relies on sphere detection, spanning a tree of variable width on the different antennas of each carrier. It automatically determines the number of nodes in the tree search as a function of the noise variance based on the channel matrix. Compared to the state-of-the-art sphere decoding with a fixed radius for each constellation symbol, the proposed approach results in a more efficient distribution of the complexity during the tree search, balancing it as a function of the needs on the different carriers and antennas. Like all sphere-detection algorithms, it is based on testing all the symbols s_i within a given radius d_i of the unsliced linear detector output ξ_i:

$$|(s_i - \xi_i)| < d_i \tag{6.7}$$

The novelty relies in the way of computing this radius parameter. It is controlled by a single high-level radius parameter R, which is used to compute the local scaled radius d_i determining the number of points to test on each carrier and antenna. This is done in the following way. The rationale is that we want to balance the probability of finding the desired symbol on all constellations from all carriers and antennas. This means that where the local noise variance is smaller, a smaller search region will be selected, while a larger region will be selected on constellations suffering from a larger noise variance.

This approach relies on computing the local noise variance on each constellation symbol. Based on the ZF linear equalizer with $N_r = N_t$, the symbols corresponding to the different antennas on a given carrier are received as the following vector $\hat{\mathbf{s}} = [\xi_1, \ldots, \xi_{N_t}]$:

$$\hat{\mathbf{s}} = \mathbf{H}^{-1}\mathbf{y} = \mathbf{s} + \mathbf{H}^{-1}\mathbf{n} \tag{6.8}$$

Assuming the noise samples to be independent and of equal variance σ_n^2, the covariance of the output noise vector $\mathbf{H}^{-1}\mathbf{n}$ can be computed as:

$$\mathbf{R}_{nn} = \mathbf{H}^{-1}(\mathbf{H}^{-1})^H \sigma_n^2 \tag{6.9}$$

When $N_r > N_t$, the equation for symbol estimation becomes the following after using the pseudo-inverse; the noise covariance matrix keeps the same expression as in (6.9):

$$\hat{\mathbf{s}} = \mathbf{s} + \left(\mathbf{H}^H\mathbf{H}\right)^{-1}\mathbf{H}^H\mathbf{n} \tag{6.10}$$

The diagonal elements of $\mathbf{R}_{nn}$ represent the equivalent noise variance on the different antennas of the considered carrier. By scaling the square root of those values—leading to the local noise standard deviation—with a constant radius R, and doing so over all the carriers, we can obtain the distance d_i used to scale the search space on all the carriers and antennas, while always keeping the property that each constellation is searched proportionally to the local noise variance.

The performance of the algorithm as function of radius is shown in Figure 6.2. Based on those results, we obtain at BER 10^{-3} a gain of 4 dB with radius 0.03, 8 dB with 0.06, and 10.5 dB with 0.1; the maximum gain at this BER is 12 dB.

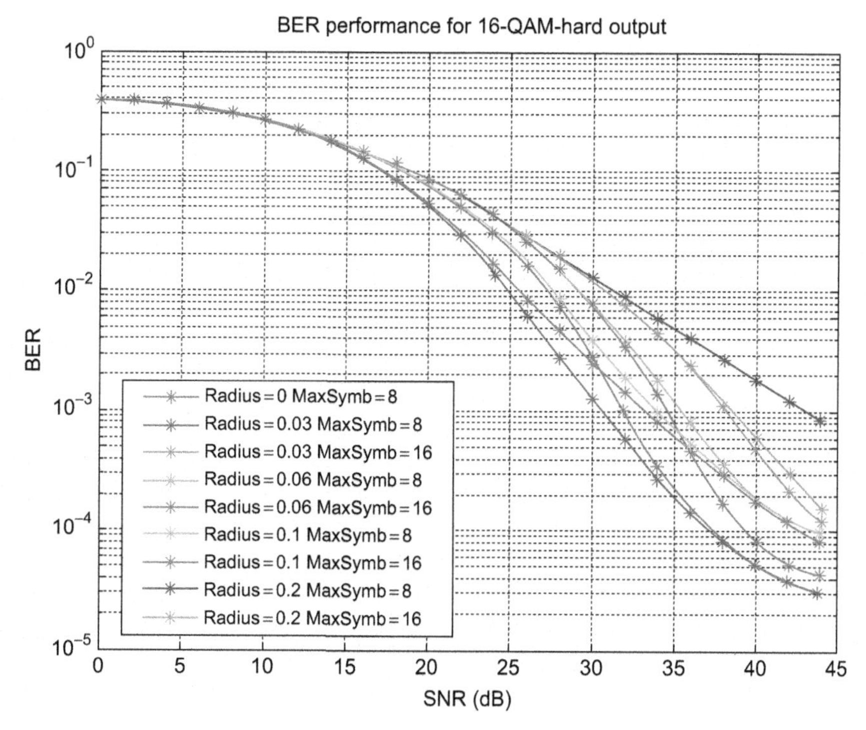

FIGURE 6.2

Performance of the radius-based detector as function of the radius parameter on a 5-MHz 2×2 LTE channel and for 16-QAM, as function of the radius parameter. A gain of 10 dB is present at BER 10^{-3} between linear ($R = 0$) and near-ML ($R \geq 0.2$). Hard-limitation to 8 out of 16 symbols was also tried as complexity-limiting additional constraints; it degrades the system performance whenever we want to reach a BER below 10^{-3}.

The latest developments around this solution have lead to a refined control strategy [LERBVdP09]. Instead of the high-level radius parameter, it is based on a probability that can be tuned at high-level, nicely scaling the detector from linear to ML performance in a more deterministic way than the radius parameter. It also extends the algorithm in order to propose LLR computations for soft-output detection. Those two extensions are not included in the implementation presented in Section 6.3. Soft-output will however be presented for the other near-ML detector described in this chapter (Section 6.4).

6.3 MAPPING OF THE RADIUS-BASED DETECTOR

After developing a scalable MIMO detection algorithm in Section 6.2, we have to map it to a digital platform in order to validate the approach and assess its real-time complexity. In this work we investigate in the ADRES CGA processor for this purpose, which is part of IMEC's software-defined radio platform [BDSV+08, BDSR+08]. The template of the processor is shown in Figure 6.3. As it can be seen, the parametrizable template consists of a Coarse Grain Array (CGA) of densely interconnected Functional Units (FUs) that have local Register Files and individual configuration memories (loop buffers). Besides, a few very large instruction word (VLIW) FUs are present. The VLIW FUs and a limited subset of the CGA FUs are connected to the global (shared) register file. This shared register file enables the exchange of data between both types of FUs. Since the VLIW FUs and the CGA FUs operate time multiplexed, two modes are available: VLIW mode and CGA mode. All FUs support SIMD (single-instruction multiple-data). For our explorations, we leverage on the DRESC C compiler framework [MLM+05]. The compiler supports both the VLIW mode and the CGA mode. In general, loops are mapped on the CGA section and the rest of the code is scheduled on the VLIW section.

The ADRES template enables the instantiation of a processor with a specific amount of instruction-level parallelism (ILP) and data-level parallelism (DLP). By changing the size of the array (number of FUs), the amount of supported ILP can be tuned. By changing the number of SIMD slots in each FU, the amount of supported DLP can be tuned. The required amount of ILP and DLP is application dependent. To determine the best combination of ILP/DLP, i.e. fulfill the performance requirements with lowest implementation complexity, extensive explorations are typically needed. In Section 6.4 we will show some exploration for another MIMO detector design. However, in this Section we map our radius-based detector to an already determined version of the platform architecture, hence having somewhat less flexibility to fine-tune the architecture to the application.

The selected instance runs at 400 MHz and consumes some 300 mW. The mapping experiment is done for a 2×2 5-MHz 3GPP-LTE system. We have selected 64-QAM as the most challenging modulation for the detector. All the simulations used for optimizing the implementation are based on a random set of 50 channels

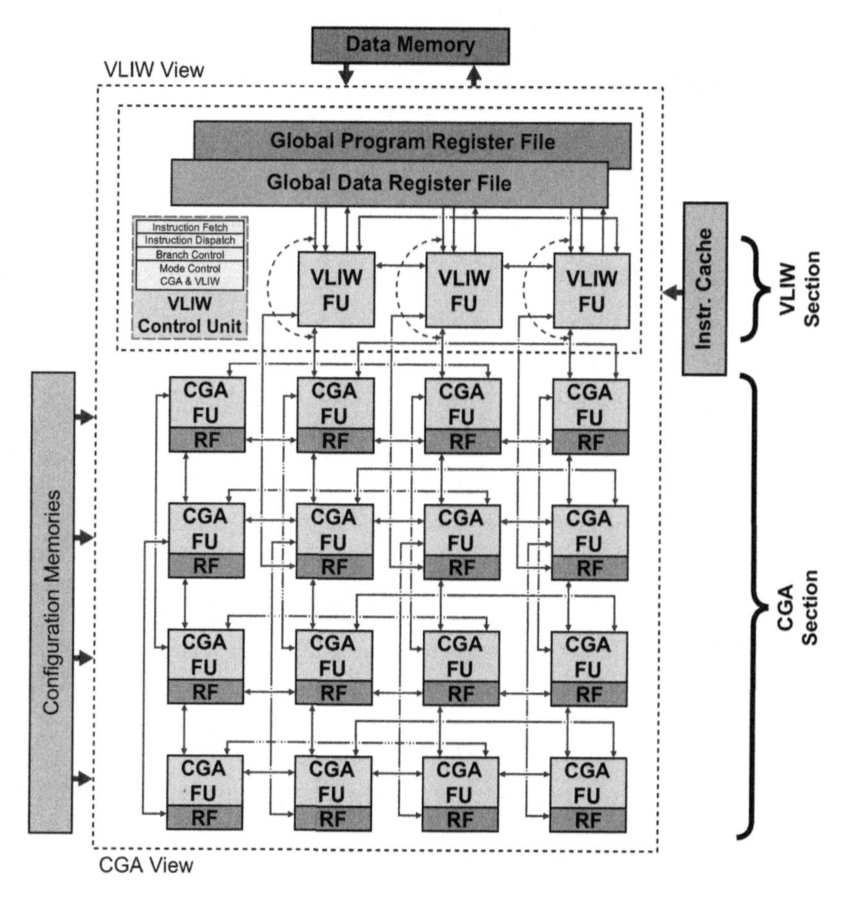

FIGURE 6.3

The ADRES template shown on a instance with 16 CGA FUs and 3 VLIW FUs.

according to the channel model of the standard, as well as realistic conditions including for example all front-end nonidealities based on the IMEC platform.

From the specific constraints of our platform, a number of modifications are made to the original solution, enabling a better implementation but sometimes at the cost of some slight performance degradation. The successive iterations between mapping and functional modifications illustrate the process leading from a theoretical solution to a practical one.

The main constraint coming from our platform is actually to exploit its parallel architecture to the maximum possible extent. Indeed, it contains 16 FUs that can be efficiently programmed in a parallel way (CGA mode), leading to the most efficient exploitation of the platform. However, if the algorithm is not suited to this architecture or not optimized in a proper way during compilation, the mapping cannot exploit correctly the parallelism, leading to inefficiencies both in throughput and in power consumption.

Another important task is quantization. More especially, in order to fit on our platform, 16-bit operations were selected as far as possible. Special tools are used first to quantize at Matlab level, and secondly for conversion into C, which is taken as source for specific compilation on our platform.

6.3.1 Direct Implementation

As a first step, we map directly the algorithm as it was defined onto the processor. The following steps are executed to fulfill the algorithm functionality:

1. Linear detection of the received symbols (ZF estimate without rounding).
2. Slicing (finding closest constellation point on each antenna).
3. Scaled radius computation based on the local channel matrix and external radius parameter.
4. Enumeration of the candidate points in a circle of computed radius.
5. Sorting of enumerated symbols (in order to enable a limitation to a maximum number of points).
6. Limitation of the number of points where needed.
7. Computation of Euclidean distance metric for all couples of candidate points on both antennas (actual ML detection).
8. Selection of the minimum-distance couple of points.
9. Demapping of the selected two symbols.

Although this implementation is functionally identical to the initial algorithm developed in MATLAB, it runs 12 times too slow on our testbench with respect to real-time, using a relatively large radius 0.2. The main reason—besides the inherent complexity of near-ML detection—is that the code is not suited to exploit parallelism: the central part of the algorithmic loops should contain a regular flow of operations, while the enumeration of symbols within circles of varying scaled radius does not enable efficient parallelization on different carriers.

6.3.2 Parallelism Exploitation and Simplifications

As was mentioned in the previous section, a straightforward implementation is not efficient at all, hence a number of modifications are needed with respect to the initial algorithm, together with some specific optimization steps in the mapping.

First of all, the general approach has to be modified in order to enable parallelization of the most inner loop in the processing. Concretely, this is achieved by computing one value of scaled radius for a group of neighboring carriers, based on the fact that the channel matrix is strongly correlated on neighboring carriers, leading to little degradation if we fix the same radius value for blocks containing several carriers. Note that the exact channel matrix interpolated from the pilots is still used for equalizing each carrier, only the scaled radius used to determine the number of

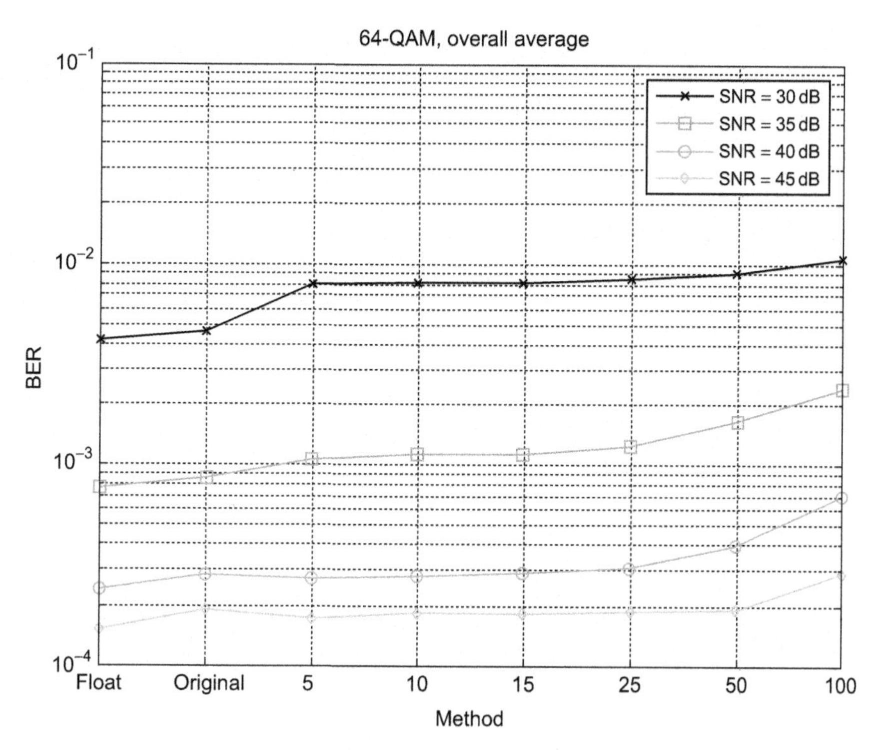

FIGURE 6.4

Impact of the block size (number of carriers) on the BER performance: comparison between floating-point, quantized (original), and modified algorithm with various block sizes between 5 and 100 carriers.

candidates is kept constant across neighboring carriers. Figure 6.4 illustrates the corresponding trade-off. Processing up to 15 or 25 neighboring carriers together with the same radius (out of the 300 carriers of the 5-MHz LTE system) does not degrade the performance.

Second, a more drastic modification is done in the selection of enumerated points. In the original algorithm, even when selecting the same radius across several carriers, this does not lead to the same number of points after enumeration, due to the dependency on the output of the linear detector. For example, if the linear detector output is close to the edge of the constellation or even outside the constellation, this will reduce the number of enumerated points with respect to a starting point in the middle of the constellation. This nondeterministic behavior again reduces the parallelization efficiency, as the determination of points within the radius distance as well as enumeration of selected points cannot be fully parallelized.

Moreover, in order to decide whether a given constellation point is within the circle of specified radius or not, we need to compute distances by squaring differences between constellation points and linear detector output along I and Q directions, requiring many multiplications as all the constellation points should be tested to check the condition, and leading to 32-bit variables which are not as effectively represented on our processor as 16-bit variables.

In order to solve those two problems, we have decided to enumerate a square sub-constellation of variable size instead of a circle of variable radius. Practically, based on a look-up table the side of the square is determined as function of the scaled radius. Moreover, if the linear detector output is close to the edge and the selected square would partly fall out of the constellation, we simply move it back to the constellation edge, making sure that the specified number of points can be enumerated. Those operations happen straightforwardly from primitives such as minimum, maximum, shift, or basic arithmetic (addition/subtraction), which are much less complex than multiplications. This approach is illustrated in Figure 6.5.

Occasionally, a few more or a few less points are enumerated, when compared to the original algorithm. However, the degradation is limited, as can be seen from Figure 6.6. If we do not use blocks larger than 25 carriers, around 1.5 dB are lost at most (in the region around BER 10^{-2}), but this loss gets down to zero when working at a BER above 10^{-1} or below 10^{-3}. The complexity is strongly reduced: an efficient parallelism is now possible and 32-bit instructions are removed (no more squaring to compute the Euclidean distances).

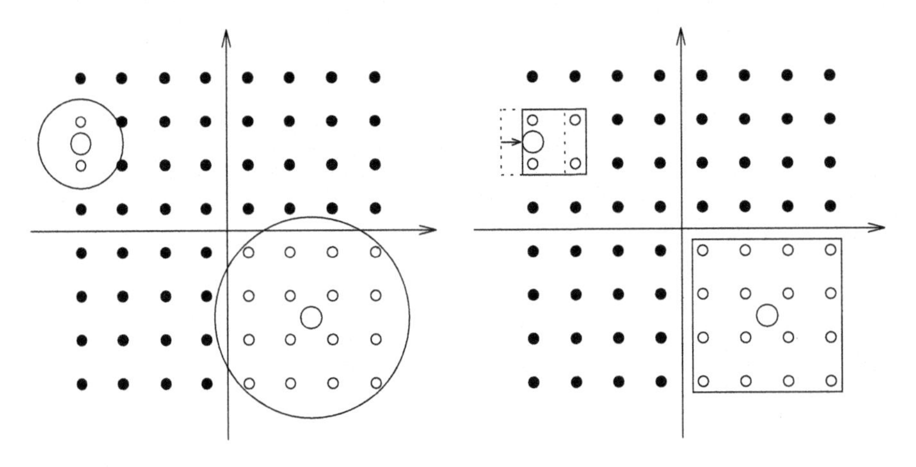

FIGURE 6.5

Modification from the original algorithm based on circles with variable radius to a more implementation-friendly version based on squares with variable size. The square partly out of the constellation (dotted line) is automatically moved to the edge of the constellation, slightly increasing the number of candidate points.

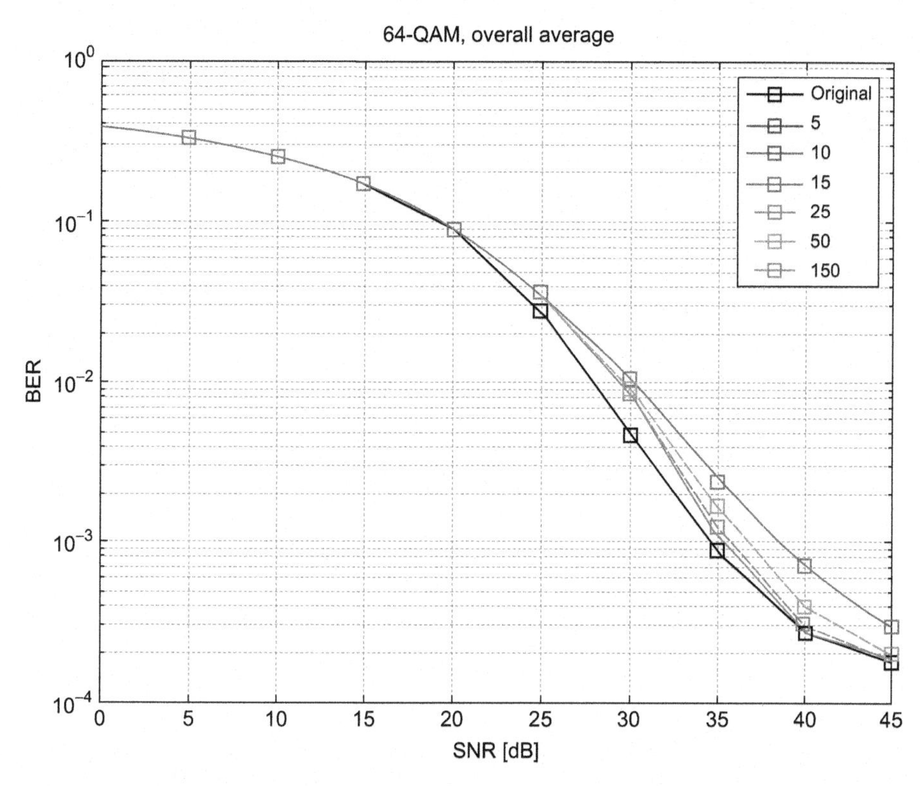

FIGURE 6.6

Comparison between the original radius-based algorithm and the simplified version based on squares of variable size. Several block sizes are illustrated as on Figure 6.4.

When looking at the typical operation of this modified algorithm, we can see that some 80% of the time only 1×1 squares are selected (Figure 6.7). This triggers a further optimization: We can introduce a switch around the main part of the enumeration and ML selection, deactivating them when a 1×1 square is selected on both antennas (no need for enumerating points, the original sliced point is the output).

Although trivial from a conceptual point of view and without changing the functionality of the algorithm, this modification saves us some extra computations. Finally, we could further quantize the square sizes, limiting them to 1×1, 2×2, 4×4, and 8×8; this approach based on powers of 2 leads to virtually no degradation (around 0.1 dB) but again improves the parallelization on our processor. Without entering in too many details, this enables the merge of several loop levels, reducing the number of prologs and epilogs when entering or quitting loops, which reduces the overhead on the processor. Optionally, the 8×8 size may be excluded as it leads to the largest complexity, but at the cost of some additional degradation in the low-BER domain.

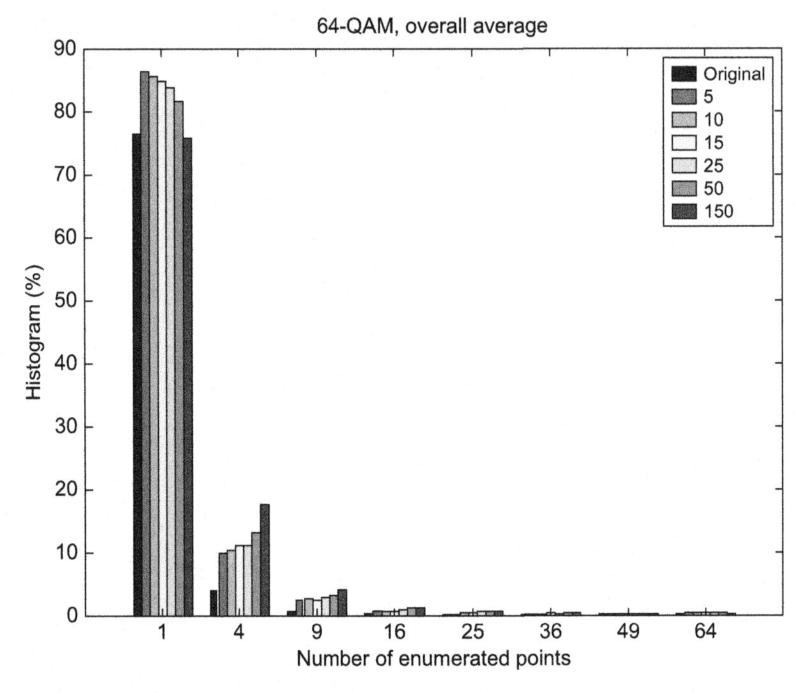

FIGURE 6.7

Size of the squares considered in typical operation of the modified algorithm (number of points), showing the interest for a switch around the 1×1 case.

6.3.3 Overall Performance and Scalability

In the previous section, we presented the complexity reductions that have been brought to the initial algorithm. The result is impressive: while the original solution was 12 times too slow with respect to real-time, the situation has improved by a factor 7. However, this means that we still cannot run the algorithm real-time with the selected parameters (a radius 0.2, which is rather at the high-performance than at the low-complexity side).

We can further reduce the load by playing with the parameters. It is worth studying in more detail the behavior of the algorithm as function of the radius as well as the channel condition. Figure 6.8 illustrates the results, with a very dynamic complexity behavior. For example, we can use a radius up to 0.08 on the average for detecting the full band in real-time. According to our results (Figure 6.2), this radius value translates into 75% of the total achievable gain.

However, the maximum radius reduces down to 0.015 on the worst channels. This means that an additional mechanism will be needed on top of radius control in order to guarantee a real-time behavior. A simple approach is to first compute the scaled radius and corresponding number of candidate points for all the carrier blocks,

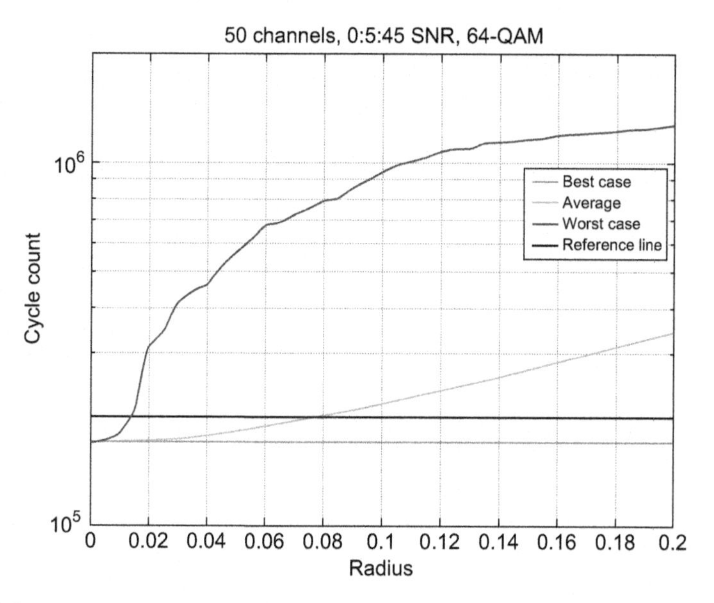

FIGURE 6.8

Illustration of the dependency of the complexity on the radius. Best, average and worst case correspond to different channel realizations. The thick horizontal line corresponds to the real-time budget on our platform (200,000 cycles for processing a 0.5-ms LTE slot).

which can be used to estimate the expected complexity (it was shown to be an affine function of the total number of enumerated points). At this point of the algorithm, if the expected total number of points to enumerate exceeds the budget, the number of points can be down-scaled on each carrier, leading to a modified value below the budget. As a final step, the near-ML enumeration runs as previously, either on original or on down-scaled values.

A last remark can be made on the real-time complexity: In this study it was always assumed that we should decode the full bandwidth in MIMO with the fastest modulation. Obviously, in a cellular standard such as LTE such extreme single-user operation will be far from realistic. Much more likely, the band will be shared among many users, and the 64-QAM will not always be used, depending on the channel conditions. By relaxing those two constraints, we can save a lot of complexity. Hence, if we do not want to implement a real-time complexity control as mentioned in the previous paragraph, a practical control alternative is for example to use the scalable radius-based detector only when a fraction of the band is allocated to the selected user, and to automatically switch back to linear detection as back-up solution in case the full band should be allocated to the considered user. This would degrade the link performance in that case, but as this scenario is unlikely to happen in a realistic cellular system this should not cause trouble in practice.

6.4 SSFE DETECTOR

In the previous section, we have seen how many modifications had to be brought to a scalable MIMO detection algorithm in order to implement it. Those two steps are common in digital design, first deriving an interesting algorithmic solution and secondly mapping it to a processor. However, in order to reduce the implementation overhead and take into account the platform constraints from the early algorithmic development, we can try to combine both steps. This algorithmic-architecture co-design approach was considered for studying another scalable MIMO detector, which is described in this section. More practically, the detection algorithm is optimized jointly with the ADRES processor described in Section 6.3, which is part of IMEC's SDR platform.

Generally speaking, the massive parallelism would enable SDR implementations of advanced wireless signal processing algorithms. Unfortunately, only simple SDR systems and algorithms have been demonstrated and reported in literature. For instance, [LLW+07, NTD08, vBHM+05] do not support Multiple-Input Multiple-Output (MIMO). Other implementations demonstrate MIMO processing [BDSR+08, EBF09, WEL09], but based on the simple Minimum Mean Square Error (MMSE) linear detection, which significantly falls behind the research progress of wireless signal processing [HtB03].

In this section, we present the near-ML Selective Spanning with Fast Enumeration (SSFE) algorithm for list generation. The SSFE algorithm was explicitly optimized for parallel architectures, and is described with more details in [LBX+08]. Contrary to the traditionally utilized K-best algorithm [GN06, CZX07, CZ07, SYK07], the SSFE algorithm results in a completely regular and deterministic dataflow structure. This is important for enabling an efficient mapping on parallel architectures. In addition, the SSFE does not require expensive memory-operations. Moreover, the SSFE algorithm is based on very simple and architecture-friendly operators such as additions, subtractions and shifts, which clearly reduces the implementation complexity. Finally, the SSFE algorithm is well-suited for scalable implementations, because it offers a parameter which determines the complexity-performance trade-off of an algorithm instance, as was the case in Section 6.2 with the radius parameter.

6.4.1 SSFE: An Implementation-Friendly Detector

A SSFE algorithm instance is characterized by a vector $\mathbf{m} = [m_1, \ldots, m_{N_t}]$, $m_i \leq M$, where M is the number of points in the selected constellation. Each entry in this vector specifies the number of scalar symbols s_i that are considered at antenna N_i. With the parameter $\mathbf{m}$, the complexity-performance trade-off point of an algorithm instance is selected. The computation of $\mathbf{s}$ can be visualized with a spanning tree. In this tree each node at level $i \in \{1, 2, \ldots, N_t\})$ is uniquely described by a partial symbol vector $\mathbf{s}^i = [s_i, s_{i+1}, .., s_{N_t}]$. Starting from level $i = N_t$, SSFE spans each node at level $i + 1$ to m_i nodes at level i.

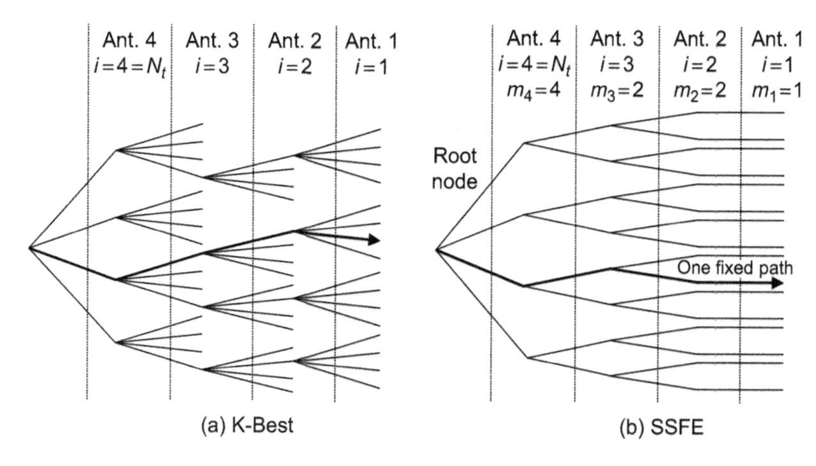

FIGURE 6.9

K-Best and SSFE search-tree topologies for 4 × 4 Quadrature Phase-Shift Keying (QPSK) modulation. Antennas are processed in decreasing order.

An example of a tree for $\mathbf{m} = [1,2,2,4]$ is shown in Figure 6.9 and compared to the state-of-the-art K-Best algorithm. K-Best first spans the K nodes at level $i+1$ to KM nodes. After spanning, K-Best sorts the KM nodes, the K best nodes are selected and the rest of the nodes are deleted. This approach results in a nondeterministic data-flow. It also requires sorting operations, as well as memory operations in order to select the best nodes at each step. In contrast, the spanned nodes in SSFE are never deleted. Therefore the dataflow in SSFE is completely regular and deterministic, as the spanning depth is predetermined at each antenna level. This is a perfect example of algorithm design with implementation constraints in mind.

The task of the SSFE during step i is to select the set of the m_i closest constellation points around the unquantized value ξ_i, which is obtained from linear detection conditioned on the values already selected for antennas $i+1$ to N_t. When $m_i = 1$, the closest constellation point to ξ_i is $p_1 = \mathcal{Q}(\xi_i)$, where $\mathcal{Q}$ is the slicing operator. When $m_i > 1$, more constellations can be enumerated based on the vector $d = \xi_i - \mathcal{Q}(\xi_i)$. Fundamentally, the technique applied here is to incrementally grow the set around ξ_i by applying heuristic-based approximations. The heuristic in SSFE is called Fast Enumeration (FE). Figure 6.10 shows an example. Compared to other schemes [GWDN05, BBW+05], the FE is independent on constellation size, so that handling 64-QAM is as efficient as handling QPSK. Moreover, the FE can be implemented with simple and architecture friendly operators, such as additions, subtractions, bit-negations and shifts. More information about the SSFE algorithm and BER performance comparisons with other schemes can be found in [LBX+08, LBL+08].

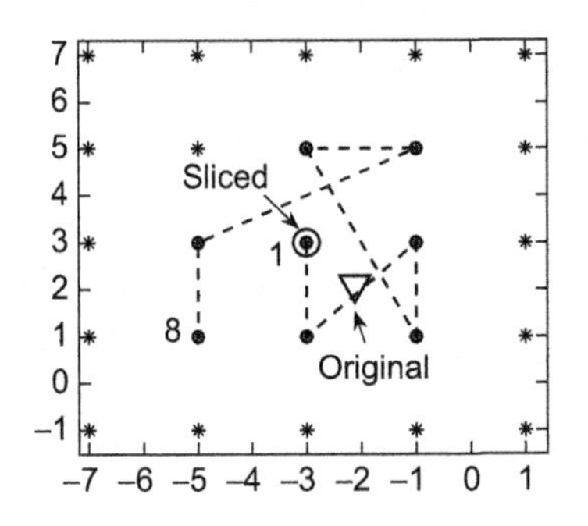

FIGURE 6.10

A fast enumeration of 8 constellation points shown on an example.

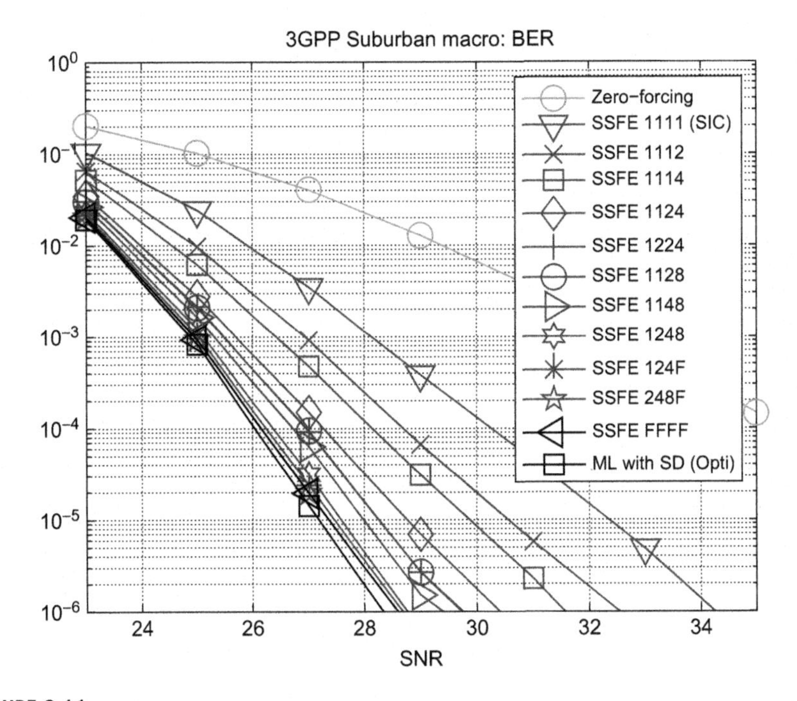

FIGURE 6.11

Illustration of the SSFE performance on a 4 × 4 LTE set-up with 64-QAM and rate 1/2 turbo coding, as function of the **m** parameter.

The performance of the SSFE algorithm is illustrated on Figure 6.11, for different values of the **m** parameter. It illustrates the available trade-off region between a single-point search, which is equivalent to an SIC detector and ML detection.

6.4.2 Soft-Output Extension

Building on the SSFE approach, we can easily provide soft-output instead of hard-output. Indeed, the list generation remains the same (SSFE tree), but the last stage is adapted in order to compute LLRs instead of simply selecting the minimum-distance branch.

The list generator provides a list of most likely candidate symbol vectors $\mathbf{s}$, denoted by $\mathcal{L}$. The task of the LLR generator is to compute the $LLR(j,b)$ for each bth bit of the jth scalar symbol in $\mathbf{s}$. This is done by considering all candidate symbol vectors $\mathbf{s}$ in $\mathcal{L}$.

For the calculation of $LLR(j,b)$ the max-log approximation can be used. It is formulated as [HtB03]:

$$LLR(j,b) = \frac{1}{2\sigma^2}\left(\min_{s\in\chi^0_{j,b}} \|\mathbf{y}-\mathbf{Hs}\|^2 - \min_{s\in\chi^1_{j,b}} \|\mathbf{y}-\mathbf{Hs}\|^2\right) \tag{6.11}$$

where $\chi^0_{j,b}$ and $\chi^1_{j,b}$ are the disjoint sets of symbol vectors that have their bth bit in their jth symbol set to 0 and 1, respectively. σ^2 is the noise variance.

The complexity of the LLR generation can be significantly reduced by a number of simplifications. First, the so-called bit-flipping strategy can be applied [WG04]. This means that instead of looking up out of the candidate vector set $\mathbf{s}$ which have value 0 and which have value 1 for the bit under consideration, we keep all the symbols of the candidate set and modify it by enforcing this bit to be 0 or 1 on all modified candidates, hence enlarging the set to keep its original size for both 0 and 1 values, instead of splitting it in two parts. This has the advantage of having a fixed number of candidates, enabling again more determinism in the flow and a better implementation when compared to splitting the set in value-0 and value-1 components. The new sets are denoted $\mathcal{L}^0_{j,b}$ and $\mathcal{L}^1_{j,b}$.

Second, applying the QR decomposition further simplifies the implementation. It not only reduces the number of multiplications to perform as only the triangular R matrix instead of having the full channel matrix will be multiplied by all the candidate symbols, but it further improves the structure by enabling reuse of partial results. Indeed, in the process of computing the distance metric over the successive antennas of the spanning trees, the rows corresponding to the successive antennas can be reused in the overall metric computation, having only a vector multiplication instead of a matrix multiplication for each new antenna.

Finally, replacing the Euclidean norm with the Manhattan (first-order) norm is also used, as was already the case in Section 6.3. This reduces both the number of multiplications and required depth of the quantization. After those three simplifications, we end up with the following expression, where $\hat{\mathbf{y}} = \mathbf{Q}^H\mathbf{y}$ and $\|\ldots\|_1$ is defined as the sum of absolute values of real and imaginary parts of a complex number:

$$LLR(j,b) = \frac{1}{2\sigma^2}\left(\min_{s\in\mathcal{L}^0_{j,b}} \|\hat{\mathbf{y}}-\mathbf{Rs}\|_1 - \min_{s\in\mathcal{L}^1_{j,b}} \|\hat{\mathbf{y}}-\mathbf{Rs}\|_1\right) \tag{6.12}$$

Some additional simplifications are possible, in order to further decrease the implementation complexity of this algorithm. We can reduce the number of required computations and memory accesses by performing algebraic simplifications at a low level data-flow. For example, the fact that Gray coding is used lets us know that flipping a bit in a symbol only changes the position in either I or Q direction, but never both at the same time. This enables a simplified implementation of the bit-flipping related computations.

We can also replace all constant multiplications by shift and add operations which enables a more efficient implementation. This is especially the case if we normalize the constellation points to small integer values, i.e., $\ldots, -3, -1, 1, 3, \ldots$ Multiplying those constants by channel-related values is easily implemented by shift-and-add.

Compared to the straightforward implementation of LLR computation, the proposed implementation with all the simplifications completely removes multiplications. Moreover, it reduces other operations (additions, shifts, and memory operations) by 50 to 90%, depending on the set-up (16 or 64-QAM, 2×2 or 4×4 MIMO, $\mathbf{m}$-parameter).

6.4.3 Corresponding Processor Architecture

The proposed algorithm has been from the beginning developed in such a way that it respects architecture and implementation constraints. This was done having in mind the ADRES processor of IMEC (see Section 6.3), but also considering implementations on a TI processor as well as an ASIC. In this sense, implementation constraints were taken into account in a generic way. We now further detail the implementation on the ADRES processor, with also propagation of constraints from the algorithm to the architecture. This can take place as it is actually a processor template that can be somewhat tuned to the application. This is somewhat different from the implementation in Section 6.3, where a fixed instance of the processor was selected and used for mapping the radius-based MIMO detector.

In mapping the SSFE, the available freedom from the processor template is used towards the algorithmic-architecture co-optimization. The template of the processor is shown on Figure 6.3. A first way to tune the processor concerns the selection of application-specific instructions. When integrated to the processor, such instructions can be used more efficiently based on packed data. This increases throughput and power efficiency, removes intermediate storage needs, and eases the compilation optimization effort. Those instructions may be rather complex. For instance, 4 specific instructions are proposed for this algorithm, 3 related to the list generation, and one for the LLR computation, which is the dominant part of the soft-output algorithm. For example, this LLR-specific instruction enables an efficient computation of absolute differences between real and imaginary parts of received signal and scaled constellation candidates. Having the flexibility to introduce such instructions strongly improves the code efficiency, both with respect to a nonflexible ADRES version and with respect to standard DSP processor such as TI's.

Based on this optimization, a throughput of 140 Mbit/s with 64-QAM 2×2 LTE was obtained on the 4×4 CGA instance of Figure 6.3 and with a $\mathbf{m}$ parameter $[1, 64]$. Although the mapping of this algorithm was not performed on the predetermined issue of the processor, it would have been much lower in throughput. This illustrates an important conclusion for the design of a specific ADRES processor instance to be integrated in a platform: A careful analysis of the dominant algorithm that will run on the platform is needed in order to select those specific instructions appropriately.

6.5 CONCLUSIONS

In this chapter, we have addressed the problem of MIMO-SDM detection algorithms and their mapping in the context of the 3GPP-LTE standard. More especially, we have considered two scalable list-based detectors, enabling the achievement of performance close to the ML solution at a reasonable complexity, but also to scale down the performance when the complexity should be reduced.

Several lessons can be drawn from those two studies. First, the gap between theoretical algorithms and practical implementations is huge. This is no new fact, but in this chapter we have looked at ways to bridge this gap. A first approach relies on transforming the initial algorithm in order to efficiently map on the selected architecture. We have shown how important this is on a parallel processor, gaining almost a factor 10 from the original implementation. Still, sphere-detection-based algorithms remain very challenging, especially given the nondeterministic complexity. A second approach was to start in a co-design philosophy between algorithm and architecture. This has led to a detector with deterministic and predictable flow, while its implementation was supported by exploiting more flexibility from an earlier stage of processor design.

A second conclusion concerns the importance of scalable or flexible solutions. Indeed, worst-case design is not an acceptable solution for complex algorithms such as MIMO detection. It becomes impossible to achieve a complexity comparable to ML detection while detecting the high-throughput streams of modern wireless standards in real-time, at least for hand-held devices with constraints of battery capacity and peak processing power. On the contrary, a scalable architecture can support fast-throughput by relying on simple algorithms when the channel is favorable, while it can reconfigure to exploit much more complex detection strategies when the channel is worse and needs them, reducing the maximum throughput at that moment. The two approaches we have proposed show such a behavior, based on a high-level parameter that can be tuned in order to select the relevant power-performance trade-off.

In the future, cellular standards will evolve further to LTE advanced. This new standard is expected to move towards more cooperative MIMO schemes, which will impact more base stations but should be more or less transparent to mobile terminals. However, there will again be a push for increasing the number of antennas and the total bandwidth through aggregation. This will definitely increase again the pressure on designing efficient implementations for hand-held terminals, and make the need for flexibility even more important.

CHAPTER

IEEE 802.11n Implementation

7

Zhipeng (Alexandre) Zhao

7.1 IEEE 802.11n PHY LAYER INTRODUCTION

Wi-Fi systems based on the IEEE 802.11 standards are widely deployed for both indoor and outdoor environments. It is well known that the Wi-Fi system suffers from the channel-fading phenomenon that degrades the link quality and limits the range of communication. Besides, the increasing demand for high-throughput brings new challenges to the existing systems.

As detailed in the preceding chapters, the studies show significant coding gain of Multiple-Input Multiple-Output (MIMO) technology compared with traditional SISO solution and many research works have been done to give a good space-time coding to make full use of the MIMO technique. Therefore, the MIMO technique has been introduced in the new IEEE 802.11n standard [IEEE P802.11n] to improve the transmission quality as well as the system spectral efficiency. The IEEE 802.11n physical layer supports data rate up to 600 Mbps in 4×4 MIMO configuration and 40 MHz bandwidth.

In this chapter, we will present the IEEE 802.11n transceiver processing to explain the operation of MIMO technique in the Wi-Fi standard. For readers' convenience, we will focus on 5.2 GHz where the 20 MHz High-Throughput (HT) mixed-mode with 800 ns guard interval duration is applied.

The MIMO technique is investigated in 3 levels in the 802.11n context: (1) using *space time block encoder* (STBC), which encodes N_{SS} spatial streams into N_{STS} space-time streams; (2) using spatial mapper for spatial expansion or beamforming, which spreads the N_{STS} space-time streams into N_{TX} transmit chains; and (3) using the receiver diversity based on multiple antennas array at the receiver's side.

7.2 IEEE 802.11n TRANSMITTER PART

The IEEE 802.11n standard (draft 2.0) gives the full description of the transmitter architecture and processing scheme. Here, we will give a quick description of the transmitter, which shows the migration from 802.11a to 802.11n. Readers are supposed to be familiar with the IEEE 802.11a physical layer.

MIMO. DOI: 10.1016/B978-0-12-382194-2.00007-1

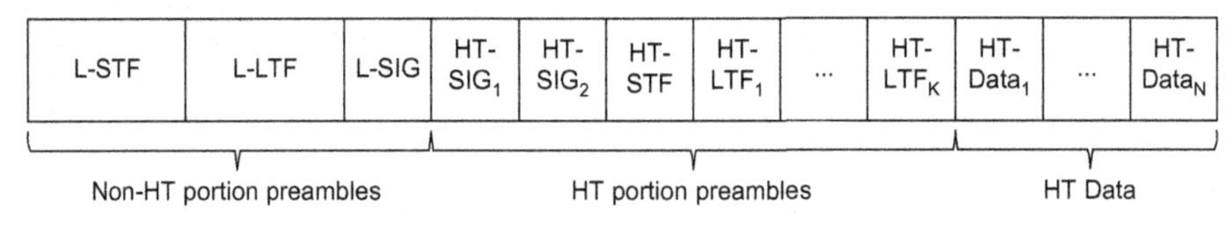

FIGURE 7.1

802.11n frame format in HT mixed format which is composed of the non-HT portion preambles, the HT portion preambles and the HT data. The N_{LTF} denotes the number of HT-LTFs and it depends on the system configuration.

7.2.1 PLCP and Frame Format

To support the MIMO transmission, a new HT *physical layer convergence procedure* (PLCP) layer is proposed as a complement to the legacy IEEE 802.11a PLCP layer. In all, IEEE 802.11n physical layer supports three types of PLCP frame formats: the *legacy* frame format, the *mixed mode*, which is compatible with the legacy devices and the *greenfield* mode, which is not compatible with the legacy devices. For the reason that the most IEEE 802.11n equipments operate in mixed-mode, we will concentrate on this mode in this chapter.

The new PLCP frame format is shown in Figure 7.1 where we see that the HT mixed format frame contains three parts: (a) the non-HT portion preambles; (b) the HT portion preamble; and (c) the HT data part.

The non-HT portion preambles is composed of the Legacy Short Training Field (L-STF), Legacy Long Training Field (L-LTF) and Legacy SIGNAL Field (L-SIG). The HT portion consists of the HT SIGNAL Field (HT-SIG), the HT Short Training Field (HT-STF) and the HT Long Training Fields (HT-LTFs).

Like in legacy mode, the L-STF is used for the Automatic Gain Controller (AGC) training and signal detection while the L-LTF is used for timing recovery (with L-STF) and legacy channel estimation. This preamble is "understandable" by legacy devices and it provides channel information to decode the following L-SIG and HT-SIG since HT-SIG contains the PLCP information including the MIMO pattern description, as illustrated in Figure 7.2. While the HT-STF helps to give a finer carrier frequency offset estimation, the HT-LTFs perform the HT channel estimation.

7.2.2 Processing

The processing is associated with the architecture illustrated in Figure 7.3, where we find several new functional blocks: Encoder Parser, Stream Parser, STBC, Cyclic Shift (CS), and Spatial Mapping. In addition, the rate of the Binary Convolutional Code (BCC) part for the Forward Error Correction (FEC) and the dimension of interleaver are both modified to meet the needs of HT transmission.

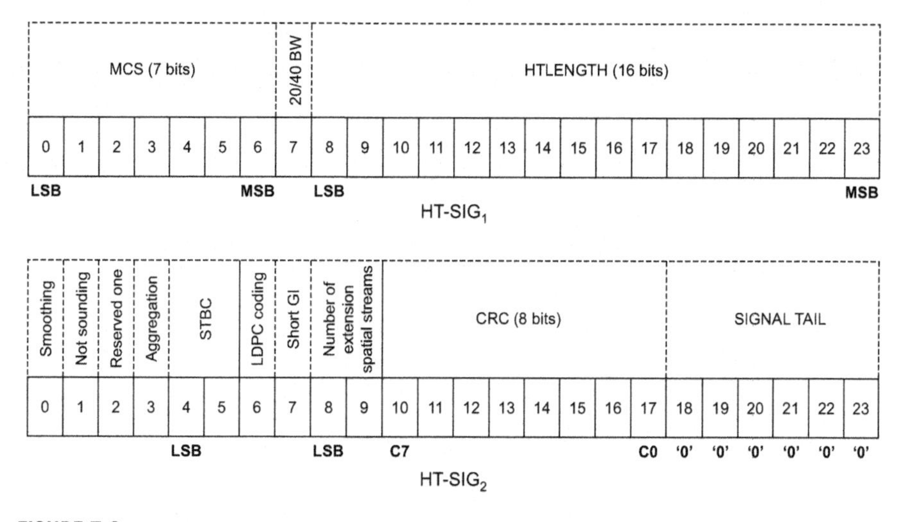

FIGURE 7.2

Format of HT-SIG: the HT-SIG is composed of 48 bits, which are sent separately in two OFDM symbols, HT-SIG$_1$ and HT-SIG$_2$. In HT-SIG$_1$: (1) MCS indicates the utilized modulation and coding scheme; (2) BW indicates the 20 MHz/40 MHz band of operation; (3) HTLENGTH indicates the length of HT *MAC Protocol Data Unit* (MPDU) in bytes. In HT-SIG$_2$: (1) Smoothing is the indicator of frequency-domain smoothing processing for the channel estimation; (2) Sounding denotes whether this frame is for sounding[1] procedure; (3) Aggregation is for the aggregation processing of MAC layer; (4) STBC indicates the use of Alamouti code, which is also the difference between N_{STS}[2] and N_{SS}[3]; (5) LDPC coding indicates the use of *low-density parity check* (LDPC) code for the *Forward Error Correction* (FEC) code; (6) Short GI indicates the use of 400 ns *guard interval* (GI) instead of the regular 800 ns one; (7) Number of extension spatial streams indicates the N_{ELTF} which is used to calculate the total number of long training fields by $N_{LTF} = N_{DLTF} + N_{ELTF}$[4].

Encoder Parser and Convolutional Encoder[5]

Briefly speaking, the scrambled binary sequence of PLCP Service Data Unit (PSDU) is fed into the encoder parser, which is activated when the data rate is greater than 300 Mbps. This is for the purpose that at high data rate, the receiver's FEC decoding procedure can be accelerated by applying two decoders for each binary stream.

[1] The sounding PLCP Protocol Data Unit (PPDU) is used to sound the MIMO channel for the receiver to obtain the channel state information for example when beamforming.
[2] N_{STS} is the number of spatial time streams.
[3] N_{SS} is the number of spatial streams, which is determined according to MCS.
[4] N_{DLTF} stands for the number of data long training fields, while N_{ELTF} stands for the number of extension long training fields. N_{LTF} is the total number of training long fields.
[5] We consider only the BCC case for FEC encoder.

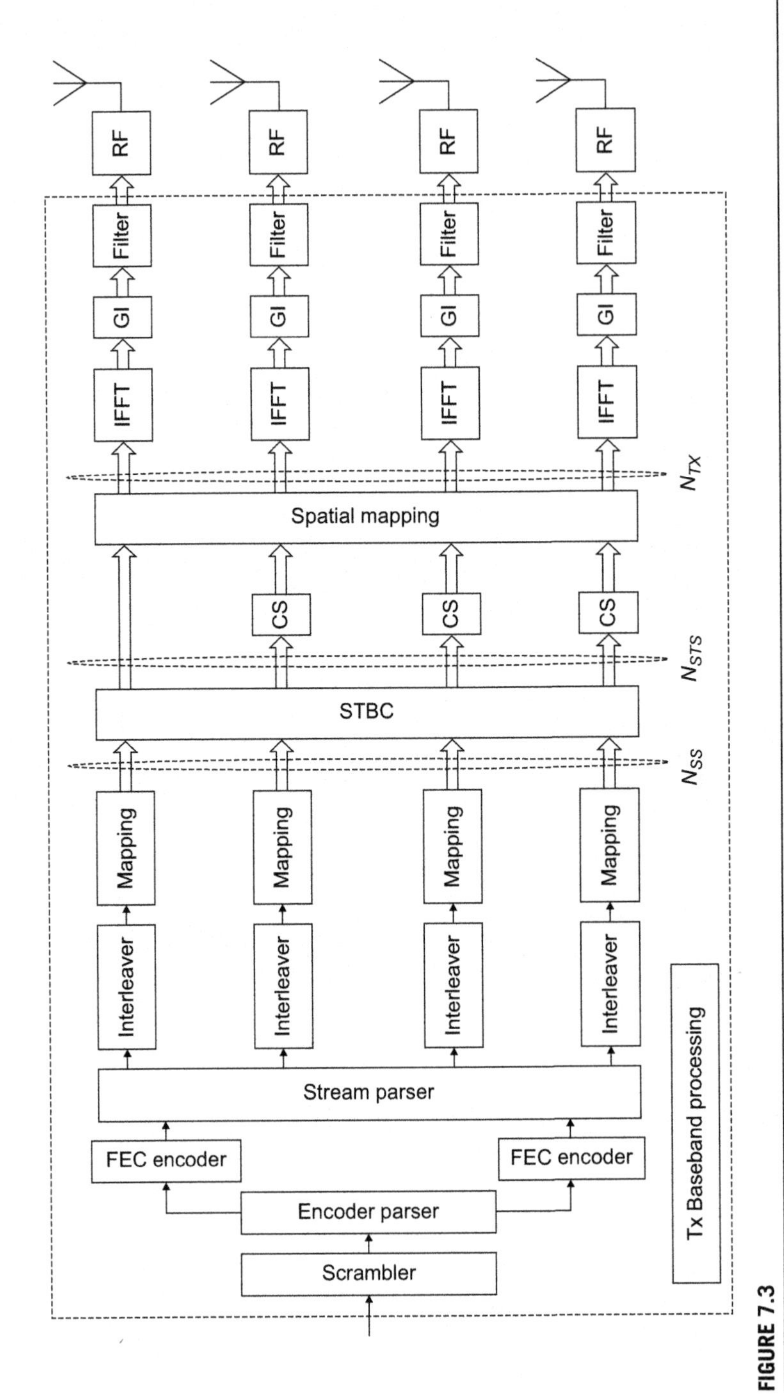

FIGURE 7.3
General architecture of IEEE 802.11n transmitter.

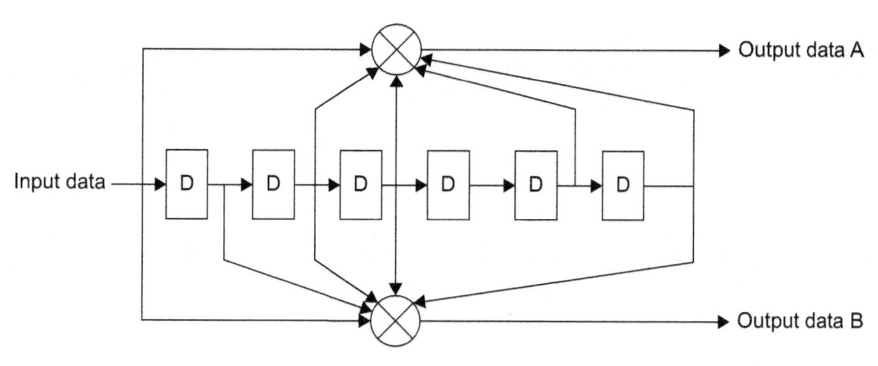

FIGURE 7.4

IEEE 802.11a/n rate-1/2 convolutional encoder with $g_1 = (133)_8$ and $g_2 = (171)_8$.

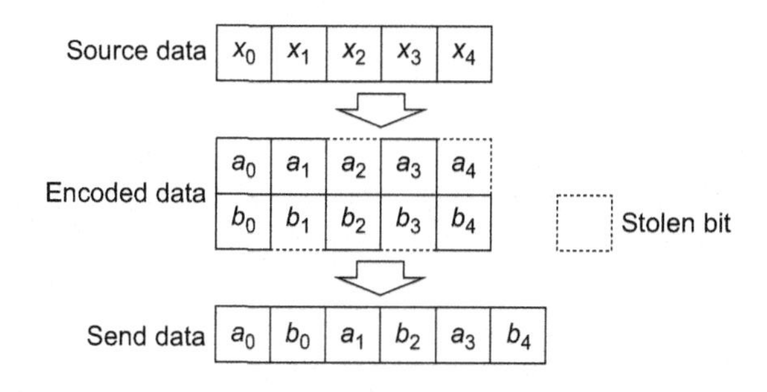

FIGURE 7.5

New puncturing mode for the coding rate 5/6.

The convolutional encoder uses the industry-standard generator polynomials with $g_1 = (133)_8$ and $g_2 = (171)_8$ with rate 1/2 as defined in 17.3.5.5 of [IEEE Std 802.11a]. The architecture of this encoder is shown in Figure 7.4, and a new puncturing mode is introduced for the coding rate 5/6 as shown in Figure 7.5.

Stream Parser

The stream parser will be active when there is more than one spatial stream. Let i_{SS} denote the spatial stream index, the consecutive blocks of $s(i_{SS})$ bits are assigned to different spatial streams in a round-robin way. The function $s(i_{SS})$ is defined as $s(i_{SS}) = \max(N_{BPSCS}(i_{SS})/2, 1)$, where $N_{BPSCS}(i_{SS})$ is the number of coded bits per single carrier of spatial stream i_{SS}. An example illustrating the stream parser is given in Figure 7.6.

Interleaver

According to 17.3.5.6 of [IEEE Std 802.11a] and 20.3.10.7.3 of [IEEE P802.11n], the interleaver works on block-by-block: for L-SIG and HT-SIG, one block is equal to 24 bits where the modulation is BPSK; for Data, the binary stream of each spatial stream

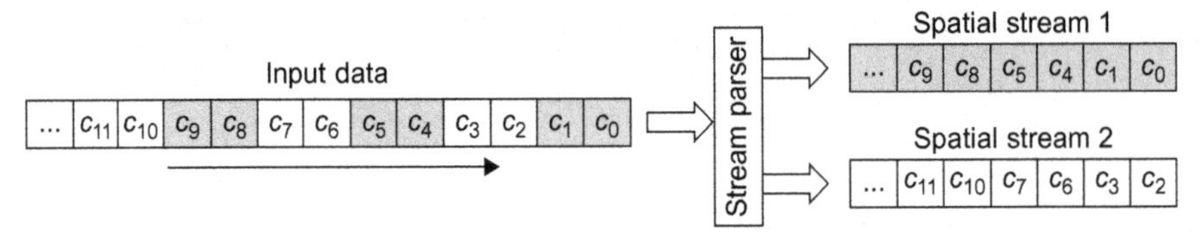

FIGURE 7.6

Stream parser operation: $N_{SS} = 2$ and 16-QAM modulation is applied on each spatial stream that $s(i_{SS}) = 2$.

are segmented in blocks of $N_{CBPSS} = 52N_{BPSCS}$[6]. The new interleaving work is realized in three steps: (a) the first permutation uses a classic rectangular interleaver; (b) the second cyclic permutation alternates the bits in the same subcarrier for 16-QAM and 64-QAM; and (c) the third permutation is a frequency rotation for $N_{SS} > 1$.

As the operation in legacy mode, the first and second permutations are accomplished with a register table of size $13 \times 4N_{BPSCS}$[7]. The third permutation processing includes the frequency rotation' which is realized by outputting the interleaved binary sequence from different bit index. This operation can minimize the impact of frequency selective fading in MIMO transmission since deep fading can occur on the same frequency index for all the spatial streams. An example of $N_{SS} = 2$ with 16-QAM is given in the following figures: The interleaving table dimension is 13×16; the input bits are written row-by-row and the second permutation for 1-bit rotation is applied to the columns 2, 4, 6, 8, 10 and 12, as shown in Figure 7.7; the output bits are read column-by-column where the bits stream is read from c_0 for $i_{SS} = 1$ and from c_{122} for $i_{SS} = 2$, as shown in Figure 7.8.

STBC

When STBC is required, STBC module will encode the N_{SS} spatial streams into N_{STS} space-time streams. The famous Alamouti code which makes full use of the transmit diversity in the configuration 2Tx – 1Tx [Ala98] is chosen in 802.11n standard as an option. The coding procedure is very simple: Let s_1 and s_2 denote the 2 information symbols to send, we send on the first space-time stream consecutively s_1 and s_2 while we send in parallel $-s_2^*$ and s_1^*. The code word can be expressed by:

$$\begin{bmatrix} s_1 & s_2 \\ -s_2^* & s_1^* \end{bmatrix} \tag{7.1}$$

For simple carrier transmission, the received signal over MISO channel is given by:

$$\begin{bmatrix} y_1 \\ y_2 \end{bmatrix} = [h_1 \quad h_2] \begin{bmatrix} s_1 & s_2 \\ -s_2^* & s_1^* \end{bmatrix} + \begin{bmatrix} W_1 \\ W_2 \end{bmatrix} \tag{7.2}$$

[6] N_{CBPSS} is the number of coded bits per OFDM symbol per spatial stream.
[7] In IEEE 802.11a, the table size is $16 \times 3N_{BPSC}$.

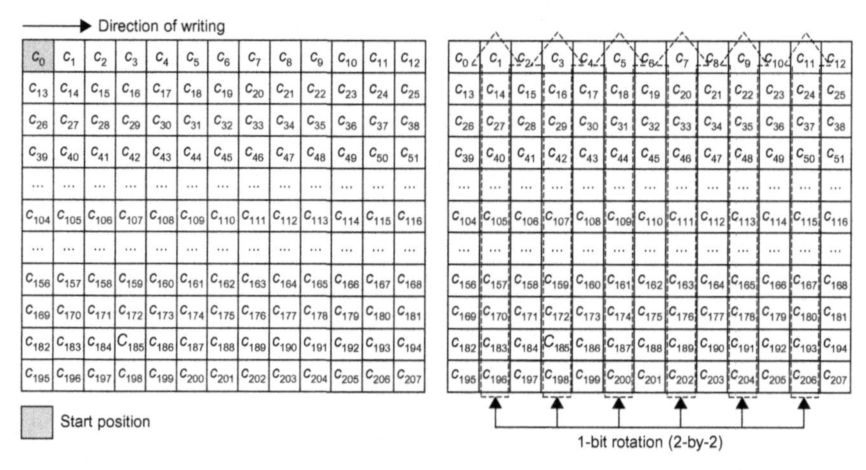

FIGURE 7.7

Bits interleaving for $N_{SS} = 2$ and 16-QAM: The left figure shows the writing in the table 13×16 and the right figure shows the 1-bit rotation.

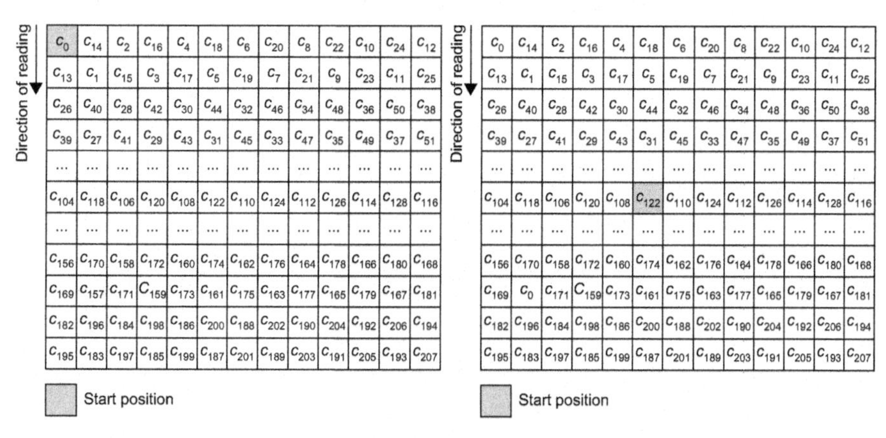

FIGURE 7.8

Bits interleaving of $N_{SS} = 2$ and 16-QAM: Left figures shows the reading for the first spatial stream and right figure shows the reading of the second spatial stream.

The w_1 and w_2 denote the Independent and Identically-Distributed (i.i.d.) Additive White Gaussian Noise (AWGN) terms on receive antennas. The system can be given in an alternative way by:

$$\begin{bmatrix} y_1 \\ y_2^* \end{bmatrix} = \begin{bmatrix} h_{11} & -h_{12} \\ h_{12}^* & h_{11}^* \end{bmatrix} \begin{bmatrix} s_1 \\ s_2^* \end{bmatrix} + \begin{bmatrix} w_1 \\ w_2^* \end{bmatrix} \tag{7.3}$$

The information symbols can be detected using Maximum-Ratio Combining (MRC)[8] as follows:

$$\begin{cases} \hat{s}_1 = \dfrac{1}{|h_1|^2 + |h_2|^2} \left(h_1^* y_1 + h_2 y_2^*\right) \\ \hat{s}_2 = \dfrac{1}{|h_1|^2 + |h_2|^2} \left(-h_2 y_1^* + h_1^* y_2\right) \end{cases} \tag{7.4}$$

Cyclic Shift Delay

The new mechanism of cyclic shift delay means to delay the space-time streams with different time reference. This technique can prevent from unintentional beamforming when the same signal is sent through different space-time streams. In [IEEE P802.11n], the cyclic shift duration is defined in Table n61 for the non-HT portion of packet, including L-STF, L-LTF, L-SIG and HT-SIG, and in Table n62 for the HT portion of packet, including HT-STF, HT-LTF and HT data. The cyclic shift can be viewed as an optimization of the MIMO communication channel and this processing is transparent to the receiver.

It is worthy to notice that in case of non-HT portion, the cyclic shift is applied directly to each transmit chain while in case of HT portion the cyclic shift is applied to each space-time stream. This implies the cyclic shift processing should be taken in frequency domain for the HT portion, in general with the spatial mapping processing. An exception is when the spatial mapping is in mode *direct mapping* that the cyclic shift can be done in the time domain.

Spatial Mapping

The spatial mapping processing maps the N_{STS} space-time streams to N_{TX} transmit chains by multiplying the spatial mapping matrix Q of dimension $N_{TX} \times N_{STS}$. This includes mainly three types of utilization: the *direct mapping*, where the spatial mapping matrix Q is an identity matrix or cyclic shift matrix; the *spatial expansion*, where the spatial mapping matrix is the product of the cyclic shift delay matrix and a spatial expansion matrix with orthogonal columns; the *beamforming steering*, where Q is a matrix which improves the transmission quality.

7.3 IEEE 802.11n RECEIVER PART

In the recommendation of IEEE 802.11n, we have a clear description of the transmitter behaviors including the processing from bits stream to transmit chain. However, there is not a standard describing the receiver implementation. It is evident that

[8]The channel matrix is orthonormal, then the maximum-likelihood performance can be obtained by *zero-forcing* (ZF) detection which is equivalent to MRC processing.

the system performance depends on the algorithms used in the receiver and the solution is not unique. We emphasize that this receiver as well as the algorithms in this section are designed and implemented for the COMSIS[9] C61000 series 802.11n System-on-Chip (SoC) IP core. The base band signal is sampled at 20 MHz.

The receiver's performance in MIMO context is discussed in the following aspects: signal detection, timing recovery (temporal synchronization), Carrier Frequency Offset (CFO) estimation, MIMO channel coefficients estimation, phase tracking and MIMO decoder. Without loss of generality, we investigate the MIMO configuration with two transmit (TX) antennas and two receive (RX) antennas, which includes SIMO (*1Tx-2Rx*), the MISO (*2Tx-1Rx*) and MIMO (*2Tx-2Rx*). The results of the other multiple-antenna cases can be derived in the same way.

7.3.1 Signal Detection

The signal detection step consists in evaluating the property of the received signal after the stabilization of AGC[10] when received signal achieves the power threshold, which is recommended in [IEEE Std 802.11a]. This signal detection algorithm is based on the legacy preamble L-STF field which is periodical of 16 samples, by examining the periodicity of 64 samples which correspond to the duration of 4 L-STFs.

Let $\{y_i^{(i_{RX})}\}, i \geq 0$ denote the samples received from the i_{RX} receive chain after the AGC training. Generally speaking, the RF switching from AC coupling to DC coupling will introduce an additional DC offset[11] to the baseband signal. Since the detection algorithm is based on the signal's autocorrelation function and intercorrelation function, the presence of DC offset may eventually degrade the performance of detection. Therefore, the baseband signal is filtered by a differential filter before entering the signal detection block.

Let $\{x_0^{(i_{RX})}, \ldots, x_{63}^{(i_{RX})}\}$ denote the 64 samples which are the inputs to the signal detection block. Then we have:

$$x_i^{(i_{RX})} = y_{i+1+t_0}^{(i_{RX})} - y_{i+t_0}^{(i_{RX})}, \tag{7.5}$$

where t_0 indicates the beginning of sliding windows, which is 0 when we start the detection. It is obvious that the filtered signal keeps the same periodicity as the input signal and that the DC offset is removed.

Let $\bar{x}_I^{(i_{RX})} = \left[x_0^{(i_{RX})}, \ldots, x_{31}^{(i_{RX})}\right]$ and $\bar{x}_{II}^{(i_{RX})} = \left[x_{32}^{(i_{RX})}, \ldots, x_{63}^{(i_{RX})}\right]$. For each antenna, we calculate the squared norm of $\bar{x}_I^{(i_{RX})}$ and $\bar{x}_{II}^{(i_{RX})}$, noted by $A_I^{(i_{RX})}$ and $A_{II}^{(i_{RX})}$, as well

[9]COMSIS is a fabless semiconductor company headquartered in Paris. It provides customized Wi-Fi solution.

[10]Identical gain is used for all the receive chains during one frame.

[11]This DC offset is antenna-dependent that we perform the SISO algorithm on each receive chain to remove this DC offset.

as their scalar product $I^{(i_{RX})}$:

$$A_I^{(i_{RX})} = \left\langle \bar{x}_I^{(i_{RX})}, \bar{x}_I^{(i_{RX})} \right\rangle = \sum_{i=0}^{31} \left| x_i^{(i_{RX})} \right|^2$$
$$A_{II}^{(i_{RX})} = \left\langle \bar{x}_{II}^{(i_{RX})}, \bar{x}_{II}^{(i_{RX})} \right\rangle = \sum_{i=32}^{63} \left| x_i^{(i_{RX})} \right|^2 \qquad (7.6)$$
$$I^{(i_{RX})} = \left\langle \bar{x}_I^{(i_{RX})}, \bar{x}_{II}^{(i_{RX})} \right\rangle = \sum_{i=0}^{31} x_i^{(i_{RX})} x_{i+32}^{(i_{RX})^*}$$

The signal periodicity is detected when:

$$\left| \sum_{i_{RX}=1}^{N_{RX}} I^{(i_{RX})} \right| > \beta \sum_{i_{RX}=1}^{N_{RX}} \left(A_I^{(i_{RX})} + A_{II}^{(i_{RX})} \right) \qquad (7.7)$$

where β depends on the implementation.

This signal detection maintains until the signal is validated for a maximal duration of 10 L-STFs. When the periodicity detection fails, the receiver will start the IEEE 802.11b signal detection since the IEEE 802.11b preamble is much longer than the OFDM case. By using the sliding-window method, this algorithm can be easily implemented in hardware level.

The MIMO technique can efficiently improve the signal detection performance. In Figure 7.9, nondetection rate ϵ_{det} is simulated in function of Signal-to-Noise Ratio (SNR) with HIPERLAN channel model A for typical office environment with NonLine-of-Sight (NLOS) conditions and channel model D for large open space with *Line-of-Sight* (LOS) conditions [MS05B]. The transmission paths are supposed independent. In the NLOS case, the simulation results reveal that the MIMO technique can effectively improve the signal detection performance: for $\epsilon_{\text{det}} = 10^{-3}$, a system gain about 5 dB is observed in the 2×2 configuration and about 3 dB is observed in the SIMO and MISO cases. In the LOS case, since the Ricean factor is $K = 10$ dB for the direct path, the fading phenomenon is less severe than the NLOS case. Therefore, the advantage of MIMO technique is less important than in NLOS case.

7.3.2 Timing Recovery

The time reference of the received frame is acquired by the timing recovery function. The following timing recovery algorithm uses the matched filters of the last 4 legacy short training symbols and the 16 first samples of L-LTF to detect the beginning of L-LTF.

Let $s_{L-STF}(n)$ denote the L-STF and $s_{L-LTF}(n)$ denote the first 16 samples of the L-LTF guard interval (GI2). The digital matched filters $h_{L-STF}(n)$ and $h_{L-LTF}(n)$ are

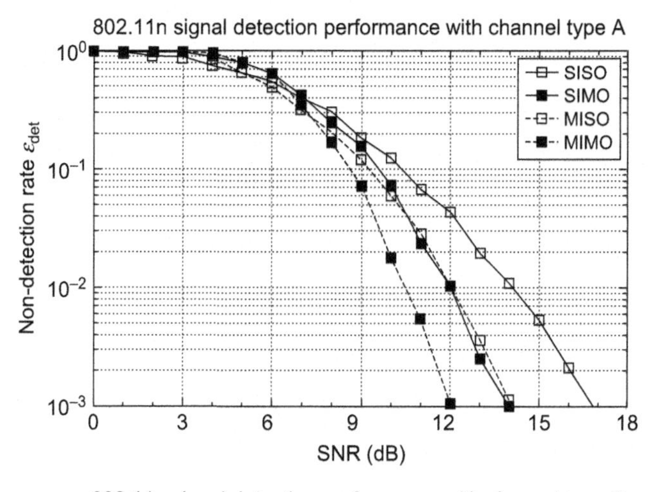

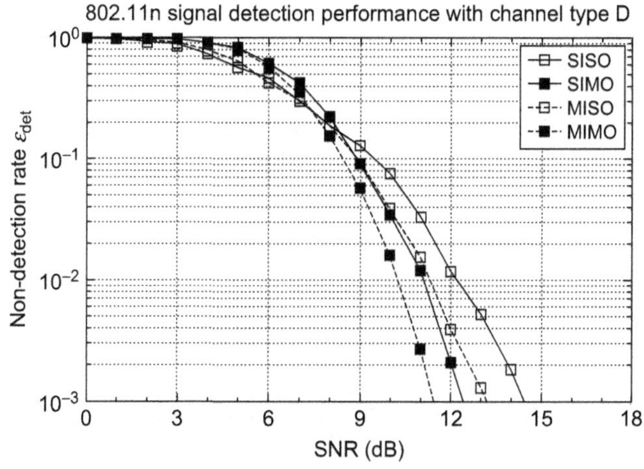

FIGURE 7.9

Signal detection performance with channel model A and channel model D: smaller ϵ_{det} means better performance.

given by:

$$\begin{cases} h_{L-STF}(n) = s^*_{L-STF}(63-n), & \text{for} \quad 0 \leq n < 64 \\ h_{L-LTF}(n) = s^*_{L-LTF}(15-n), & \text{for} \quad 0 \leq n < 16 \end{cases} \tag{7.8}$$

Let $y_{i_{RX}}(n)$ denote the received signal from i_{RX} receive chain where n is the time index. The received signal is filtered by the L-STF matched filter and the L-LTF matched filter respectively.

$$\begin{cases} x_{i_{RX}}^{(L-STF)}(n) = y_{i_{RX}}(n) * h_{L-STF}(n) = \sum_{t=0}^{63} y_{i_{RX}}(n-t)\, s_{L-STF}^{*}(t) \\ x_{i_{RX}}^{(L-LTF)}(n) = y_{i_{RX}}(n) * h_{L-LTF}(n) = \sum_{t=0}^{15} y_{i_{RX}}(n-t)\, s_{L-LTF}^{*}(t) \end{cases} \tag{7.9}$$

The powers of $x_{i_{RX}}^{(L-STF)}$ and $x_{i_{RX}}^{(L-LTF)}$, denoted respectively by $E_{i_{RX}}^{(L-STF)}$ and $E_{i_{RX}}^{(L-LTF)}$, are then defined by:

$$\begin{cases} E_{L-STF}^{(i_{RX})}(n) = \left| x_{i_{RX}}^{(L-STF)}(n) \right|^2 \\ E_{L-LTF}^{(i_{RX})}(n) = \left| x_{i_{RX}}^{(L-LTF)}(n) \right|^2 \end{cases} \tag{7.10}$$

For multiple receive chains, the sum of all the $E_{i_{RX}}^{(L-STF)}$ is then calculated as well as the sum of $E_{i_{RX}}^{(L-LTF)}$:

$$\begin{cases} E_{L-STF}(n) = \sum_{i_{RX}=1}^{N_{RX}} E_{i_{RX}}^{(L-STF)}(n) \\ E_{L-LTF}(n) = \sum_{i_{RX}=1}^{N_{RX}} E_{i_{RX}}^{(L-LTF)}(n) \end{cases} \tag{7.11}$$

The energy is given by the accumulation of E_{L-STF} and E_{L-LTF} for duration $M = 6$ that[12]:

$$\begin{cases} S_{L-STF}(n) = \sum_{m=0}^{M-1} E_{L-STF}(n-m) \\ S_{L-LTF}(n) = \sum_{m=0}^{M-1} E_{L-LTF}(n-m) \end{cases} \tag{7.12}$$

The sum defined in (7.13) of these accumulations is examined:

$$\begin{aligned} S(n) = S_{L-LTF}(n) + S_{L-STF}(n-16) + S_{L-STF}(n-32) \\ + S_{L-STF}(n-48) + S_{L-STF}(n-64) \end{aligned} \tag{7.13}$$

The synchronization algorithm for the detection of L-LTF is based on the observation of $S(n)$: during each 16 samples, the maximum $S_{\max}$ and the sample index $n_{\max}$ are memorized as well as the previous pair, noted by $S'_{\max}$ and $n'_{\max}$, an illustration is

[12]The configuration of $M = 6$ is corresponding to the accumulation time of 300 ns, which is a trade-off between the system latency and the channel response duration.

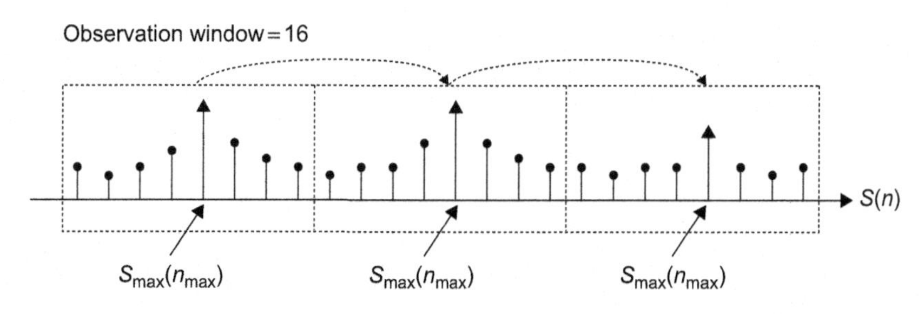

FIGURE 7.10

Observation of energy evolution for the detection of $(S_{\max}, n_{\max})$ and $(S'_{\max}, n'_{\max})$.

shown in Figure 7.10; the sample index $n'_{\max}$ will be taken as the beginning of L-LTF when the following condition is met:

$$S_{\max} < \rho \cdot S'_{\max}, \tag{7.14}$$

where ρ is a constant which is recommended to be 0.85 in the implementation.

This algorithm works by examining the peaks of energy since the autocorrelation function of L-STF and L-LTF as well as their intercorrelation function are nearly *Dirac* functions. This good property allows us to detect the waveform of L-STF and L-LTF by evaluating the energy of the filtered signal: the result of filtering the legacy preamble is given in Figure 7.11 to show this property.

For the multipath channels, the channel response's duration is essential to the design of the timing recovery function. The HIPERLAN channel model A is with 50 ns root mean square (rms) delay spread and the channel model D is with 140 ns rms delay spread [WE04]. In both cases, the channel's response duration is considered smaller than the accumulation time which is 300 ns for $M = 6$[13]. An example of the channel response for the channel model A is given in Figure 7.12. We apply the proposed algorithm with this channel response. The SAs Acoarse time synchronization can be given by examining the drop of the maximum $S_{\max}$ while precise time synchronization is provided by $n_{\max}$: The $S_{\max}$-based time recovery method with the channel response shown in Figure 7.12 is given in Figure 7.13.

Before the synchronization performance, let us introduce the *perfect synchronization time* which is denoted by T_{perf}. Let T_0 be the start of transmission at transmitter's side and $h(i_{TX}, i_{RX}, n)$ be the channel response from transmit antenna i_{TX} to receive antenna i_{RX}, with consideration of the cyclic shift, then T_{perf} is given by:

$$T_{perf} = T_0 + T_{opt} \tag{7.15}$$

[13] The maximal WLAN channel delay spread is achieved with the channel model E, which is 250 ns in rms.

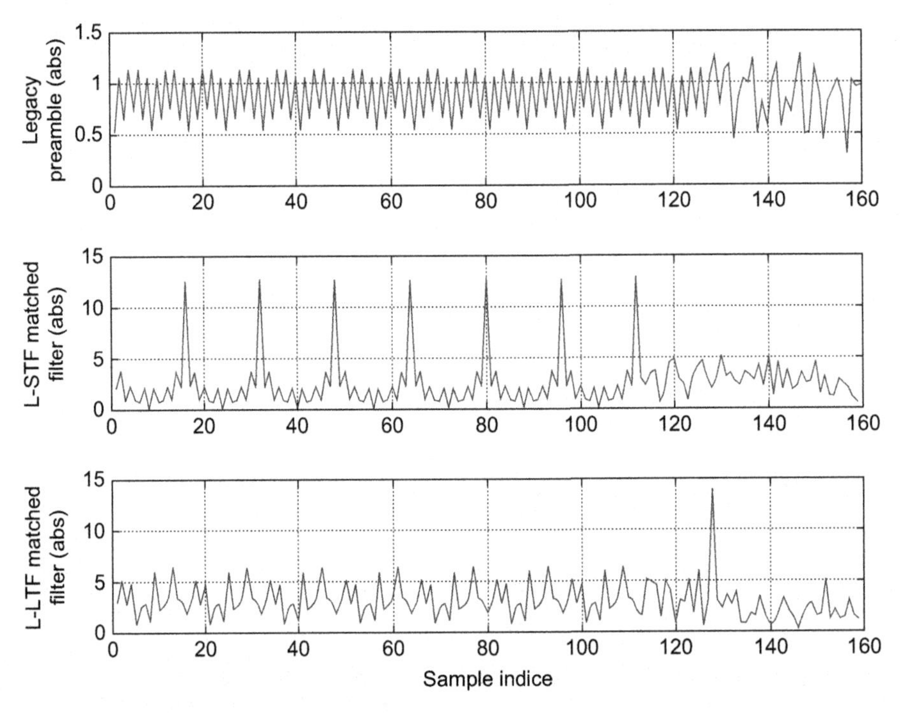

FIGURE 7.11

Filtering the legacy IEEE 802.11a preamble with L-STF and L-LTF matched filters: The peaks of L-STF and L-LTF matched filters are observed when the signal is synchronized. The L-STF filter peak is periodic of 16 samples and it disappears when filtering the GI2 of L-LTF. At the same time, the L-LTF filter peak appears.

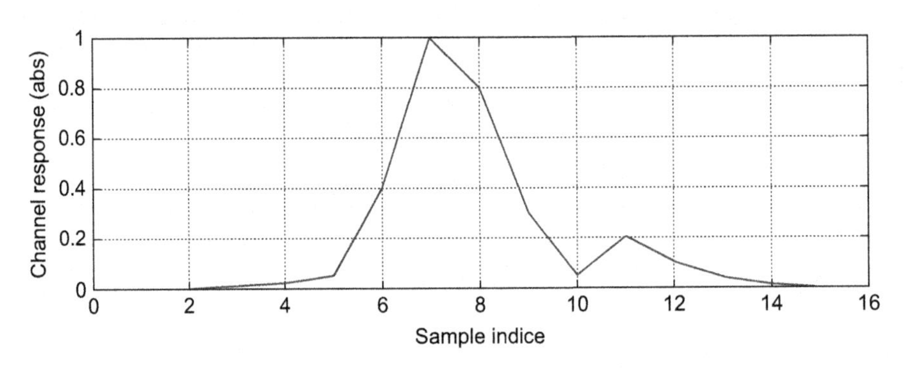

FIGURE 7.12

An example of channel response: The channel response is concentrated from index 7 to 12.

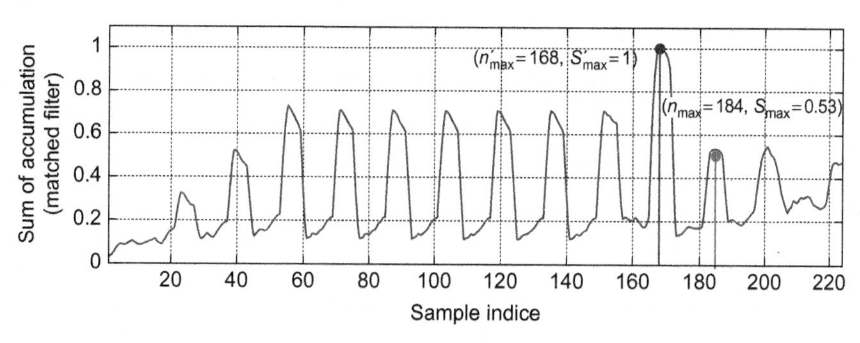

FIGURE 7.13

Detection of $S_{\max}$ and $n_{\max}$: (1) the received frame starts at index 0 and the channel response is shown in the above figure without noise; (2) the total duration of L-STF field is 160 samples; (3) the peak marked by $(n_{\max}, S_{\max})$ is corresponding the point satisfying the condition in equation (7.14) and the peak marked by $(n'_{\max}, S'_{\max})$ indicates the synchronized beginning of L-LTF.

where T_{opt} is defined as:

$$T_{opt} = \arg\max_{T} \sum_{i_{RX}=1}^{N_{RX}} \sum_{n=0}^{15} \left| \sum_{i_{TX}}^{N_{TX}} h(i_{TX}, i_{RX}, T+n) \right|^2 \tag{7.16}$$

The T_{opt} is the moment when the total energy of channel response accumulated over the GI is maximized with full channel information. This moment is optimal in terms of the Inter-Symbol Interference (ISI). However, it is necessary to point out that T_{opt} is not the same for non-HT portion and HT portion due to different cyclic shift processing. We take into account only the T_{opt} of the non-HT portion to analyze the timing recovery function's performance.

For the simulation results, the performance is evaluated by the *nonsynchronization rate*, noted by ϵ_{sync}, where we consider a frame is not synchronized temporally when the gap between the T_{perf} and the synchronized time index T_{sync} exceeds 8 samples which is the half duration of GI. The simulation results for both channel model A and channel model D are given in Figure 7.14.

Let us first see the performance in the SISO and SIMO case: The multiantenna array configuration at the receiver side efficiently improves the timing recovery performance so that the gain is about 5 dB $\sim$ 5.5 dB for both channel model A and channel model D. Analyzing the performance with multiple-antennas at the transmitter's side is a little slightly more difficult: In MISO case, the synchronization performance is worse than the one in SISO case and the same behaviour is observed for the SIMO and MIMO case. This is essentially because the channel response is delayed by the related cyclic shift so that the channel response seen at each receive

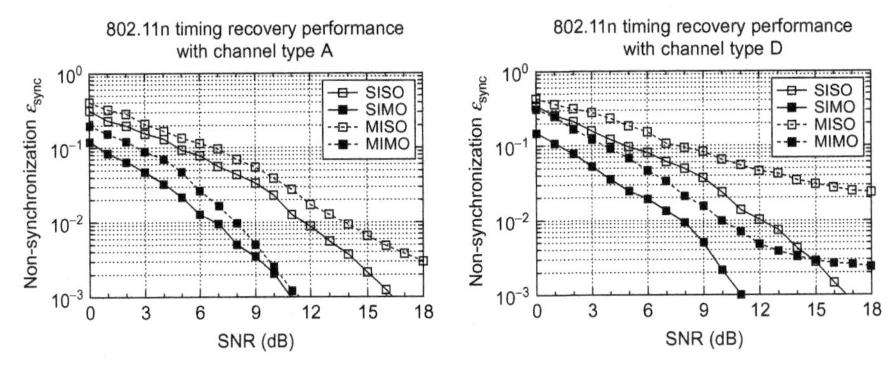

FIGURE 7.14

Timing recovery function performance with channel model A and channel model D: the frame is considered nonsynchronized when the synchronization time is outside of interval $[T_{perf} - 8, T_{perf} + 8]$.

antenna is spread. The extended length of the channel response brings some difficulties to the present timing recovery function. The synchronization function is based on the observation of $S_{\max}$ by assuming the channel delay spread is small, however the cyclic shift processing will delay the channel response. For non-HT portion, the maximum cyclic shift delay is 200 ns. In case of the channel model A, the rms delay spread is 50 ns, which is trivial so that the cyclic shift impact is not less significant than in case of the channel model D where the channel rms delay spread is 140 ns. Thus, the configuration of the multiple transmitter antennas will introduce difficulties for the timing recovery function for the environments with long channel response.

Briefly speaking, the timing recovery function in MISO/MIMO performance is degraded compared with the SISO/SIMO configuration. It is true that the performance can be improved by enlarging the accumulation window or by using more adaptive algorithms, but in general these solutions cost more latency or calculation complexity. At the same time, the channel model plays another important role in the performance: In case of indoor environment where the channel response is short, the impact of MISO/MIMO is less important than in case of long channel response, which is associated with typical outdoor environment.

It is worthy to notice that the measurement is based on the temporal difference. The case considered nonsynchronized does not mean a total loss of time information but more ISI. The impact of ISI on the transmission performance is hard to give since it depends on the other parameters of transmission.

7.3.3 Carrier Frequency Offset (CFO) Estimation

Like all the OFDM systems, the IEEE 802.11a/n is very sensitive to the carrier frequency offset (CFO) which is typically caused by the mismatch of the transmitter and receiver oscillators. In order to ensure the orthogonality of the subcarriers, accurate frequency synchronization must be performed before processing in frequency domain. This task involves the CFO estimation and the pre-FFT phase correction. The CFO is estimated by using L-STF and L-LTF.

Let $s(n)$ denote the transmitted signal, then the received signal is expressed by:

$$r_{i_{RX}}(n) = e^{j\frac{2\pi\Delta f_c n}{T_s}} \sum_{i_{TX}=1}^{N_{TX}} h(i_{TX}, i_{RX}, n)^* s(n) + w_{i_{RX}}(n), \tag{7.17}$$

where $h(i_{TX}, i_{RX}, n)$ denotes channel response from i_{TX} transmit antenna to i_{RX} receive antenna and $w_{i_{RX}}(n)$ denotes the noise received at i_{RX} receive antenna terms which is i.i.d AWGN. The operation [*] denotes the time domain discrete convolution. T_S is the sampling time which is 50 ns. The CFO is represented by $\Delta f_c = f_c^{(RX)} - f_c^{(TX)}$.

Instead of estimating Δf_c, the CFO can be measured in using the phase-shift per sample which is defined by:

$$\phi = 2\pi\Delta f_c / T_S \tag{7.18}$$

First estimation result is based on the L-STF, which is periodic with a 16 samples period. By using the last 4 legacy short training symbols, which present 64 samples, the L-STF based CFO estimation is given by:

$$\phi_{L-STF} = \frac{1}{32}\arg\left[\sum_{i_{RX}=1}^{N_{RX}} \sum_{t=0}^{31} r_{i_{RX}}(t_0 - t - 1)\, r_{i_{RX}}^*(t_0 - 33 - t)\right], \tag{7.19}$$

where t_0 is the time index of the first sample of L-LTF and the L-LTF field is phase-corrected by ϕ_{L-STF}:

$$\bar{r}_{i_{RX}}(n - t_0) = e^{j(n-t_0)\phi_{L-STF}}\, r_{i_{RX}}(n - t_0), \quad \text{for} \quad t_0 \le n < t_0 + 160 \tag{7.20}$$

The phase-corrected L-LTF is used to give the non-HT channel estimation while the second CFO estimation based on the noncorrected L-LTF is taken at the same time. By using the L-LTF field, which is periodic of 64 samples, the second estimation is expressed by:

$$\phi_{L-LTF} = \frac{1}{64}\arg\left[\sum_{i_{RX}=1}^{N_{RX}} \sum_{t=0}^{63} r_{i_{RX}}(t_0 + 159 - t)\, r_{i_{RX}}^*(t_0 + 95 - t)\right] \tag{7.21}$$

The new L-LTF based estimation result is combined with φ_{L-STF} in order to give a more accurate CFO estimation:

$$\widehat{\phi} = \alpha\phi_{L-STF} + (1-\alpha)\,\phi_{L-LTF} \tag{7.22}$$

where α is the weighting factor which is defined in the implementation.

This result is applied in the CFO correction for the rest of the frame[14] that we have:

$$\bar{r}_{i_{RX}}(n - t_0) = e^{j(n-t_0)\widehat{\phi}}\, r_{i_{RX}}(n - t_0), \quad \text{for} \quad n > t_0 + 160 \tag{7.23}$$

[14]The HT-STF can also be employed in the CFO estimation procedure by using the same estimation algorithm as L-STF.

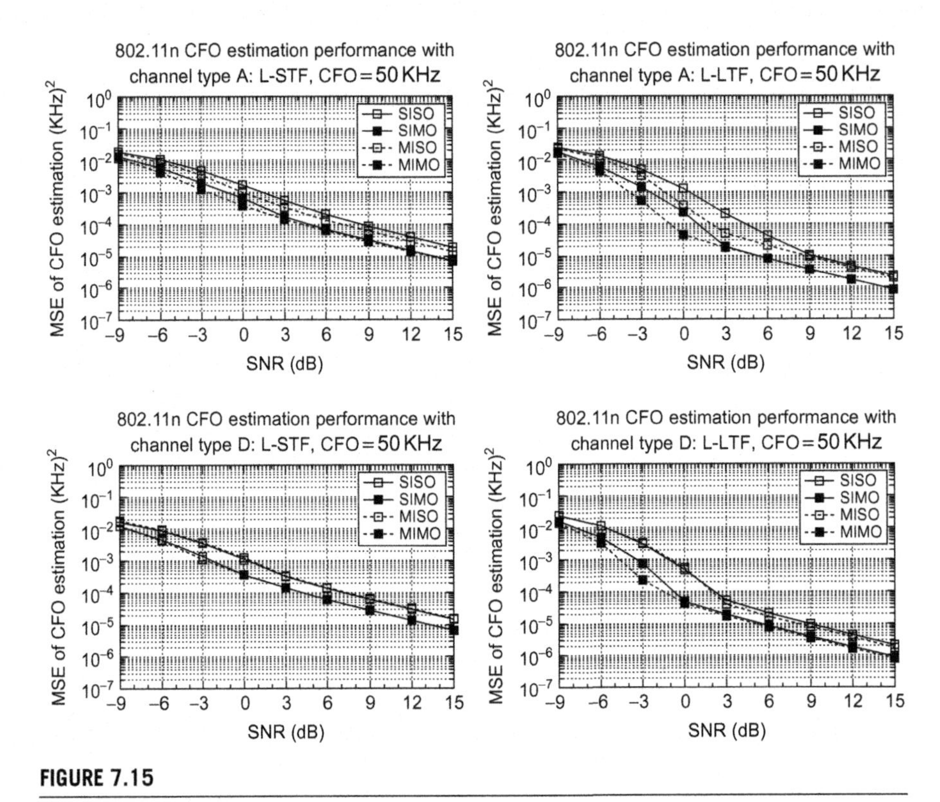

FIGURE 7.15

L-STF and L-LTF based CFO estimation: the CFO is fixed to 50 KHz and the estimation performance is reported in MSE of the results based on L-STF and L-LTF, respectively.

The residual CFO phase error and Sampling Frequency Offset (SFO) are tracked during the reception of data by using the pilots, which will be detailed in Section 7.3.5. The estimation of CFO can be optimized provided the information of SFO and DC offset[15]. However, since the impact of SFO is much smaller than CFO, we will not consider the SFO impact during one or two OFDM symbols for the sake of simplicity of hardware.

The performance in IEEE 802.11n context is illustrated in Figure 7.15. We assume a perfect timing recovery for the CFO estimation function and the performance is reported in Mean Squared Error (MSE) in all configurations. For the L-STF based estimation, the performance can be improved by the use of multiple receive antennas in case SIMO and MIMO while the multiple transmit antennas have less importance on the performance since there is nearly no gain of MISO system. For the L-LTF

[15]In the implementation, the DC offset is also estimated with L-STF. Since this impact is trivial to the result of CFO estimation and it is independent from the MIMO technique, it is not presented in this article for the simplicity.

based estimation, both SIMO and MIMO show the great advantages to SISO or MISO system especially in the SNR region.

7.3.4 Channel Coefficients Estimation

The channel matrix estimation is handled in two parts: For the non-HT portion, the L-LTF is used to give the legacy channel estimation. We neglect the residual CFO error in this section and the received signal during L-LTF time is given by:

$$r_{i_{RX}}(n) = \sum_{i_{TX}=1}^{N_{TX}} h(i_{TX}, i_{RX}, n)^* s(n) + w_{i_{RX}}(n), \quad \text{for} \quad t_0 + T_{\mathrm{GI2}} \le n < t_0 + 2T_{\mathrm{OFDM}} \tag{7.24}$$

where T_{GI2} is the length of guard interval of L-LTF field and T_{OFDM} is the length of one OFDM symbol which is 64 in this case.

We take into account the average of received two L-LTFs for the legacy channel estimation that:

$$r_{i_{RX}}(n) = \frac{1}{2}\left[r_{i_{RX}}(n - t_0 - T_{\mathrm{GI2}}) + r_j(n - t_0 - T_{\mathrm{GI2}} + T_{\mathrm{SYM}})\right], \quad \text{for} \quad 0 \le n < T_{\mathrm{SYM}} \tag{7.25}$$

Let $X(k)$ be the DFT of $x(n)$ and $M_u^{(\text{non-HT})}$ denote the set of all used sub-carriers in non-HT portion, then we have:

$$R_{i_{RX}}(k) = \sum_{i_{TX}=1}^{N_{TX}} H_{i_{RX}}(k) S_{L-LTF}(k) + W_{i_{RX}}(k), \quad \text{for} \quad k \in M_u^{(\text{non-HT})} \tag{7.26}$$

where $W_{i_{RX}}(k)$ is the frequency domain noise term.

It is remarkable that the long training symbols are normalized in frequency domain, $S_{L-LTF}(k) \in \{-1, +1\}$, so the legacy channel coefficients are given by:

$$\widehat{H}_{i_{RX}}(k) = R_{i_{RX}}(k) S_{L-LTF}(k) \tag{7.27}$$

In MIMO case, the HT-LTF field is used to estimate the MIMO channel matrix. Without loss of generality, we discuss the case when the spatial mapping matrix is identity matrix[16] and the number of HT-LTF is equal to the number of space-time stream and the number of transmit antennas: $N_{HTLTF} = N_{STS} = N_{TX}$. Let $M_u^{(\text{HT})}$ denote the set of all used sub-carriers in HT portion and we have:

$$\mathbf{R}_{N_{RX} \times N_{STS}}(k) = \mathbf{H}_{N_{RX} \times N_{STS}}(k) \mathbf{S}_{N_{STS} \times N_{STS}}^{HT-LTF}(k) + \mathbf{W}_{N_{RX} \times N_{STS}}(k) \quad \text{for} \quad k \in M_u^{(\text{HT})} \tag{7.28}$$

[16]The spatial mapping processing is transparent to the channel estimation.

With

$$\begin{cases} R_{N_{RX}\times N_{STS}}(k) = \begin{bmatrix} R_{1,1}(k) & \cdots & R_{1,N_{STS}}(k) \\ \vdots & \ddots & \vdots \\ R_{N_{RX},1}(k) & \cdots & R_{N_{RX},N_{STS}}(k) \end{bmatrix} \\ H_{N_{RX}\times N_{STS}}(k) = \begin{bmatrix} H_{1,1}(k) & \cdots & H_{1,N_{STS}}(k) \\ \vdots & \ddots & \vdots \\ H_{N_{RX},1}(k) & \cdots & H_{N_{RX},N_{STS}}(k) \end{bmatrix} \\ S_{N_{STS}\times N_{STS}}(k) = \begin{bmatrix} S_{1,1}(k) & \cdots & S_{1,N_{STS}}(k) \\ \vdots & \ddots & \vdots \\ S_{N_{STS},1}(k) & \cdots & S_{N_{STS},N_{STS}}(k) \end{bmatrix} \\ W_{N_{RX}\times N_{STS}}(k) = \begin{bmatrix} W_{1,1}(k) & \cdots & W_{1,N_{STS}}(k) \\ \vdots & \ddots & \vdots \\ W_{N_R,1}(k) & \cdots & W_{N_R,N_{STS}}(k) \end{bmatrix} \end{cases} \tag{7.29}$$

According to the generation of HT-LTF, $\mathbf{S}^{HT-LTF}_{N_{STS}\times N_{STS}}(k)$ can be written as:

$$\mathbf{S}^{HT-LTF}_{N_{STS}\times N_{STS}}(k) = S_{HT-LTF}(k)\,\mathbf{P}_{N_{STS}\times N_{STS}} \tag{7.30}$$

where $S_{HT-LTF}(k) \in \{-1,+1\}$ and $\mathbf{P}_{N_{STS}\times N_{STS}}$ is the first N_{STS} rows and first N_{STS} columns of the Hadamard matrix:

$$\mathbf{P}^{Hadamard}_{4\times 4} = \begin{bmatrix} 1 & 1 & 1 & 1 \\ 1 & -1 & 1 & -1 \\ 1 & 1 & -1 & -1 \\ 1 & -1 & -1 & 1 \end{bmatrix} \tag{7.31}$$

Therefore, we have:

$$\mathbf{P}_{N_{STS}\times N_{STS}}\mathbf{P}^T_{N_{STS}\times N_{STS}} = N_{STS}\mathbf{I}_{N_{STS}\times N_{STS}} \tag{7.32}$$

The estimation of $\mathbf{H}(k)$ is given by:

$$\begin{aligned} \hat{\mathbf{H}}_{N_{RX}\times N_{STS}} &= \frac{S_{HT-LTF}(k)}{N_{STS}}\mathbf{R}_{N_{RX}\times N_{STS}}(k)\,\mathbf{P}^T_{N_{STS}\times N_{STS}} \\ &= \mathbf{H}_{N_{RX}\times N_{STS}} + \frac{S_{HT-LTF}(k)}{N_{STS}}\mathbf{W}_{N_{RX}\times N_{STS}}(k)\,\mathbf{P}^T_{N_{STS}\times N_{STS}} \end{aligned} \tag{7.33}$$

We notice that the noise term remains i.i.d AWGN for each estimated channel coefficient so the estimation performance is the same as the one in SISO configuration.

7.3.5 Fractional Time Delay (FTD) Estimation and Post-FFT Phase Tracking

The Sampling Frequency Offset (SFO) leads to phase error when the sampling time is sufficiently long. The precision of local oscillator is about 20 ppm, this offset is not as significant as the CFO problem; however when the frame is long enough, the accumulation of time offset will cause an important time shift of sampling, namely Fractional Time Delay (FTD) which turns to be a phase error for the coherent detection.

Let us denote the FTD by Δt. The FTD is accumulated during the reception and it turns to be a phase error in frequency domain in function of the reception time. In the proposed receiver architecture, both the estimation and correction of this time shift is performed in frequency domain in a post-FFT manner.

The SFO estimation starts from the first OFDM data symbol and uses the pilots mapped in corresponding subcarrier positions. Let $y^{(ref)}_{iRX,n}(t)$ denote the nth OFDM symbol observed at i^{th}_{RX} antenna[17] without sampling offset which corresponds to $Y^{(ref)}_{iRX,n}(f)$ in frequency domain. Supposing a time offset Δt_n occurs, the observed OFDM signal can be presented by $y_{iRX,n}(t, \Delta t_n) = y^{(ref)}_{iRX,n}(t + \Delta t_n)$. After the FFT processing, the frequency domain signal could be written as:

$$Y_{iRX,n}(f, \Delta t_n) = DFT\left(y_{iRX,n}(t, \Delta t_n)\right) = e^{2\pi f \Delta t_n} Y^{(ref)}_{iRX,n}(f) \tag{7.34}$$

In the meanwhile, the residual carrier frequency offset can co-exist with the sampling frequency offset. The phase error due to CFO is mainly estimated and compensated as explained in Section 7.3.3. However, the residual frequency error accumulates and for long frame transmission, it harms both the QAM detection and FTD estimation. In order to simplify the post-FFT phase correction, this CFO residual phase error is modeled as a constant $\Delta\phi_{n,0}$ during one OFDM symbol by:

$$y_{iRX,n}\left(t, \Delta t_n, \Delta\phi_{n,0}\right) = e^{j\Delta\phi_{n,0}} y_{iRX,n}(t, \Delta t_n) \tag{7.35}$$

In frequency domain, the signal is expressed by:

$$Y_{iRX,n}\left(f, \Delta t_n, \Delta\phi_{n,0}\right) = \mathrm{DFT}\left(e^{j\Delta\phi_{n,0}} y_{iRX,n}(t, \Delta t_0)\right) = e^{j(2\pi \Delta t_n f + \Delta\phi_{n,0})} Y^{(ref)}_{iRX,n}(f) \tag{7.36}$$

Let $\Delta\phi_n(f) = 2\pi f \Delta t_n + \Delta\phi_{n,0}$ denote the total phase error in function of f. It can be illustrated in the phase-error/frequency plan, as shown in Figure 7.16.

In fact, with respect to the channel estimation results, $\Delta\phi_{n,0}$ is a function of the frequency offset $\Delta f_c = \Delta f_c - \Delta f_c$ which could be expressed by:

$$\Delta\phi_{n,0} \approx 2\pi n T_{OFDM} \Delta f_c \tag{7.37}$$

Let $f_e^{(ref)}$ denote the reference sampling rate and $f_e^{(local)}$ denote the local sampling rate, then the sampling offset is defined by $\Delta f_e = f_e^{(local)} - f_e^{(ref)}$. The interval of sampling is therefore noted by $T_e^{(ref)} = 1/f_e^{(ref)}$ and $T_e^{(local)} = 1/f_e^{(local)}$ by:

$$\Delta T_e = T_e^{(ref)} - T_e^{(local)} \approx -T_e^{(ref)}\left(\Delta f_e / f_e^{(ref)}\right) \tag{7.38}$$

[17] We consider the beginning of GI as the reference time.

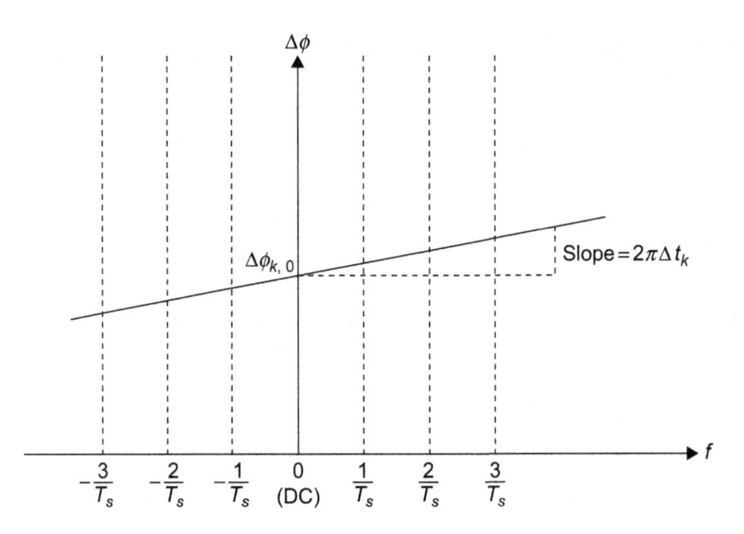

FIGURE 7.16

Phase-error in frequency domain: This phase error is a linear function of f where the slope is equal to $2\pi t_k$ and $\Delta\phi_{k,0}$ is the phase-intercept.

For the nth data symbol, the time error is expressed by:

$$\Delta t_n = n\left(N_{FFT} + N_{GI}\right)\Delta T_e = -nT_{OFDM}\Delta f_e / f_e^{(ref)} \tag{7.39}$$

It is easy to see that the total phase error is linear in time. To estimate this phase error, the pilots embedded in the data OFDM symbols are compared with the pilots in the preamble in order to track the phase variation.

By (7.37) and (7.39), the phase error at each pilot subcarrier is:

$$\Delta\phi_{pilot}(n,k) = -2\pi nT_{OFDM}\left(f_k\Delta f_e / f_e^{(ref)} + \Delta f_c\right), \quad \text{for} \quad k \in M_p \tag{7.40}$$

where M_p denotes the set of pilot subcarriers' index.

Let $\bar{P}_{n,i_{RX},k}$ denote the received pilot of nth OFDM data symbol at i_{RX} antenna at frequency index k. It can be expressed by:

$$\bar{P}_{n,i_{RX},k} = e^{\Delta\phi_{pilote}(n,k)} \sum_{i_{STS}=1}^{N_{STS}} H_{f_k,i_{RX},i_{STS}} P_{i_{STS},k}^{(n)} + W_{n,i_{RX},k} \tag{7.41}$$

Where $H_{f_k,i_{RX},i_{STS}}$ denotes the i_{RX}-row i_{STS}-column entry of channel matrix on pilot subcarrier f_k and W_{n,i_{RX},f_k} represents the complex AWGN term that $w_{n,i_{RX},f_k} \sim CN(0,N_0)$.

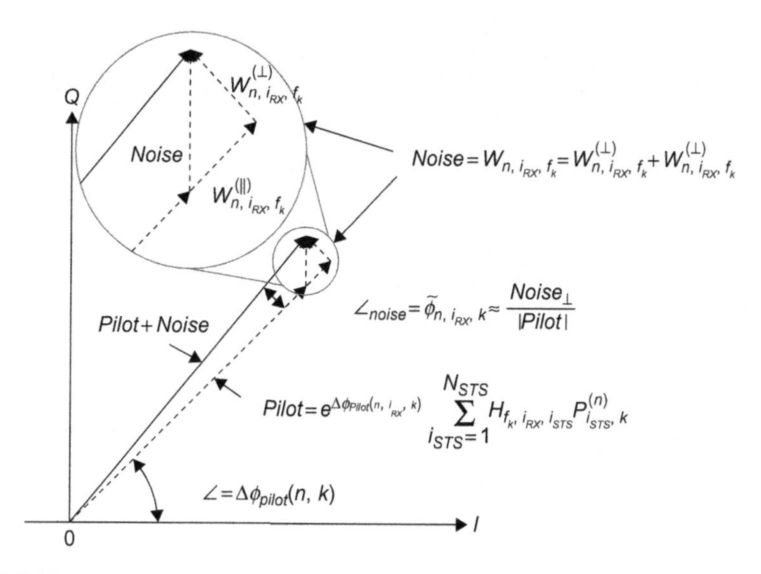

FIGURE 7.17

Phase noise.

The phase error observed at receive antenna i_{RX} can be modeled by:

$$\begin{aligned}\Delta\bar{\phi}_{pilot}(n,i_{RX},k) &= \Delta\phi_{pilot}(n,k)+\bar{\phi}_{n,i_{RX},k}\\ &= -2\pi nT_{OFDM}\left(f_k\Delta f_e/f_e^{(ref)}+\Delta f_c\right)+\bar{\phi}_{n,i_{RX},k} \end{aligned} \tag{7.42}$$

Here, $\bar{\varphi}_{n,i_{RX},k}$ is the phase estimation error which could be considered as an AWGN noise that $\bar{\phi}_{n,i_{RX},k} \sim N\left(0,\sigma^2_{n,i_{RX},k}\right)$. The AWGN property of this estimation error can be explained with Figure 7.17 where the complex noise term W_{n,i_{RX},f_k} is decomposed into $W^{(\perp)}_{n,i_{RX},f_k}$ and $W^{(//)}_{n,i_{RX},f_k}$: only $W^{(\perp)}_{n,i_{RX},f_k}$ introduces the phase estimation error.

Since $W_{n,i_{RX},k} \sim CN(0,N_0)$, we have $W^{(\perp)}_{n,i_{RX},f_k} \sim N(0,N_0/2)$. The phase estimation error can then be approximated by $\bar{\phi}_{n,i_{RX},k} \approx Noise_{\perp}/|Pilot|$, which is AWGN that $\bar{\phi}_{n,i_{RX},k} \sim N\left(0,\sigma^2_{n,i_{RX},k}\right)$ with:

$$\sigma^2_{n,i_{RX},k} = \frac{N_0}{2}\left|\sum_{i_{STS}=1}^{N_{STS}} H_{k,i_{RX},i_{STS}}P^{(n)}_{i_{STS},k}\right|^2 \tag{7.43}$$

For the nth OFDM symbol, the phase error is given by the average of the phase errors based on the observation of last L data OFDMs:

$$\Delta\phi_{n,k} = \frac{N}{N-L}arg\left[\sum_{i_{RX}=1}^{N_{RX}}\sum_{q=1}^{L}\bar{P}^*_{i_{RX},k,n-L+q}\bar{P}_{i_{RX},k,q}\right], \quad \text{for} \quad k\in M_p \tag{7.44}$$

The parameter L depends on the system configuration. The average pilots' phase errors are then used to estimate the residual CFO phase error $\Delta\varphi_{0,n}$ and FTD Δt_n:

$$\Delta\hat{\phi}_{n,k} = \frac{1}{|M_p|}\sum_{k\in M_p}\Delta\phi_{n,k}$$
$$\text{slope}_n = \frac{1}{\sum_{k\in M_p} k^2}\sum_{k\in M_p} k\Delta\phi_{n,k} \tag{7.45}$$

where $|M_p|$ denotes the number of elements in set M_p which is 4 in the 20 MHz bandwidth.

We remark that the variance of the estimation noise $\sigma^2_{n,i_{RX},k}$ is a function of the symbol index n, the receive antenna index i_{RX} and the pilot frequency index k. The direct summing result is a sub-optimal solution in order to reduce the system complexity. More accurate results could be obtained when taking consideration of this fact[18].

Besides, the noise variance $\sigma^2_{n,i_{STS},k}$ is periodic with respect to n since the pattern of $\{P^{(n)}_{i_{STS},k}\}$ is periodic for all pilot subcarriers, $P^{(n)}_{i_{STS},k} = P^{(n+N_{period})}_{i_{STS},k}$.

Let take the case of 20 MHz and $N_{STS} = 2$, for example. The transmitted pilot pattern on sub-carrier $k = -21$ is given by:

$$\left\{\begin{bmatrix} P_{n,1,k=-21} \\ P_{n,2,k=-21} \end{bmatrix}\right\} = \left\{\begin{bmatrix} 1 \\ 1 \end{bmatrix}_{n=1}, \begin{bmatrix} 1 \\ -1 \end{bmatrix}_{n=2}, \begin{bmatrix} -1 \\ -1 \end{bmatrix}_{n=3}, \begin{bmatrix} -1 \\ 1 \end{bmatrix}_{n=4}, \ldots\right\}, \quad \text{period} = 4 \tag{7.46}$$

The variance is periodic with period 2:

$$\left\{\begin{bmatrix} \sigma^2_{n,1,k=-21} \\ \sigma^2_{n,2,k=-21} \end{bmatrix}\right\} = \frac{N_0}{2}\left\{\begin{bmatrix} |H_{k=-21,1,1} + H_{k=-21,1,2}|^2 \\ |H_{k=-21,2,1} + H_{k=-21,2,2}|^2 \end{bmatrix}, \begin{bmatrix} |H_{k=-21,1,1} + H_{k=-21,1,2}|^2 \\ |H_{k=-21,2,1} + H_{k=-21,2,2}|^2 \end{bmatrix}, \ldots\right\}, \quad \text{period} = 2 \tag{7.47}$$

In practice, the length of the observation window is a multiple of the variance period and the phase error estimation results are updated periodically in order to avoid the different noise variance. The processing shown in equation (7.44) takes advantage of this periodicity by putting an observation window of length $2L_{\max}$ with $L_{\max} = 6$. When the number of receiving OFDM symbols is less than $2L_{\max}$, the estimation is taken on the actually received OFDM symbols when the index is a multiple of the 2 periods. This procedure is illustrated in Figure 7.18: the estimation of SFO and residue CFO is taken each 4 OFDM time symbols; when the number of

[18]When estimating the phase error, it is preferred to use a pre-calculated weighting function to refine the estimation result.

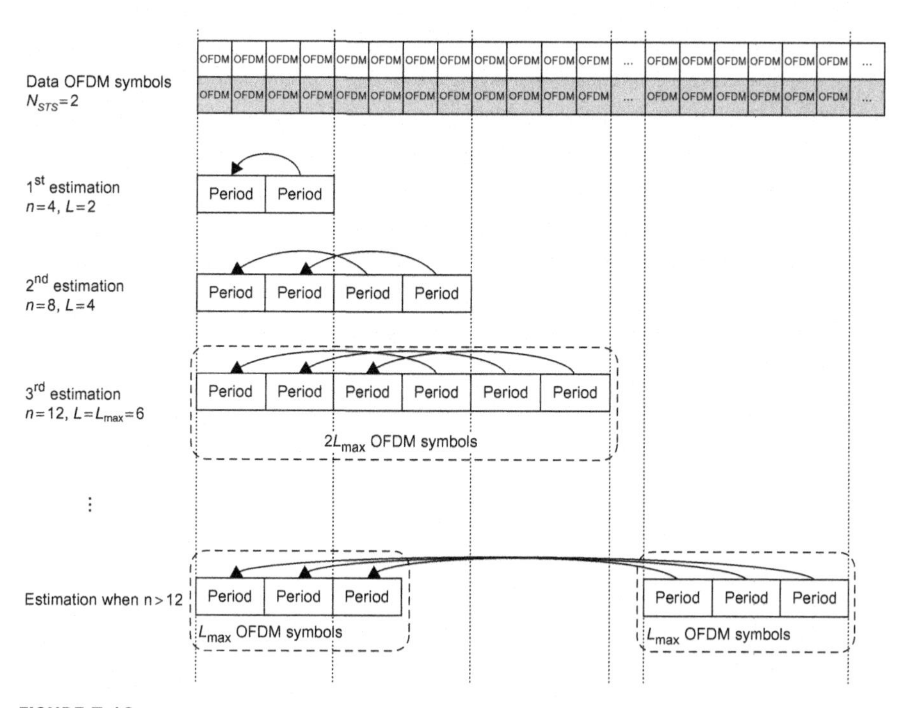

FIGURE 7.18

Example of phase tracking procedure with $N_{STS} = 2$ and $BW = 20$ MHz: The phase noise variance's period is 2 and the observation window's size is $2L_{\max} = 12$.

received symbols is greater than 12, the estimation is based on the comparison of the first received 6 OFDM symbols and the last received 6 OFDM symbols.

7.3.6 MIMO Decoder

The MIMO decoder is the key factor to the MIMO transmission performance, especially when multiple spatial streams are used for multiplexing gain. The implementation at hardware level can be very different from the soft methods due to the latency constraint or the circuit complexity. The architecture of the MIMO decoder as well as the selected algorithm can greatly vary according to the demand of transmission quality. Many researches are focusing on the Maximum-Likelihood (ML) or suboptimal algorithms in order to give a good trade-off between the performance and the complexity. We present in this section a VHDL-implementable 2×2 BLAST decoder which achieves ML performance. This decoder operated for both $N_{SS} = 1$ and $N_{SS} = 2$ cases with $N_{RX} = 2$. The general architecture is illustrated in Figure 7.19.

The data is processed by the "Y Projection & H Scalar Product" blocks, before being decoded. The controller block operates in two modes: For the data with $N_{SS} = 1$,

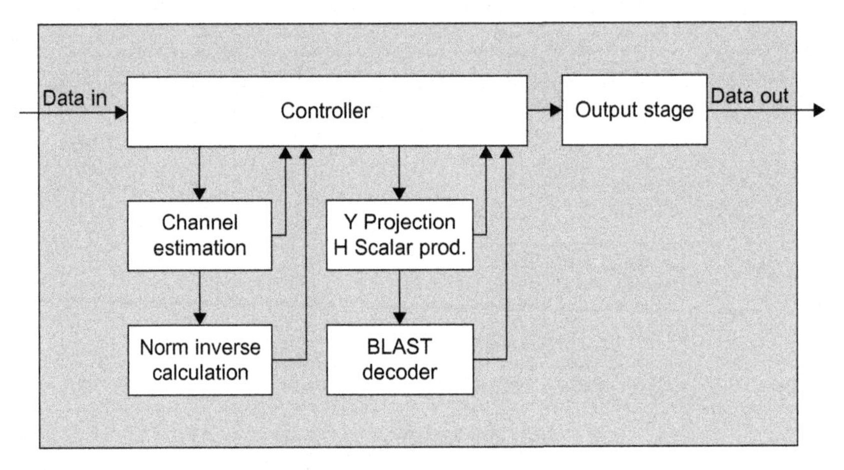

FIGURE 7.19

2 × 2 BLAST MIMO decoder general architecture.

it hard-decodes the received data via the Maximum-Ratio Combining (MRC) (inside the controller); for the data with $N_{SS} = 2$, the dedicated "BLAST Decoder" block will be used for ML decoding.

The channel estimation gives the channel matrix, denoted by H which is a complex matrix defined as follows:

For $N_{SS} = 1$,

$$H = \begin{bmatrix} h_{1,1} & 0 \\ h_{2,1} & 0 \end{bmatrix}$$

For $N_{SS} = 2$,

$$H = \begin{bmatrix} h_{11} & h_{12} \\ h_{21} & h_{22} \end{bmatrix}$$

Where $h_{i,j}$ is the complex channel coefficient at a given data subcarrier between the transmit antenna j and the receive antenna i.

The sequence "Data In" represents the stream of received words, denoted by Y, which is a complex vector of dimension 2:

$$Y = \begin{bmatrix} y_1 \\ y_2 \end{bmatrix}$$

Where y_i is the complex signal received by the receive antenna i.

The "Norm Inverse Calculation" block receives the estimated 2 or 4 channel coefficients. From the channel matrix, both column squared-norms N_1 and N_2 are calculated and compared. If N_1 is greater than N_2, then the columns are swapped.

$$\begin{cases} N_1 = |h_{11}|^2 + |h_{21}|^2 \\ N_2 = |h_{21}|^2 + |h_{22}|^2 \end{cases} \tag{7.48}$$

An output signal SWAP keeps track of whether the swap occurred or not: this signal is one bit variable defined as follow:

$$\mathrm{SWAP} = \begin{cases} 0 & \text{if} \quad N_2 > N_1 \\ 1 & \text{if} \quad N_2 > N_1 \end{cases} \tag{7.49}$$

The greater norm is then passed to the inverse calculator in order to get:

$$N_{\mathrm{max}}^{-1} = \frac{1}{N_{2-\mathrm{SWAP}}} \tag{7.50}$$

Both the norm inverse and the swap indicator are sent back to the controller.

The "Controller" receives the coding mode, which takes value from 0 to 3, indicates which QAM constellation is used for decoding and the value of N_{BPSC}, which is detailed in the Table 7.1.

This value is communicated with "Output Stage" and "BLAST Decoder" blocks.

The constellation normalization factor, denoted by K, depends on the coding mode. "Controller" block sends K and the inverse of the constellation normalization factor, denoted by K^{-1}, to the "Y Projection & H Scalar Product" block. K and K^{-1} are both real positive constants, which are provided according to Table 7.2.

This treatment simplifies the notation of Re(x) and Im(x) candidates. The possible values of Re(x) and Im(x) are shown in Table 7.3 as a function of coding mode/constellation.

The "Y Projection block & H Scalar Product" block use the following input signals: the received symbol Y, the corresponding channel coefficients H, the norm inverse N_{max}^{-1}, the inverse of the normalization factor K^{-1} and the SWAP parameter.

Table 7.1 Coding Mode and Constellation Correspondence

Coding Mode	**0**	**1**	**2**	**3**
Constellation	BPSK	QPSK	16-QAM	64-QAM
N_{BPSC}	1	2	4	6

Table 7.2 Coding Mode and Constellation Normalization Factor

Coding Mode	**0**	**1**	**2**	**3**
K	$8/7$	$2/\sqrt{2}$	$4/\sqrt{10}$	$8/\sqrt{42}$
K^{-1}	$7/8$	$\sqrt{2}/2$	$\sqrt{10}/4$	$\sqrt{42}/8$

Note: with this convention a QAM symbol s can be written $s = K.\, x$ with $-1 < \mathrm{Re}(x) < 1$ and $-1 < \mathrm{Im}(x) < 1$.

Table 7.3 Coding Mode and QAM Values

Coding Mode	Constellation	Re(x)	Im(x)
0	BPSK	$-7/8, 7/8$	0
1	QPSK	$-1/2, 1/2$	$-1/2, 1/2$
2	16-QAM	$-3/4, -1/4, 1/4, 3/4$	$-3/4, -1/4, 1/4, 3/4$
3	64-QAM	$-7/8, -5/8, -3/8, -1/8,$ $1/8, 3/8, 5/8, 7/8$	$-7/8, -5/8, -3/8, -1/8,$ $1/8, 3/8, 5/8, 7/8$

Three calculations are processed with this block:

1. z, the received symbol Y projected by the channel coefficients and corrected by the scaling factor K^{-1}. The calculation equation is described as below:

$$z = N_{\max}^{-1} \cdot K^{-1} \cdot \left[h_{1,2-swap}^{*} \quad h_{2,2-swap}^{*} \right] \cdot Y \tag{7.51}$$

2. P, the scalar product of the channel coefficient matrix. The calculation equation is:

$$P = N_{\max}^{-1} \cdot \left(h_{1,1+SWAP} \cdot h_{1,1-SWAP}^{*} + h_{2,1+SWAP} \cdot h_{2,2-SWAP}^{*} \right) \tag{7.52}$$

By construction, we have $|P| < 1$; thus $-1 < Re(P) < 1$ and $-1 < Re(P) < 1$.

3. H', the channel coefficients scaled by factor K:

$$H' = K \cdot \begin{bmatrix} h_{1,1+SWAP} & h_{1,2-SWAP} \\ h_{2,1+SWAP} & h_{2,2-SWAP} \end{bmatrix} \tag{7.53}$$

z is sent to both "Controller" for MRC decoding and the "BLAST Decoder" block, while P and H' are sent only to the "BLAST Decoder."

The "Blast Decoder" browses an entire space-time layer, deriving for each possible coordinate the closest corresponding coordinate in the second layer (by orthogonal projection). Then the Euclidean metric between received symbol and each computed point is calculated. By providing the candidate pair (x_1, x_2), this metric is defined as:

$$M(x_1, x_2) = \left| y_1 - h'_{1,1}x_1 - h'_{1,2}x_2 \right|^2 + \left| y_1 - h'_{2,1}x_1 - h'_{2,2}x_2 \right|^2 \tag{7.54}$$

The point featuring the smallest metric of all is elected to be the decoded codeword.

The output module outputs $N_{SS} \times N_{BPSC}$ bits. These bits match with the Gray Code of the N_{SS} QAM symbols decoded by the ML decoder. By convention, the first N_{BPSC} bits match with the QAM symbol of the first spatial stream.

The whole decoding algorithm is composed of 2 parts: the channel processing and the received word processing.

1. Channel processing (when channel coefficients are ready):
   Determine K and K^{-1};
   Compute N_1 and N_2;

Determine *SWAP*;
Compute N_{max}^{-1};
If $N_{SS} = 1$, compute P and H'.

2. Word processing (when data are ready)
Compute z;
If $N_{SS} = 1$, round Re(z) and Im(z) to get the closest QAM point x and output the binary Gray code of x;
If $N_{SS} = 2$, for all the possible value of x_1 in the QAM constellation:
Compute $t = z - P \cdot x_1$;
Round Re(z) and Im(z) to get the closest QAM point x_2;
Compute the metric $M(x_1, x_2)$;
Select the couple (x_1, x_2) which possesses the minimum metric;
If *SWAP*, swap x_1 and x_2;
Output the binary Gray code of x_1 and then the binary Gray code of x_2.

7.4 SIMULATION RESULTS

We give the performance of simulation in the IEEE 802.11n for the configurations SISO, SIMO, MISO and MIMO[19]. The number of spatial stream N_{SS} is 2 in MIMO case and 1, otherwise. The simulation is taken with hypothesis of perfect synchronization condition and perfect channel information. In order to give a good comparison, the transmission of the frame at length 1000 bytes is investigated for data throughput of 13 Mbps (MCS = 1 and MCS = 8), 26 Mbps (MCS = 3 and MCS = 9) and 52 Mbps (MCS = 5 and MCS = 11). Please see Table 20-29 and Table 20-30 of [IEEE P802.11n] for details of MCS definition. Both the Frame Error Rate (FER) and the efficient throughput are reported.

7.4.1 Without STBC

The simulation results without use of STBC are given in Figure 7.20 for data rate 13 Mbps, Figure 7.21 for data rate 26 Mbps and Figure 7.22 for data rate 52 Mbps.

For the SIMO system, the multiple receive antennas system shows significant gain in SNR thanks to the receive diversity gain in both indoor and outdoor environments. Compared with the SISO case, this gain is about 10.0 dB at FER $= 10^{-2}$ in all configurations.

In addition, we see the channel model D introduces more ISI so that there are obvious performance degradations in SISO and MISO systems for data rate 26 Mbps and 52 Mbps. A floor of FER about 5% is observed in case of data rate 26 Mbps as shown in Figure 7.20 for both SISO and MISO systems. The performance with channel model D becomes worse in case of data rate 52 Mbps that about 70% frames are lost for SISO system and 90% for MISO system. With multiple receive

[19] The antenna settings are 1Tx-1Rx, 1TX-2RX, 2Tx-1Rx and 2Tx-2Rx, respectively.

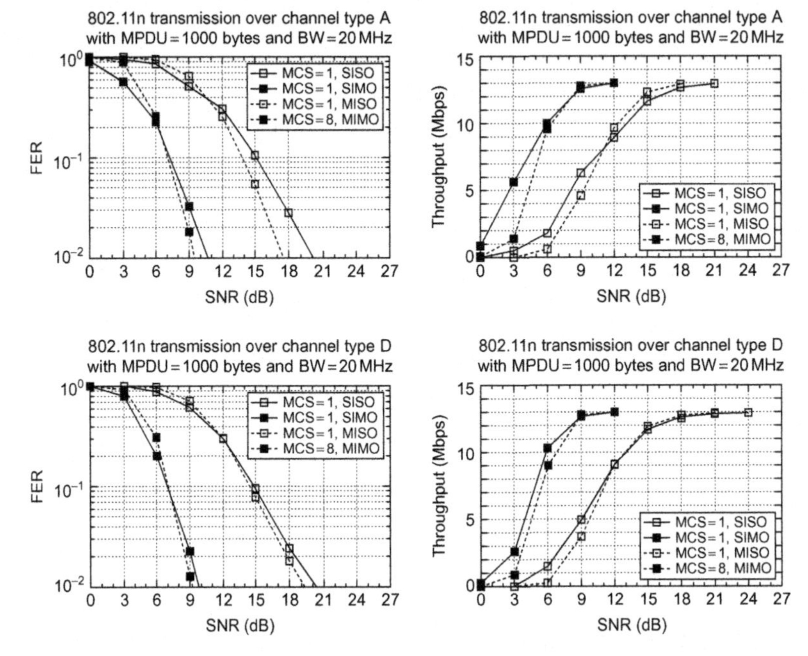

FIGURE 7.20

Simulation results of channel model A and channel model D with MPDU = 1000 bytes at 13 Mbps: For SISO, SIMO and MISO cases, the frame is sent with MCS = 1; for MIMO case, the frame is sent with MCS = 8.

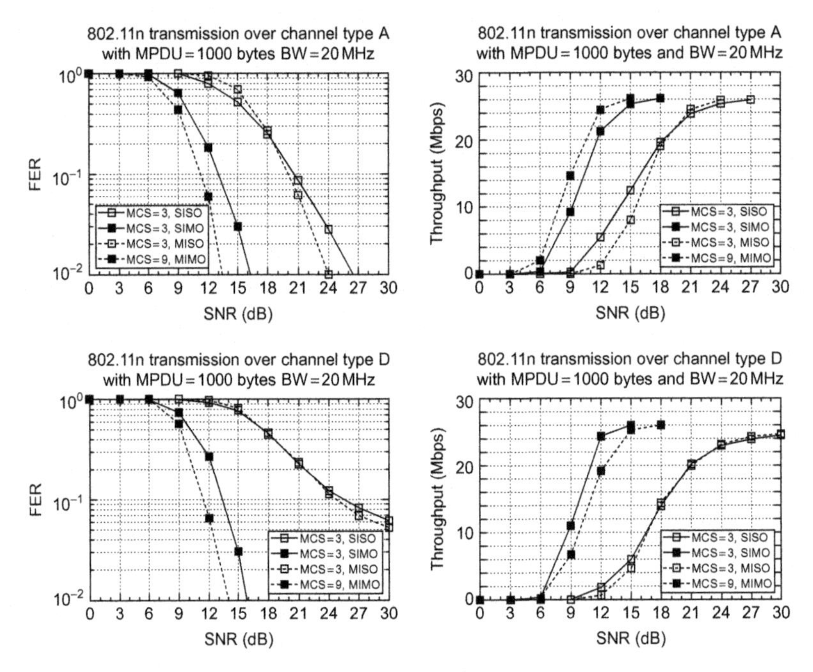

FIGURE 7.21

Simulation results of channel model A and model D with MPDU = 1000 bytes at 26 Mbps: for SISO, SIMO and MISO cases, the frame is sent with MCS = 3; for MIMO case, the frame is sent with MCS = 9.

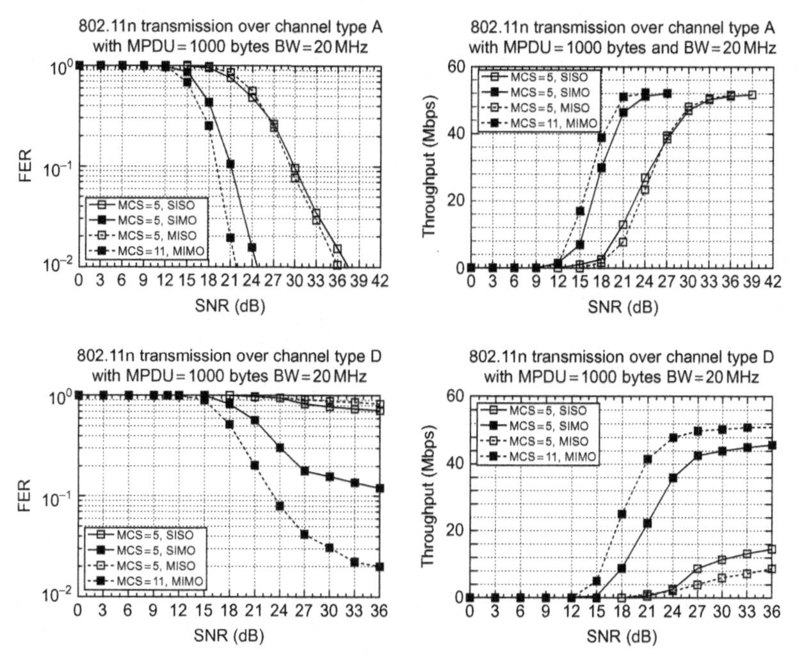

FIGURE 7.22

Simulation results of channel model A and model D with MPDU = 1000 bytes at 52 Mbps: for SISO, SIMO and MISO cases, the frame is sent with MCS = 5; for MIMO case, the frame is sent with MCS = 11.

antennas, like the SIMO and MIMO systems, the ISI impact can be diminished thus the throughput is efficiently improved.

In terms of throughput, the multiple receive antennas system brings great improvements. Let us compare the SISO and SIMO systems for example. We find in case of data rate 13 Mbps at SNR = 6 dB, the throughput gain can be up to 8 Mbps with channel model A and up to 9 Mbps with channel model D; in case of data rate 26 Mbps, the throughput gain is more than 14 Mbps at SNR = 12 dB with channel model A and more than 18 Mbps at SNR = 15 dB with channel model D; in case of data rate 52 Mbps, a throughput gain about 32 Mbps is reported at SNR = 21 dB with channel model A and at SNR = 27 dB with channel model D.

For the MISO system, the use of spatial expansion technique gives an important gain in typical indoor environment with channel model A. By comparing with SISO system performance at FER = 10^{-2}, we report a gain about 2.5 dB in cases of data rate 13 Mbps and 26 Mbps, and a gain about 1.5 dB in case of data rate 52 Mbps. However, in the outdoor environment with channel model D, this gain is less important since the propagation channel is approaching Ricean channel which limits the diversity gain of spatial expansion. In terms of throughput, MISO system fails to give considerable improvement and its performance with data rate 52 Mbps and channel model D is even worse than SISO system.

For the MIMO configuration, besides of the receive diversity gain, the *multiplexing gain* plays also an important role in the performance. The MIMO system allows the data to be transmitted with several spatial streams that in each spatial stream we can use a simpler constellation. For example, in case of 13 Mbps, we can use BPSK modulation (MCS = 1) instead of QPSK (MCS = 8) for a more robust transmission. Therefore, the MIMO system can outperform the SISO/SIMO/MISO systems. This advantage is more obvious with high data rate. Compared with SIMO system, in case of data rate 26 Mbps, MIMO system gives a gain about 2.5 dB with channel model A and 2.0 dB with channel model D at FER = 10^{-2}. In case of data rate 52 Mbps, MIMO system gives a gain about 3.0 dB with channel model A and it improves largely the throughput with channel model D: we report an improvement more than 16 Mbps at SNR = 21 dB.

7.4.2 With STBC

The use of STBC can greatly improve the performance of MISO system. As shown in Figure 7.23 (MCS = 1, 13 Mbps), Figure 7.24 (MCS = 3, 26 Mbps) and Figure 7.25

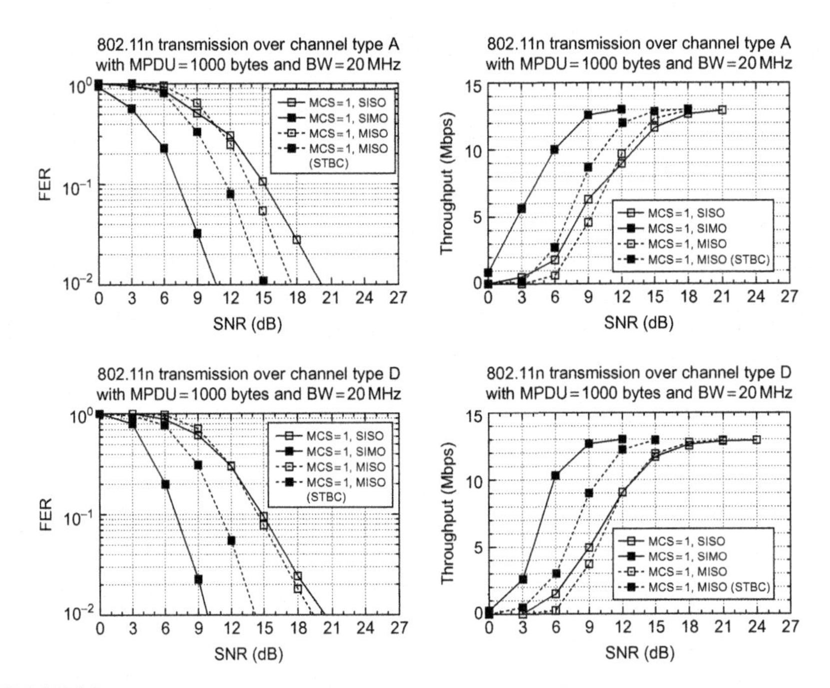

FIGURE 7.23

Simulation results of channel model A and channel model D with MPDU = 1000 bytes at 13 Mbps (MCS = 1).

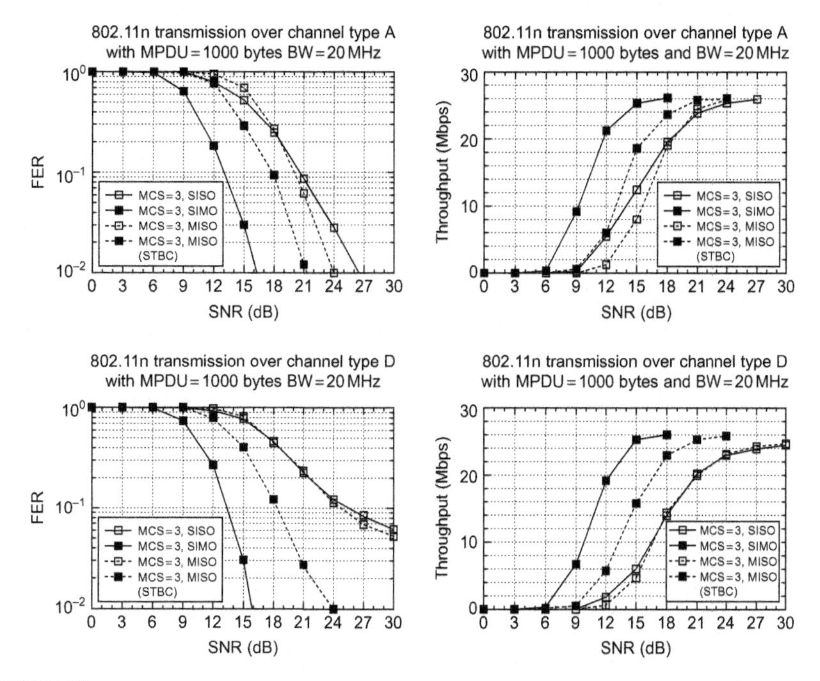

FIGURE 7.24

Simulation results over channel model A and model D with MPDU = 1000 bytes at 26 Mbps (MCS = 3).

(MCS = 5, 52 Mbps), we compare the performances of SISO system, SIMO system, MISO system without STBC and MISO system with STBC. The MISO system with STBC gives a performance between the SISO and SIMO performance. The use of STBC is much advantageous with channel model D where STBC exploits more efficiently the transmitter diversity.

Compared to the MISO system without STBC, at FER $= 10^{-2}$ the use of STBC for MCS = 1 achieves a coding gain about 2.5 dB with channel model A and 5.0 dB with channel model D.

For MCS = 3, STBC brings a coding gain of 2.5 dB at FER $= 10^{-2}$ is reported with channel model A and it gives an improvement in throughput about 10 Mbps with channel model D at SNR = 15 dB.

For MCS = 5, a coding gain of 6.0 dB is observed with channel model A. The impact of ISI with channel model D is slightly restrained such that the throughput is generally increased by $4 \sim 6$ Mbps.

The STBC can be applied in the network with asymmetric antenna configuration. An example is that the AP is equipped with 2 antennas while the users are equipped with only one. The uplink is therefore a SIMO system and the downlink is

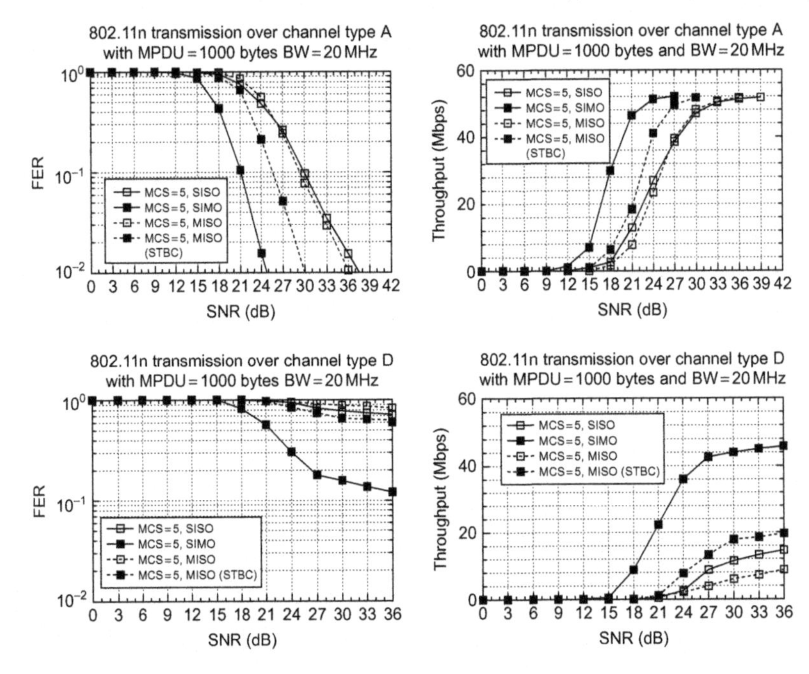

FIGURE 7.25

Simulation results over channel model A and model D with MPDU = 1000 bytes at 52 Mbps (MCS = 5).

a MISO system. This configuration with STBC support can improve the quality of transmission in both directions with very low complexity.

7.5 CONCLUSION

The simulation results show the great advantage of MIMO technology in the present Wi-Fi system. This technology protects the wireless system from the fading problem that the system can operate with either larger cover range or less transmission power. The multiple spatial stream technique allows the system to achieve higher data throughput within the same bandwidth, which appears to be good solution to meet the increasing demand of data transfer.

CHAPTER

WiMAX Implementation

8

Serdar Sezginer, Guillaume Vivier, Bertrand Muquet,
Ambroise Popper, and Amélie Duchesne

8.1 INTRODUCTION[1]

Multiple Input Multiple Output MIMO techniques have been incorporated in all of the recently developed wireless communications standards including IEEE 802.11n, IEEE 802.16e, and Long-Term Evolution (LTE). In this chapter, we focus on the schemes from current and future IEEE 802.16 specifications. First, we describe the schemes existing in current IEEE 802.16e specifications and, then, we discuss briefly the solutions proposed for IEEE 802.16m. Last, we provide the benefits of UL-MIMO together with a new method, namely, tile-switched diversity (TSD), which benefits from transmit diversity without any channel knowledge at the transmitter side.

8.2 EXISTING SCHEMES IN IEEE 802.16e

8.2.1 Description

IEEE 802.16e specifications [IEE09a] include several MIMO profiles for two, three, and four transmit antennas. They provide transmit diversity, Spatial Multiplexing (SM) or combine the advantages of both. Most of the MIMO schemes included in the IEEE 802.16e specifications are based on two schemes which are defined for two transmit antennas. The first one, called Matrix A in the specifications, is based on the Space-Time Block Code (STBC) proposed by Alamouti for transmit diversity [Ala98]. This code achieves a diversity order that is equal to twice the number of antennas at the receiver, but it is only rate-1 code since it only transmits two symbols using two time slots. The other profile, defined as Matrix B, provides Spatial Multiplexing (SM) and uses two transmit antennas to transmit two independent data streams. This scheme is a rate-2 code, but it does not benefit from any diversity gain at the transmitter, and, at best, it provides a diversity order equal to the number

[1] This section has been adapted from contribution from the authors into [WIN+D17].

MIMO. DOI: 10.1016/B978-0-12-382194-2.00008-3

of receive antennas. Furthermore, these two schemes are the only options defined for Uplink (UL) transmission. These two schemes have also been included in the WiMAX Forum specifications as two mandatory profiles for use on the downlink.

It is believed that these schemes using two transmit antennas will be two basic profiles of most future standards, such as the IEEE 802.16 m for mobile WiMAX evolutions and the LTE-Advanced of the 3GPP. However, there may be a need to include new codes combining the respective advantages of the Alamouti code and the SM while avoiding their drawbacks. Such a code actually exists in the IEEE 802.16e specifications as Matrix C. This code is a variant of the Golden code [BRV05] (see also [YW03] and [DV05] for other variants), which is known to the 2×2 STBC code achieving the diversity-multiplexing trade-off with the highest coding gain. [TV07].

In the following, we provide these three schemes of IEEE 802.16e, while retaining the notations of [IEE09a] (i.e., Matrix A, B, and C), previously introduced:

$$A=\begin{bmatrix} s_1 & -s_2^* \\ s_2 & s_1^* \end{bmatrix}, \quad B=\begin{bmatrix} s_1 \\ s_2 \end{bmatrix}, \quad C=\frac{1}{\sqrt{1+r^2}}\begin{bmatrix} s_1+jr\cdot s_4 & r\cdot s_2+s_3 \\ s_2-r\cdot s_3 & jr\cdot s_1+s_4 \end{bmatrix}, \tag{8.1}$$

where $r=\left(-1+\sqrt{5}\right)/2, s_k$ the kth transmitted symbol.

As the number of transmit antennas increases, the complexity of full-rate full-diversity codes increases exponentially and, therefore, for higher number of antennas only the combination of Alamouti and SM is preferred to improve the performance while keeping the detection complexity reasonable. This is, in fact, the case with IEEE 802.16e specifications and the existing schemes mainly use Alamouti code and SM given in equation (8.1) for three and four transmit antennas. Particularly, for three transmit antennas, the main schemes are defined as:

$$A=\begin{bmatrix} \tilde{s}_1 & -\tilde{s}_2^* & 0 & 0 \\ \tilde{s}_2 & \tilde{s}_1^* & \tilde{s}_3 & -\tilde{s}_4^* \\ 0 & 0 & \tilde{s}_4 & \tilde{s}_3^* \end{bmatrix}, \quad B=\begin{bmatrix} \sqrt{\frac{3}{4}} & 0 & 0 \\ 0 & \sqrt{\frac{3}{4}} & 0 \\ 0 & 0 & \sqrt{\frac{3}{2}} \end{bmatrix}\begin{bmatrix} \tilde{s}_1 & -\tilde{s}_2^* & \tilde{s}_5 & -\tilde{s}_6^* \\ \tilde{s}_2 & \tilde{s}_1^* & \tilde{s}_6 & \tilde{s}_5^* \\ \tilde{s}_7 & -\tilde{s}_8^* & \tilde{s}_3 & -\tilde{s}_4^* \end{bmatrix}, \quad C=\begin{bmatrix} s_1 \\ s_2 \\ s_3 \end{bmatrix}, \tag{8.2}$$

Here the complex symbols to be transmitted are taken as x_1,x_2,x_3,x_4, which take values from a square QAM constellation, and we have $s_i=x_ie^{j\theta}$ for $i=1,2,\ldots,8$, where $\theta=\tan^{-1}(1/3)$. Then, the matrix elements are obtained as $\tilde{s}_1=s_{1I}+js_{3Q}$; $\tilde{s}_2=s_{2I}+js_{4Q}$; $\tilde{s}_3=s_{3I}+js_{1Q}$; $\tilde{s}_4=s_{4I}+js_{2Q}$ where $s_i=s_{iI}+js_{iQ}$.

The first two matrices, namely, Matrix A and B, benefit from transmit diversity exploited by means of Alamouti code. Moreover, in these two schemes, the coordinate interleaved notion [J01] is also added over the phase-rotated symbols to increase the transmit diversity. Indeed, it can be easily seen that for any combination of the complex symbols x_1,x_2,x_3,x_4—where at least one symbol takes nonzero value—Matrix A of three transmit antennas have always full raw rank and this makes it a full diversity matrix. Both Matrix A and Matrix B are defined as space-time-frequency codes (i.e., they are transmitted over two time slots and two subcarriers) and exploit the orthogonality of the Alamouti code for complexity reduction. Matrix C with three transmit antennas is the pure SM.

Similar to the three transmit antenna case, for four transmit antennas, we have the following matrices:

$$A = \begin{bmatrix} s_1 & -s_2^* & 0 & 0 \\ s_2 & s_1^* & 0 & 0 \\ 0 & 0 & s_3 & -s_4^* \\ 0 & 0 & s_4 & s_3^* \end{bmatrix}, \quad B = \begin{bmatrix} s_1 & -s_2^* & s_5 & -s_7^* \\ s_2 & s_1^* & s_6 & -s_8^* \\ s_3 & -s_4^* & s_7 & s_5^* \\ s_4 & s_3^* & s_8 & s_6^* \end{bmatrix}, \quad C = \begin{bmatrix} s_1 \\ s_2 \\ s_3 \\ s_4 \end{bmatrix}. \quad (8.3)$$

Again, the first two matrices benefit from transmit diversity by means of Alamouti scheme, and defined as rate-1 and rate-2 options, respectively. We have also pure SM as a rate-4 option. In addition to the mentioned open-loop schemes, closed-loop schemes have also been included in [IEE09a]. However, none of these schemes have been included in the WiMAX profile and implemented in the existing products.

8.2.2 Simulated Performance

Although IEEE 802.16e specifications include the above mentioned matrices and different permutations of them, WiMAX profile only includes the two simple ones: Matrix A and Matrix B with two transmit antennas. During the standardization period, there have been discussions to include Matrix C with two transmit antennas in order to benefit from transmit diversity while maximizing the transmission rate. However, because of its high decoding complexity it has not been included in the final WiMAX profile.

Indeed, the optimal Maximum Likelihood (ML) detection complexity grows exponentially with the modulation order and number of transmitted symbols in the code matrix and this prevents the usage of Matrix C especially for high constellation sizes. However, because of the rapid change in wireless technologies, it is evident that the implementation of more complex decoders will be possible. Therefore, such optimum codes are still thought to be strong candidates for future standards. In this retrospect, it is meaningful to investigate the performance of these codes in real environments.

We now present the performance of Matrix C in a real WiMAX environment, where advanced WiMAX features such as frequency permutation, namely, Partial Usage of Subcarriers (PUSC) as defined in IEEE 802.16e, and Convolutional Turbo Codes (CTCs) are used. We particularly compare Matrix B and Matrix C in a 2×2 downlink MIMO WiMAX system for different coding rates, while using a soft-output Schnorr-Euchner decoder based on the single tree search algorithm [SBB06]. FFT size is chosen equal to 1024, which corresponds to a system bandwidth of 10 MHz. Both uncoded and coded cases (with CTC having coding rates of 1/2 and 3/4) are treated in order to demonstrate the effect of channel encoding on space-time codes using QPSK modulation. At the receiver, following the soft-output Schnorr-Euchner decoder, we have a soft-input CTC decoder together with a bit de-interleaver. In the simulations, Jake's channel model is used in a Pedestrian B environment at a speed of 3 km/h. The Pedestrian B test environment parameters are given in Table 8.1.

Table 8.1 Pedestrian B Test Environment Tapped-Delay-Line Parameters.

Tap	Relative Delay (ns)	Average Power (dB)	Doppler Spectrum
1	0	0	Classic
2	200	−0.9	Classic
3	800	−4.9	Classic
4	1200	−8.0	Classic
5	2300	−7.8	Classic
6	3700	−23.9	Classic

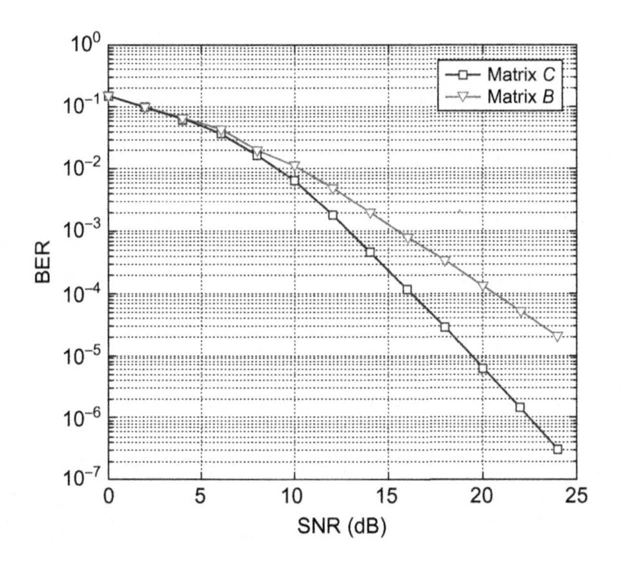

FIGURE 8.1

Uncoded BER, Matrix B and Matrix C, QPSK.

Figure 8.1 shows the Bit Error Rate (BER) performance of Matrix B and Matrix C for an uncoded QPSK signal constellation as a function of Signal-To-Noise Ratio (SNR), where SNR is defined as the ratio of the received signal energy per antenna to noise spectral density. In the uncoded case, as expected, Matrix C outperforms Matrix B with a transmit diversity advantage which can be clearly observed above 10 dB SNR. On the other hand, as shown in Figure 8.2(a), in the presence of channel coding with a rate of 3/4, Matrix C still performs better than Matrix B. However, the gap between the BER curves is dramatically reduced compared to the uncoded case and both schemes exploit essentially the same diversity order. When we further decrease the coding rate to 1/2 (see Figure 8.2(b), Matrix C only becomes closer to Matrix B below the BER value of 10^{-4} which requires an SNR value above 5 dB. These figures simply show that BER performance is dominated more by the diversity

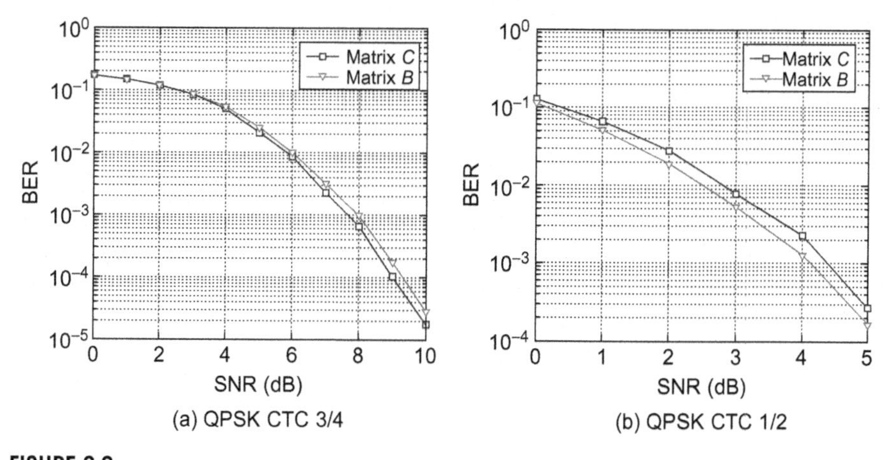

FIGURE 8.2

Coded BER, Matrix B and Matrix C.

exploited by channel coding than the diversity exploited by STBC. In other words, channel codes may recover the diversity loss that Matrix B suffers from. A theoretical investigation of diversity brought by Golden codes in coded systems has been proposed in [MROU10].

Another interesting observation is that Matrix B outperforms Matrix C in the SNR range of interest with a much lower ML detection complexity. Similar results obtained also for higher modulation sizes and presented in [KSB09]. As a conclusion, these results along with the long and complex decoding of Matrix C compared to Matrix B, do not justify its use at least in the current WiMAX systems.

8.2.3 Requirements on Signaling and Measurements

In the current WiMAX profile there is only one MIMO feedback option which allows a switch between Matrix A and Matrix B with two transmit antennas and requires a 6-bit feedback. Within this 6 bit, we have both MIMO mode and the permutation information.

8.2.4 Requirements on Architecture and Protocols

Multiple transmit and receive antennas are needed in order to facilitate the MIMO transmission and reception of multiple spatial layers.

8.3 MIMO CANDIDATES FOR IEEE 802.16m

8.3.1 Description

In this section, we focus on the MIMO schemes included in the System Description Document (SDD) [IEEE16mSDD] of the IEEE 802.16m. Generally speaking, the

SDD document includes the MIMO schemes (or at least the general descriptions of the schemes) which are decided to be included in the amendment document.

Concerning the IEEE 802.16m standardization period, the schemes considered up to now are modified versions of the ones existing in the IEEE 802.16e specifications with two and four transmit antennas based on Alamouti and SM. The main concern is to find the codes that provide the best trade off between performance/rate/complexity. Recent attempts mainly focus on this issue and try to find different and better alternatives to the existing solutions. Despite the existence of numerous STBCs that provide interesting performance, they are not included in the SDD as they are all thought to be too complicated in terms of implementation.

In SDD, MIMO schemes are divided into two main groups: open-loop MIMO schemes and closed-loop MIMO schemes. Currently, closed-loop and open-loop schemes use the same codebooks (or subset of these codebooks).

In open-loop schemes, rate-1 schemes are collected in transmit diversity modes while the higher rate schemes have been put in spatial multiplexing modes. In particular, the transmit diversity modes are defined as:

2Tx rate-1: For $M = 2$, SFBC with precoder, and for $M = 1$, a rank-1 precoder
4Tx rate-1: For $M = 2$, SFBC with precoder, and for $M = 1$, a rank-1 precoder
8Tx rate-1: For $M = 2$, SFBC with precoder, and for $M = 1$, a rank-1 precoder

where M denotes the number of symbols at a given time and SFBC refers to space-frequency block code. The precoding matrix will be based on the selected codebook and defined in the amendment document.

On the other hand, the spatial multiplexing modes include:
Rate-2 spatial multiplexing modes:

2Tx rate-2: rate 2 SM with precoding
4Tx rate-2: rate 2 SM with precoding
8Tx rate-2: rate 2 SM with precoding

Rate-3 spatial multiplexing modes:

4Tx rate-3: rate 3 SM with precoding
8Tx rate-3: rate 3 SM with precoding

Rate-4 spatial multiplexing modes:

4Tx rate-4: rate 4 SM with precoding
8Tx rate-4: rate 4 SM with precoding

In closed-loop MIMO, unitary codebook based precoding is supported for both Frequency-Division Duplex (FDD) and Time-Division Duplex (TDD) systems. In TDD systems, sounding based precoding will also be supported. For codebook based precoding, two types of codebook are currently discussed in the SDD: 802.16e codebook and Discrete Fourier Transform (DFT) codebook.

8.3.2 Concluding Remarks

Although IEEE 802.16e specifications include many MIMO schemes, only the well known Alamouti scheme and SM have been included in WiMAX profile for two transmit antennas. In order to have a 16m amendment closer to implementation, 16m working groups are trying to define minimum number of MIMO schemes (both mandatory and optional ones) which provide the best trade off between performance/rate/complexity. The optional ones will be included only if they provide significant improvements compared to mandatory ones.

8.4 UL-MIMO SCHEMES IN WiMAX SYSTEMS

8.4.1 Introduction

In order to fulfill the WiMAX promises, WiMAX semiconductor platforms are providing solutions for all types of WiMAX equipment makers and service providers in different markets. However, to meet the requirements of extended coverage, high data rate, and low power consumption, silicon solutions need to deliver some challenging trade offs presented by these key constraints.

In WiMAX systems, as in other wireless systems, the uplink channel becomes the limiting factor for coverage. Improving the uplink performance yields benefits for both operators and end users; it lowers infrastructure costs and improves user experience. Typically, in current WiMAX systems, MIMO is implemented mostly on the downlink channel (see previous section). The implementation of dual transmit channels in a single user terminal, is one of the WiMAX capabilities that can improve uplink performance.

Moreover, if an appropriate algorithm is used to implement the second transmit channel, substantial improvement can be achieved with little or no incremental cost to the mobile station and no cost at all to the base station.

Another way to see UL MIMO scheme, still relying on terminal having single transmit antenna is to consider collaboration between two terminals producing a collaborative UL MIMO scheme to the base station. This scheme is included in the WiMAX standard under the name of "Collaborative Spatial Multiplexing". This scheme enables an operator to spatially multiplex two different users in the uplink. This does not double the instantaneous user data rate but increases the cell capacity in the uplink. It is currently part of the profile since it does not require the mobile station to be equipped with two Tx. This scheme is illustrated on the Figure 8.3.

In the following, we provide the benefits of UL-MIMO, when the user terminal has two Tx enabled. We describe a new method, namely, Tile-Switched Diversity (TSD), which benefits from transmit diversity without any channel knowledge at the transmitter side. This method is mainly based on the transmit diversity concept for clustered OFDM systems and was first introduced in [CDS96]. The idea has been developed for WiMAX systems and its advantages have been shown based on both

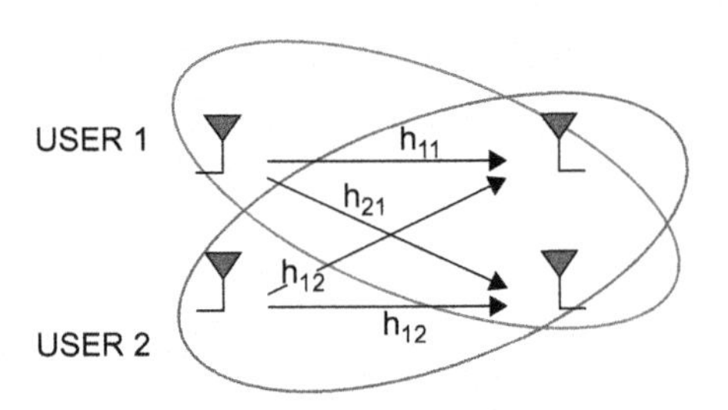

FIGURE 8.3

Collaborative MIMO: spatial multiplexing of two separates users in the UL.

performance analysis and implementation aspects. As described in the sequel, the proposed technique is completely transparent and it does not require any additional processing at the base station side. This makes the algorithm very attractive for mobile stations from the implementation point of view as it could be immediately deployed in the field with no change on the base station side.

8.5 CYCLIC DELAY DIVERSITY (CDD)

The principle of UL CDD is to transmit different cyclic delayed versions of the original signal on different transmit antennas. One antenna will transmit the regular waveform and the second antenna will transmit OFDMA symbols that have been cyclically shifted in time with the corresponding cyclic prefix. This technique does not add any inter symbol interference and as a result does not require any particular equalizer at the receiver side. The receiver could still use Maximum Ratio Combing if it has multiple receive antennas. Therefore, a conventional BS is still able to demodulate terminals using UL CDD.

CDD introduces strong frequency selective component on the channel seen by the receiver. The frequency selectivity depends on the amount of delay introduced between the two antennas. The downside of CDD is that it can actually degrade performance in line-of-sight, or near line-of-sight environments since the artificial phase shift between the two antennas can create deep fade in such environment. Indeed, the optimal delay depends on the channel.

CDD is indeed a well know diversity technique. It is already considered on the BS side for downlink transmission.

8.6 TILE-SWITCHED DIVERSITY (TSD)

The presented technique uses the basic idea of antenna selection and benefits from the frequency selectivity of the channel by means of channel coding. It requires the implementation of two transmit antennas at the mobile station but is fully transparent to the base station like CDD transmission. However, compared to CDD transmission, it has better or equal performance in fading environments and significantly better performance in line-of-sight environments.

In a typical WiMAX system [IEE09a], tiles are defined as the minimum resource allocation unit for uplink transmission. A tile consists of four subcarriers and extended over three OFDMA symbols, as shown in Figure 8.4. In TSD, tiles from data slots are split between the two transmit antennas of the mobile station. In particular, each uplink slot (defined as the group of six tiles) is split in two groups of three tiles. Each group of tiles goes to a different transmit antenna, and, therefore, each group of tiles is affected by a different channel, as illustrated by Figure 8.5. This transmission technique is completely transparent to the base station, as the channel estimation is done on a slot-by-slot basis. It is worth noting that the presented TSD scheme may be applied to PUSC (namely, partial usage of subchannels) permutation in the UL. Specifically, PUSC is defined as the permutation type in the specification which benefits from the frequency diversity of the channel together with the convolutional turbo code (CTC) used for channel encoding.

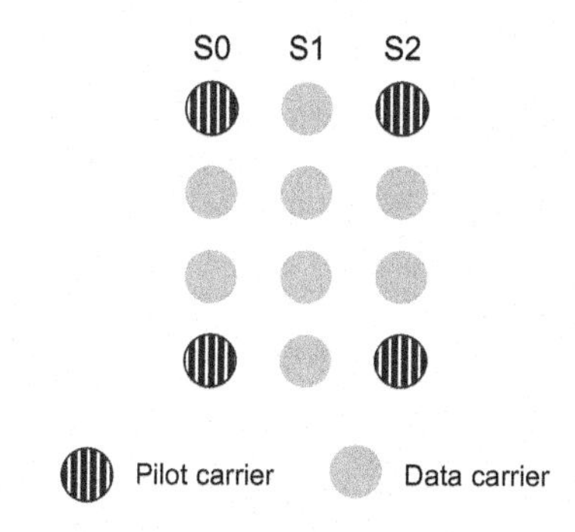

FIGURE 8.4

Tile structure defined for UL PUSC transmission in WiMAX systems [IEE09a].

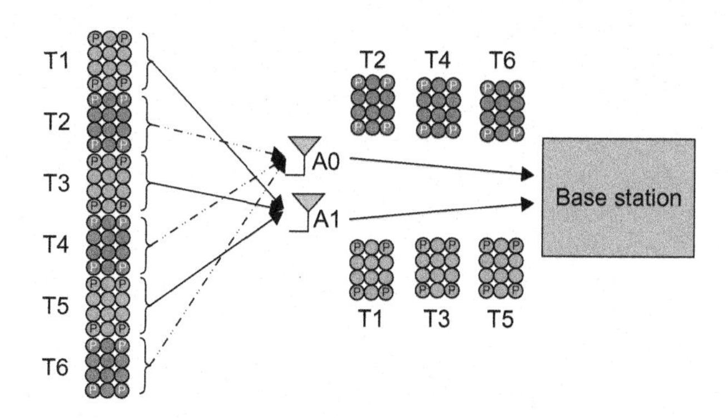

FIGURE 8.5

Allocation of tiles over two transmit antennas.

In WiMAX systems, channel coding is performed on a slot-by-slot basis. Therefore, the diversity introduced by the proposed TSD on the tiles will be directly exploited by the convolutional encoder to reduce the error probability. The exact order of diversity is a function of the coding order. Furthermore, a TSD scheme can be further enhanced by the use of MRC at the base station.

As mentioned before, TSD provides superior performance compared to CDD. However, the benefits of TSD are not limited to this. More specifically, unlike CDD, no interference phenomenon is created since a tile is never transmitted simultaneously by the two Tx antennas. Moreover, TSD does not raise synchronization ambiguities as CDD does where the OFDMA signal is transmitted with different delays on the two Tx antennas, making synchronization more difficult. Finally, TSD does not decrease the coherence bandwidth of the actual channel and, therefore, it does not incur any channel estimation performance degradation. All these favor the usage TSD for UL-MIMO transmission.

8.7 PERFORMANCE

In this section, we will describe the performance benefits of TSD together with other possible UL MIMO transmission schemes based on some results obtained in a real WiMAX environment. In multiple transmit antenna schemes, due to the physical combination of the transmit signals from both antennas over the air, the resulting power of the transmitter is equal to the sum of the power transmitted on each antenna. Thus, if the power transmitted on each antenna is the same, UL-MIMO schemes provide a 3 dB combining gain in the link budget. This combining gain is present with whatever uplink MIMO scheme is used.

In the sequel, we first describe briefly the scenarios that are simulated on the WiMAX environment and, then, present some results showing the performance comparison between TSD and conventional UL transmission schemes. Results are obtained for modulation scheme QPSK with a coding rate of 1/2 using convolution turbo coding where the FEC block size is taken as defined in the IEEE 802.16e specifications [IEE09a]. Both the transmitter and receiver parts are in floating point and antenna correlation is not taken into account. An FFT size of 1024 has been used together with a CP length equal to 1/8 of OFDM symbol duration. At the receiver side perfect timing and frequency synchronization have been assumed and a soft output equalizer with a bit de-interleaver and a soft-input CTC decoder are used. For AWGN channel the perfect channel estimation corresponds to the case where the channel estimation module is disabled. In the figures, both the perfect CSI results and results with channel estimation are presented. The results are depicted as packet error rate (PER) vs. C/N where PER simply corresponds to burst errors containing several FEC blocks and C/N denotes the received signal-to-noise ratio per receive antenna.

Figure 8.6 depicts the performance comparison on AWGN. In particular, we have the performances of MRC alone (i.e., two antennas at the receiver side), and STC

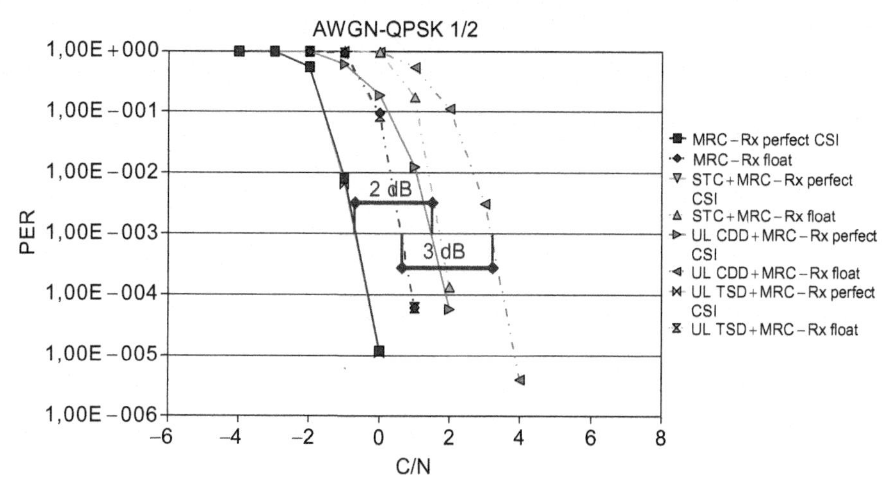

FIGURE 8.6

PER comparison in AWGN channel.

(i.e., Alamouti scheme) with MRC, UL TSD with MRC and UL CDD with MRC with a cyclic delay of six samples. It is worth noting that in the following figures, the delay has been chosen as the one giving the overall best performance. We can see that when the channel information is perfectly available at the receiver, MRC only, STC and TSD provide the same performance as expected. However, CDD has 2 dB performance degradation compared to the others for a PER of 10^{-3}. This simply implies that TSD does not degrade the performance in AWGN channel, which is not the case for CDD transmission. On the other hand, when we enable the channel estimation module CDD has additional 1 dB degradation compared to MRC alone and TSD. Moreover, since the number of pilots are half in STC transmission that of TSD, STC suffers from 1 dB performance degradation.

Figure 8.7 compares the performances of MRC only, STC, TSD and CDD on the ITU pedestrian channel model B. It can be observed that both TSD and CDD bring an important diversity gain that will noticeably enhance the system performance when the channel is perfectly known at the receiver. Indeed, they provide 2 dB gain at the PER of 10^{-3} compared to MRC only transmission with a slight complexity increase at the transmitter side. However, the performance of TSD and CDD seem not as efficient as that of STC. This is actually the case with perfect CSI and when we include the channel estimation errors TSD and CDD provide almost the same performance as STC and better than MRC only transmission. Consequently, in fading channels the principle gain of both transparent techniques, that is, TSD and CDD, is better than MRC only transmission and remains similar to STC transmission which requires specific implementation at the receiver side. Moreover, on a typical vehicular channel representing mobile terminals, the conclusions remain unchanged.

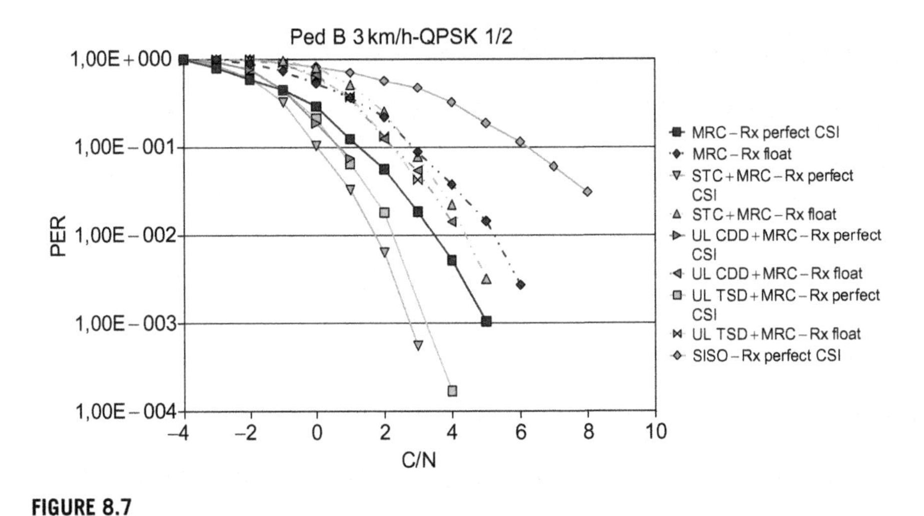

FIGURE 8.7

PER comparison in ITU Pedestrian B channel at 3 km/h.

Presented UL-MIMO schemes TSD and CDD do not require additional signaling. In addition, the TSD method does not require any specific decoder mechanism at the base station side and it is completely a transparent scheme for UL transmission.

8.8 POTENTIAL IMPACTS ON ARCHITECTURE

Figure 8.8 illustrates a typical mobile station RF front-end design. It is worth noting that passive components for power supply decoupling are omitted for the sake of clarity. The blocks depicted light gray are the main components of a 1Tx/2Rx mobile station. On the other hand, the blocks in dark gray are the additional components for a 2Tx/2Rx mobile station. In particular, the require components to realize the second transmit antenna path are simply a second power amplifier, ordinary filters and duplexer.

The additional size required for these components is approximately 50 mm^2. Actually, this may vary depending on the components chosen but the order gives the size which can be achieved without considering the recent advances in the component design. Indeed, in the future, this could even be smaller, as integrated Front-End Modules (FEM) are used. Note that some FEM vendors are planning integrated dual-FEM designs in a single package. This process would further reduce the impact of the second transmitter. Similarly, it is difficult to provide exact price estimates due to variable vendor pricing. Nevertheless, we estimate the added cost of the second transmitter in this case to be about $2 in 500 K volumes.

To go further in assessing the impact of using a two Tx device, we should compare designs with for equivalent performances. We have seen from the performance curves

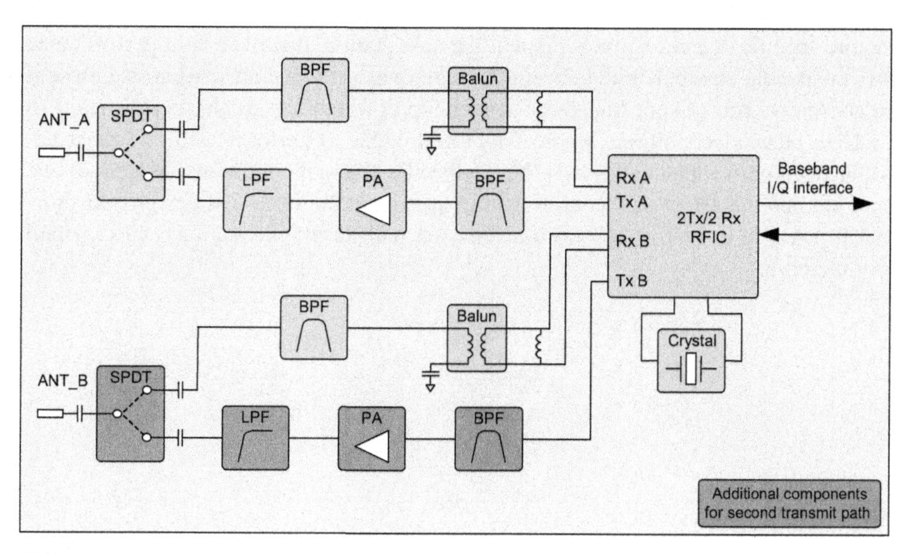

FIGURE 8.8

Typical mobile station RF front end.

that roughly speaking the two Tx scheme can bring 2 to 3 dBs gain from diversity gains. In addition to this gain the two Tx scheme obviously transmit with two Tx and thus benefit from 3 dB gains from doubling the transmitting power. As a result, a 2-Tx scheme using 23 dBm transmit power should be compared to a 1-Tx scheme with a 23 + 2 dB diversity gain +3 dB combining gain = 28 dBm.

By analyzing data for a wide range of power amplifier available on the market, we evaluate the power consumption reduction for a 2-Tx scheme at about 800 mW. This is mainly due to the fact that much efficient power amplifier could be found for 23 dBm output power compared to 28 dBm.

On the other hand, using 2-Tx scheme require additional computation at the baseband side estimated at 50 mW. As a result, in overall, the 2-Tx scheme presents a 750 mW gain compared to an equivalent (performance wise) scheme at 1Tx.

At last, various RF requirements such as transmit Error Vector Magnitude (EVM), spectral masks or control of spurious emissions, can be more easily met with the 2-Tx approach, greatly simplifying the design of the terminal.

8.9 CONCLUSIONS

UL-MIMO techniques are introduced in the standards to provide additional gains on the uplink link budget compared to single transmit antenna schemes. In addition to the STC based on Alamouti scheme, the current WiMAX specifications include CDD as the transparent UL transmission technique. Although this technique does not require specific detection mechanism at the base station, it suffers from performance loss in specific channels like near line of side ones. In this contribution, we present an alternative transparent transmission technique, which has similar performance to CDD in channel conditions where CDD is thought to perform well and does not suffer from performance loss as CDD in specific channel conditions. Moreover, we have studied the effect of second transmit antenna in the front-end design and show that it possible to realize second transmission path in mobile stations with a small cost increase.

CHAPTER

LTE and LTE-Advanced

9

Thomas Derham

LTE and LTE-Advanced represent the recent efforts of the 3rd Generation Partnership Project (3GPP) to define the evolution of cellular communications beyond 3rd Generation (3G) CDMA-based technology. The existing family of 3G standards ranging from WCDMA to HSPA+ has achieved strong commercial success and, in the latter case, can provide peak downlink speeds in excess of 20 Mbps. For LTE, the emphasis is not only on achieving further significant enhancements to data rates, but also reducing the total cost-per-bit, optimizing for cellular traffic that is increasingly dominated by multimedia data, and implementing a simplified All-IP network architecture including packet-switched, rather than circuit-switched, voice services [STB09].

It is expected that a wide range of mobile services will be enabled or enhanced by the deployment of LTE, including streaming video-on-demand, video conferencing, high quality VoIP, high-speed upload of user-generated multimedia content, consistent low-latency online gaming, paid dynamic content such as E-newspapers, video-based mobile advertising, on-demand music download and storage, application sharing and cloud-based services [UMT08]. LTE supports peak downlink speeds up to 300 Mbps with maximum channel bandwidth of 20 MHz.

The LTE standard was essentially completed in December 2008 and encapsulated in 3GPP Release 8 [3GP09a], with limited commercial roll-out beginning during 2009. The first large-scale deployments of LTE are planned during 2010, and it has achieved very wide support from operators across the global cellular industry.

Compared to the continued refinement of CDMA-based technologies in HSPA+, LTE has benefited from not requiring backward compatibility at the physical layer. One key example is the introduction of OFDM-based access technology, which provides a convenient basis for the implementation of advanced MIMO schemes as a fundamental part of the LTE system design. In addition, OFDMA-based multiple access results in a high degree of orthogonality between the signals at the receiver associated with different users within the cell, and therefore represents a rather different philosophy to the multiuser interference averaging approach of CDMA. In particular, in LTE there is more flexibility for the scheduler to maximize multiuser gain by adaptive fine-grain allocation of the shared channel resources to users in the frequency, time and spatial domains.

MIMO. DOI: 10.1016/B978-0-12-382194-2.00009-5

LTE-Advanced (LTE-A) is a logical extension of LTE that is due for completion in 2011 to be encapsulated in 3GPP Release 10. The transition from LTE to LTE-A will be relatively straightforward since LTE-A maintains the same basic OFDM transmission structure and is fully back-compatible with LTE at the physical layer. However, LTE-A contains several significant enhancements to LTE, which are primarily aimed at achieving the requirements defined by the Radiocommunications sector of the International Telecommunications Union (ITU-R) for IMT-Advanced [ITU-R08]. In addition, 3GPP has defined its own performance requirements that are considerably more ambitious, with a peak data rate of up to 1 Gbps for low mobility, and particular attention paid to improving cell-edge and uplink throughput [3GP10b]. A summary of these performance targets in terms of spectral efficiency (for given MIMO configurations) and latency is shown in Table 9.1 [TR308]. The major enhancements in LTE-A to achieve this performance are carrier aggregation, support for relays, and enhanced MIMO schemes.

The MIMO schemes implemented in LTE and LTE-A represent a distillation of the large collection of literature on MIMO theory that has arisen from the research community over recent years. However, the remit of the standardization process is to not only consider theoretical optimality, but also practical issues such as complexity, cost, power consumption, hardware impairments, testability, backwards compatibility and feedback channel overhead. This final issue is particularly important since it is now well understood that the greatest MIMO gains are achieved by techniques that require knowledge of Channel State Information at the Transmitter (CSIT). In practice, particularly for FDD systems, obtaining CSIT requires feedback from the receiver, which imposes an overhead that reduces the effective spectral efficiency of the reverse link. Therefore, much of the MIMO-related studies within 3GPP relate to the proposal and assessment of schemes that provide a good trade-off between pure throughput and reverse link overhead. This issue has become more complex in the context of LTE-A, where efforts are being focused particularly on providing enhanced support for MU-MIMO as well as Cooperative MultiPoint (CoMP) techniques. In these cases, the MIMO schemes are dependent on CSIT information from multiple users, which may even be associated with different cells.

Table 9.1 Performance Targets for LTE-A

	LTE Performance	ITU-R Requirement	LTE-A Target
Peak (DL)	15 bps/Hz (4×4)	15 bps/Hz (4×4)	30 bps/Hz (8×8)
(UL)	3.6 bps/Hz (1×2)	6.75 bps/Hz (2×4)	15 bps/Hz (4×4)
Cell average (DL)	1.87 bps/Hz (4×2)	2.2 bps/Hz (4×2)	2.6 bps/Hz (4×2)
(UL)	0.74 bps/Hz (1×2)	1.4 bps/Hz (2×4)	2.0 bps/Hz (2×4)
Latency (control)	100 ms	<100 ms	<100 ms
(user)	5 ms	<10 ms	<5 ms
Bandwidth	≤20 MHz	≥40 MHz	≤100 MHz

9.1 TRANSMISSION STRUCTURE

9.1.1 LTE Downlink

The LTE transmission structure[1] for the downlink between the base-station (eNB) and users (UE) is shown in Figure 9.1 [3GPP07]. One slot (0.5 ms) consists of seven consecutive OFDM symbols, while one subframe (1 ms) comprises two consecutive slots and corresponds to the Transmission Time Interval (TTI). One radio frame (10 ms) comprises 10 subframes, which for FDD operation (paired spectrum) are all assigned to the downlink. For TDD operation (unpaired spectrum), subframes may

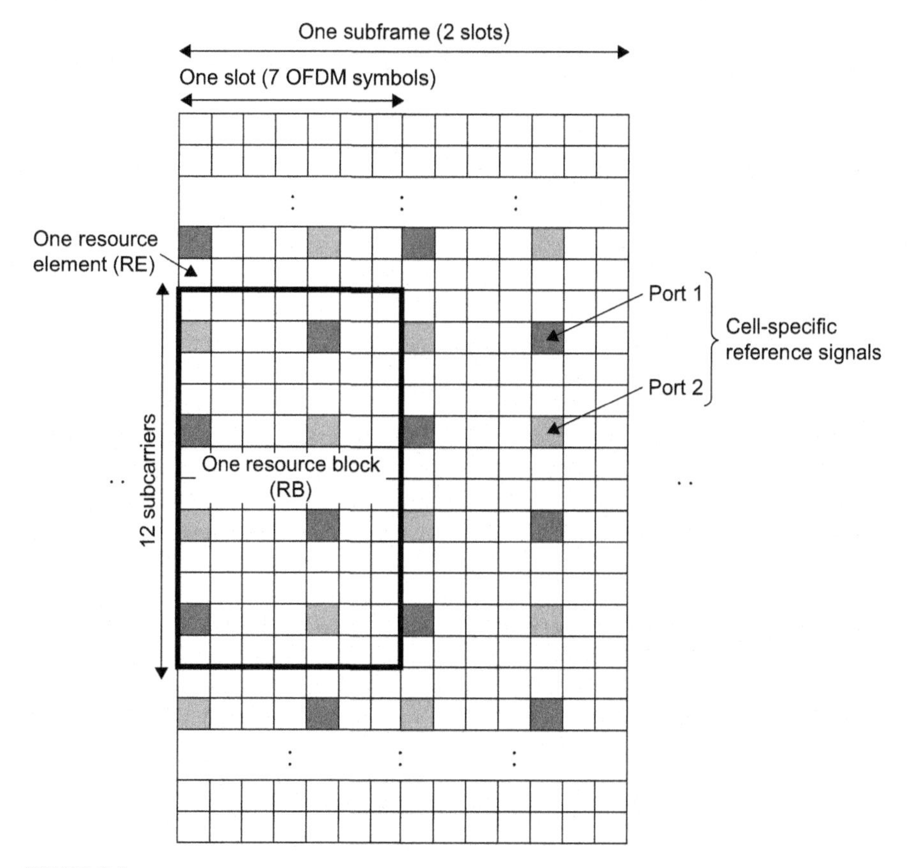

FIGURE 9.1

LTE downlink structure.

[1]For the standard parameters of normal cyclic prefix and 15 kHz subcarrier spacing.

Table 9.2 Logical LTE Downlink Data Channels

Logical Channel	Description
Physical Broadcast Channel (PBCH)	Broadcasts basic configuration information. Fixed mapping so can be detected by UE without knowledge of system bandwidth.
Physical Multicast Channel (PMCH)	Used for Multimedia Broadcast and Multicast Services (MBMS). Transmitted exclusively in subframes dedicated to PMCH.
Physical Downlink Shared Channel (PDSCH)	Main data transporting channel.

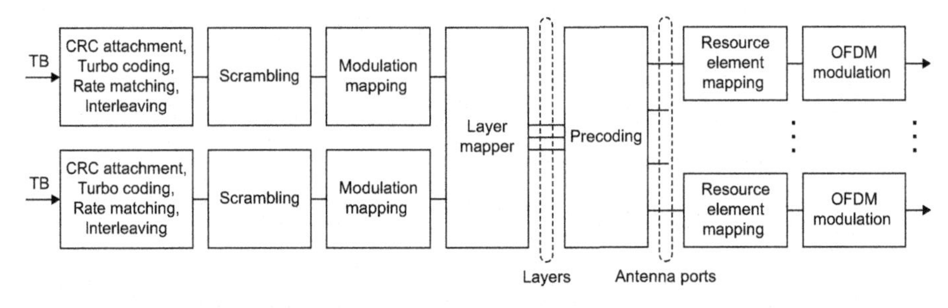

FIGURE 9.2

LTE downlink transmission structure.

be assigned to either the downlink or uplink, and special frames including a guard period are used for switching between downlink and uplink transmission.

A Resource Element (RE) is a single subcarrier on one OFDM symbol. A Resource Block (RB) consists of 12 consecutive subcarriers over one slot. The unit for allocating resources to UEs (scheduling) is a Scheduling Block (SB), which comprises two RBs consecutive in time over one subframe. System bandwidths ranging from 1.08 MHz (6 RBs) to 19.8 MHz (110 RBs) are supported.

There are three types of logical data transporting channels which share these resources, as described in Table 9.2. The broadcast channel (PBCH) is used for basic system configuration information, while the multicast channel (PMCH) is used for MBMS Single Frequency Network (MBSFN) transmissions that deliver services such as mobile TV over the LTE infrastructure. The downlink shared channel (PDSCH) is the main data transporting channel, and is used to transmit blocks of data, known as Transport Blocks (TB), which are passed down from the MAC layer.

Up to two TBs may be transmitted simultaneously per UE and subframe using RBs which need not be contiguous in the frequency domain. Each TB is independently coded and modulated, and the resulting modulation symbols are mapped to up to four layers (spatial streams), as shown in Figure 9.2. Precoding maps the layers to antenna ports, where RE mapping and OFDM modulation is performed to obtain the transmitted signal. A TB is transmitted during a single TTI, and retransmission of incorrectly received TBs is dealt with by the Hybrid Automatic Repeat reQuest (HARQ) functionality.

The logical control channels, primarily the Physical Downlink Control Channel (PDCCH), are allocated to the control channel region in a configurable number of OFDM symbols (1, 2 or 3) at the start of each subframe. In addition, certain REs are dedicated to the transmission of reference signals, which are described in Section 9.2.6.

9.1.2 LTE Uplink

The LTE transmission structure for the uplink between the UEs and the eNB is shown in Figure 9.3. The main structural components—slot, subframe, RB and RE—are the same as the downlink. In order to minimize the PAPR[2] of the signals transmitted by UEs, the uplink is based on DFT-spread OFDM (otherwise known as SC-FDMA), as shown in Figure 9.4.

Each UE can transmit one TB per subframe on the uplink shared channel (PUSCH) using RBs allocated by the eNB which must be contiguous in the frequency domain. Control information is transmitted on the uplink control channel (PUCCH), which is allocated to a configurable number of RBs at the edge of the system bandwidth. Code Division Multiplexing (CDM) is used to allow multiple UEs to transmit control information on the PUCCH simultaneously. In addition, certain symbols in each subframe are dedicated to the transmission of reference signals, which are described in Section 9.2.6.

9.2 LTE MIMO SCHEMES

The focus of efforts during the standardization of LTE was on specifying efficient schemes for downlink SU-MIMO—that is, the MIMO link between the eNB and a single UE—for various system architectures and channel conditions.

On the other hand, the LTE uplink was developed on the assumption that the UE transmitter hardware comprises just a single signal chain. Therefore, the extent of uplink SU-MIMO support is limited to adaptive transmit antenna switching. Uplink MU-MIMO (sometimes known as virtual MIMO) is not precluded and is left to implementation—the scheme resolves to a classical spatial Multiuser Detection (MUD) problem at the eNB [Ver98] for which little explicit support in the standard is required.

Therefore, this section focuses on the downlink MIMO implementation for PDSCH. There are a total of five SU-MIMO modes, as shown in Table 9.3, ranging from transmit diversity to closed-loop spatial multiplexing, plus one mode that provides a basic MU-MIMO implementation.

[2] In 3GPP, instead of PAPR, the related Cubic Metric (CM) is generally used, which better correlates to the power amplifier back-off required to meet the distortion requirements.

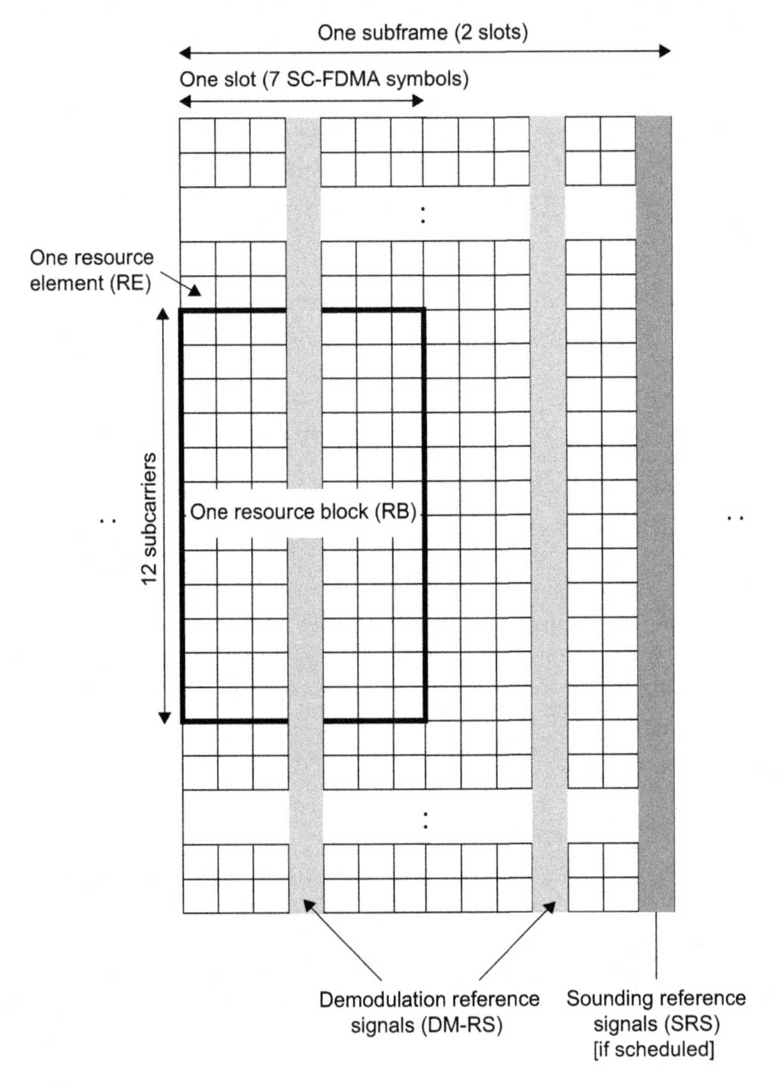

FIGURE 9.3

LTE uplink structure.

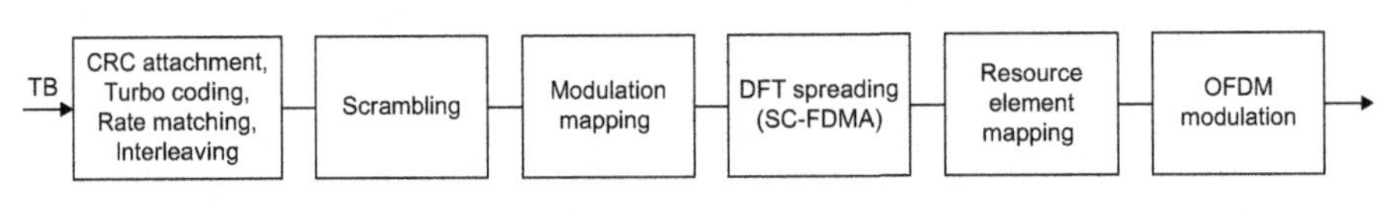

FIGURE 9.4

LTE uplink transmission structure.

Table 9.3 MIMO Transmission Modes in LTE Downlink PDSCH

Mode	Description
2	Transmit diversity
3	Open-loop spatial multiplexing
4	Closed-loop spatial multiplexing
5	MU-MIMO
6	Closed-loop rank-1 precoding
7	Transmission using UE-specific reference signals

9.2.1 MIMO System Model and LTE Feedback Scheme

It is useful at this stage to briefly review the classical MIMO downlink system model, which can be written for a narrowband channel matrix $\mathbf{H}_{(k,l)}$ (corresponding to the RE on the kth subcarrier and lth OFDM symbol) as:

$$\mathbf{y}_{(k,l)} = \mathbf{H}_{(k,l)}\mathbf{W}_{(k,l)}\mathbf{x}_{(k,l)} + \mathbf{n}_{(k,l)} \tag{9.1}$$

where $\mathbf{y}_{(k,l)}$ is the column vector of signals at the receive antennas, $\mathbf{W}_{(k,l)}$ is the precoding matrix, $\mathbf{x}_{(k,l)}$ is the column vector of transmitted modulation symbols mapped to this RE on each layer, and $\mathbf{n}_{(k,l)}$ is the column vector of noise and interference at the receive antennas. The transmit antennas are located at the eNB, while the receive antennas may be located at one or multiple UEs – thus both SU-MIMO and MU-MIMO are supported by this model.

The feedback scheme (from a UE to the eNB) in LTE comprises three components—Channel Quality Indicator (CQI), Precoding Matrix Indicator (PMI), and Rank Indicator (RI). CQI is the index of the preferred Modulation and Coding Scheme (MCS) applied to each TB. LTE supports a total of 15 MCS corresponding to QPSK, 16QAM and 64QAM modulation with various coding rates, as specified in Table 9.4. In general, the CQI reported by the UE corresponds to the maximum MCS that, it is estimated, can be applied to the corresponding TB while maintaining the BLER below a threshold, typically 0.1. PMI is the index of the preferred precoding matrix $\mathbf{W}_{(k,l)}$ from a predefined codebook, while RI indicates the preferred number of layers to be transmitted to the UE.

9.2.2 Transmit Diversity Mode

The transmit diversity scheme (mode 2) makes use of all antennas at both transmitter and receiver to mitigate the effect of multipath fading. It is based on the classical Alamouti code, and supports either two or four transmit antennas. While the maximum data rate is limited since no spatial multiplexing gain is achieved, it may be preferred in several scenarios such as low SNR, or when CSI feedback is not possible (e.g., MBMS transmissions)—neither PMI nor RI are required. It may also be employed in the PDCCH control channel.

Table 9.4 LTE Modulation and Coding Schemes

CQI Index	Modulation	Coding Rate	Effective Bits per Modulation Symbol
0	–	–	0
1	QPSK	78/1024	0.1523
2	QPSK	120/1024	0.2344
3	QPSK	193/1024	0.3770
4	QPSK	308/1024	0.6016
5	QPSK	449/1024	0.8770
6	QPSK	602/1024	1.1758
7	16QAM	378/1024	1.4766
8	16QAM	490/1024	1.9141
9	16QAM	616/1024	2.4063
10	64QAM	466/1024	2.7305
11	64QAM	567/1024	3.3223
12	64QAM	666/1024	3.9023
13	64QAM	772/1024	4.5234
14	64QAM	873/1024	5.1152
15	64QAM	948/1024	5.5547

Table 9.5 Transmit Diversity Mapping Scheme for Two Antennas

Antenna	Subcarrier k_1	Subcarrier k_2
1	x_1	x_2
2	$-x_2^*$	x_1^*

In transmit diversity mode, only one TB may be transmitted per UE and subframe. This mode is unique in that one coded modulation symbol is mapped to more than one RE. In the case where two transmit antennas are used, Alamouti coding maps two consecutive modulation symbols x_1, x_2 from the coded data stream to a pair of adjacent subcarriers k_1, k_2—a basic Space-Frequency Block Coding (SFBC) scheme—as shown in Table 9.5. In terms of the MIMO model in Equation 9.1, this scheme can be written as:

$$\mathbf{W}_{(k,l)} = \begin{pmatrix} 1 & 0 \\ 0 & 1 \end{pmatrix} \tag{9.2}$$

$$\mathbf{x}_{(k_1,l)} = \begin{bmatrix} x_1 & -x_2^* \end{bmatrix}^T \tag{9.3}$$

$$\mathbf{x}_{(k_2,l)} = \begin{bmatrix} x_2 & x_1^* \end{bmatrix}^T \tag{9.4}$$

In the case where four transmit antennas are used, SFBC is combined with Frequency Switched Transmit Diversity (FSTD). Specifically, two Alamouti codes are

Table 9.6 Transmit Diversity Mapping Scheme for Four Antennas

Antenna	Subcarrier k_1	Subcarrier k_2	Subcarrier k_3	Subcarrier k_4
1	x_1	x_2	0	0
2	0	0	x_3	x_4
3	$-x_2^*$	x_1^*	0	0
4	0	0	$-x_4^*$	x_3^*

mapped to the pairs of antennas (1,3) and (2,4) over four adjacent subcarriers k_1, k_2, k_3, k_4 as shown in Table 9.6, with the equivalent model given by:

$$\mathbf{W}_{(k,l)} = \begin{pmatrix} 1 & 0 & 0 & 0 \\ 0 & 1 & 0 & 0 \\ 0 & 0 & 1 & 0 \\ 0 & 0 & 0 & 1 \end{pmatrix} \tag{9.5}$$

$$\mathbf{x}_{(k_1,l)} = \begin{bmatrix} x_1 & 0 & -x_2^* & 0 \end{bmatrix}^T \tag{9.6}$$

$$\mathbf{x}_{(k_2,l)} = \begin{bmatrix} x_2 & 0 & x_1^* & 0 \end{bmatrix}^T \tag{9.7}$$

$$\mathbf{x}_{(k_3,l)} = \begin{bmatrix} 0 & x_3 & 0 & -x_4^* \end{bmatrix}^T \tag{9.8}$$

$$\mathbf{x}_{(k_4,l)} = \begin{bmatrix} 0 & x_4^* & 0 & x_3^* \end{bmatrix}^T \tag{9.9}$$

In both cases, estimates of the transmitted modulation symbols can be obtained using a simple linear receiver implementation [HMG05]—if the MIMO channel over the adjacent subcarriers is constant, the diversity streams are orthogonal and the linear receiver provides the optimal SNR. The actual diversity gain achieved depends on the spatial correlation of the MIMO channel—large separation and/or different polarization of the eNB transmit antennas result in greater gains.

9.2.3 Spatial Multiplexing Modes

Modes 3 and 4 perform SU-MIMO spatial multiplexing of up to four layers to a single UE using open-loop and closed-loop techniques, respectively. In principle, the number of layers transmitted is determined from feedback of the channel rank[3] in the RI. Except for the case of rank-one transmission, two independently coded TBs (code words) are mapped to these layers as specified in Table 9.7. In the case of rank 3 and 4 transmissions, modulation symbols from one code word are alternately mapped across two layers. Therefore, the UE should take into account the strength of both layers when determining the corresponding preferred CQI for the corresponding TB.

In LTE, receiver algorithms are in general not standardized and are left to implementation. However, for a conventional linear SU-MIMO receiver where the signals

[3]Note that the term *rank* in LTE does not imply the mathematical rank (i.e., number of nonzero singular values) of the channel matrix.

Table 9.7 Mapping of Code Words to Layers in LTE

Channel rank	Code word 1	Code word 2
1	Layer 1	–
2	Layer 1	Layer 2
3	Layer 1	Layer 2 + Layer 3
4	Layer 1 + Layer 2	Layer 3 + Layer 4

at all receive antennas can be jointly decoded, an estimate $\hat{\mathbf{x}}_{(k,l)}$ of the transmitted signals is found as:

$$\hat{\mathbf{x}}_{(k,l)} = \mathbf{D}_{(k,l)}\mathbf{y}_{(k,l)} \tag{9.10}$$

where $\mathbf{D}_{(k,l)}$ is the decoding matrix, which is commonly determined using the MMSE criterion [Die08]. The limited number of transmitted code words (compared to the number of layers) helps reduce the complexity of the Turbo decoder implementation at the UE.

In principle, the precoding matrix $\mathbf{W}_{(k,l)}$ should be designed to maximize throughput. Physically speaking, the columns of $\mathbf{W}_{(k,l)}$ are the beamforming vectors for each layer, and the purpose of precoding is both to ensure that the layers can be effectively separated by the decoder, and also preferably to optimally *direct* energy towards the receiver taking into account the (multipath) channel. The basic difference between modes 3 and 4 is how the precoding matrix $\mathbf{W}_{(k,l)}$ is determined.

Open-Loop Spatial Multiplexing

The open-loop technique used in mode 3 does not require PMI feedback, and so is particularly suited to the case of high UE velocity where the CSI feedback achievable with reasonably low overhead would be unreliable. However, since the transmitter has no knowledge of the channel $\mathbf{H}_{(k,l)}$, it cannot design $\mathbf{W}_{(k,l)}$ to allow optimal decoding of the layers at the UE.

It is well-known that, for the case where a random channel is unknown at the transmitter, the optimal spatial multiplexing scheme in terms of ergodic capacity is to set the precoder to the scaled identity matrix and transmit independent data streams from each transmit antenna with equal power allocation [Gol05]. However in LTE, since each TB is mapped over just a single subframe and a finite bandwidth, ergodic capacity is not necessarily the best optimization criterion because the channel over these REs may be highly mutually correlated. Therefore, when the transmitted rank is greater than one, mode 3 uses a precoding scheme based on Cyclic Delay Diversity (CDD), which has been shown to give throughput gains in system-level simulations compared to identity precoding [3GP07].

In CDD, a predefined series of precoding matrices is applied to each RE, which have the effect of cyclically progressing the beam angle for each layer over the subcarriers. As a result, frequency-domain diversity is enforced even when the actual channel over the REs is flat-fading. Mathematically, for transmission of v layers

$(2 \leq v \leq 4)$, $\mathbf{W}_{(k,l)}$ is determined from the product of three matrices as follows:

$$\mathbf{W}_{(k,l)} = \hat{\mathbf{W}}(k)\mathbf{D}(k)\mathbf{U} \tag{9.11}$$

where $\mathbf{D}(k)$ is a $v \times v$ diagonal matrix where the (n,n)th element equals $\exp(-j2\pi(n-1)k/v)$, which phase shifts the signals on each antenna according to the subcarrier index k to produce the required beam angle. In addition, $\mathbf{U}$ is the $v \times v$ DFT matrix, which orthogonalizes the precoding vectors for each layer; in other words, it optimally separates the beam angles of each layer for a given subcarrier. In the case of two transmit antennas, $\hat{\mathbf{W}}(k)$ is always the 2×2 identity matrix. However, in the case of four transmit antennas, $\hat{\mathbf{W}}(k)$ is a $4 \times v$ matrix cyclically selected from a predetermined codebook, which results in improved performance when the MIMO channel is spatially correlated [3GPP07].

When the channel rank is one, instead the transmit diversity scheme is used.

Closed-Loop Spatial Multiplexing

The closed-loop technique used in mode 4 provides the highest possible throughput in LTE, but requires reliable CSIT feedback in the form of the PMI. Hence, it is best suited to cases of low to medium UE mobility with medium to high SNR.

It is well-known that, for the case where the channel is perfectly known at the transmitter, the optimal spatial multiplexing scheme is Singular Value Decomposition (SVD) precoding [Gol05], which spatially decomposes the MIMO channel into mutually orthogonal virtual channels onto which independently coded data streams are transmitted. The precoding matrix for the SVD scheme is given by:

$$\mathbf{W}^{\mathrm{SVD}}_{(k,l)} = \mathbf{V}^{(1)}{}_{(k,l)}\mathbf{P}_{(k,l)} \tag{9.12}$$

where $\mathbf{V}^{(1)}{}_{(k,l)}$ is obtained from the SVD of $\mathbf{H}_{(k,l)}$ given by:

$$\mathbf{H}_{(k,l)} = \mathbf{U}_{(k,l)}\Sigma_{(k,l)}\left[\mathbf{V}^{(1)}{}_{(k,l)} \quad \mathbf{V}^{(0)}{}_{(k,l)}\right]^{H} \tag{9.13}$$

where $\mathbf{V}^{(1)}{}_{(k,l)}$ contains the right singular vectors corresponding to nonzero singular values, and $\mathbf{P}_{(k,l)}$ is a diagonal matrix representing the power allocation for each layer, which is determined by the waterfilling algorithm based on the singular values in $\Sigma_{(k,l)}$ and the SNR.

However, there are several differences between the assumptions on which the optimality of SVD precoding is based and the fundamental transmission scheme in LTE:

- In practice, particularly for FDD systems, obtaining (almost) perfect knowledge of $\mathbf{H}_{(k,l)}$ for every RE would imply a massive feedback overhead.
- In general, the transmit power constraint in LTE is *per antenna* summed over transmissions to all served UEs (in order to maximize power amplifier efficiency), rather than total transmitted power for a single UE.
- The transmitted modulation symbols in LTE are derived from a finite set of constellations (modulation schemes), rather than being Gaussian random variables.

- For rank 3 and 4 transmission, the data streams on two layers are not independent since they are mapped from the same code word.

The scheme implemented in mode 4 may be considered as an adaptation of the SVD precoding scheme to take into account these practical constraints.

It can be seen from Equations 9.12 and 9.13 that the SVD precoding matrix is dependent only on the SNR and the channel matrix $\mathbf{H}_{(k,l)}$, which in practice is estimated at the receiver from reference signals. While in principle $\mathbf{H}_{(k,l)}$ could be directly quantized and fed back to the transmitter, given the limited feedback rate it is preferable to feedback a quantized form of $\mathbf{W}_{(k,l)}$ in the form of PMI, since the important spatial information contained in the *eigenstructure*[4] of the channel is better preserved. Specifically, PMI can be considered as feedback of approximations to the dominant right singular vectors of the channel matrix.

In LTE, the PMI codebook for two transmit antennas comprises three matrices, while the four antenna codebook comprises 16 matrices (the definitions of which depend on the transmission rank), each of which has the constant-modulus property [3GPP07]. This property means that the transmit power from each antenna will be equal. However, it also means there is no way to indicate the power allocation per layer as determined by waterfilling. In fact, in LTE the transmit power allocation per layer is equal, and the CQI and RI are calculated accordingly to adapt the MCS to the resulting strength of each layer. In addition, to further reduce feedback overhead, a single PMI is fed back for multiple adjacent subcarriers (typically around five RBs).

The algorithm used to determine the PMI is implementation specific. However, since the codebook size is quite small, it is not necessary to explicitly calculate the SVD of $\mathbf{H}_{(k,l)}$. Instead, for example, an exhaustive search of all PMI and CQI[5] combinations may be carried out by estimating the effective SINR of each layer (averaged over the subcarriers) to determine the greatest total throughput to the UE. This approach has the advantages that no further feedback or control channel overhead is required to indicate the power per layer, different receiver capabilities can be transparently supported in the mapping from effective SINR to CQI, and also that it enables simple adaptation of the transmit scheme to the channel using the finite set of MCS that are supported in LTE.

The Role of Space-Time/Frequency Coding in LTE

It may be surprising that the implementation of STBC/SFBC in LTE is seemingly limited to the basic transmit diversity scheme in mode 2, despite continuing to be a major topic of interest in the research community. One reason for this is that LTE is optimized for low-mobility (although it supports high mobility too) and therefore is designed from the ground up to support efficient closed-loop CSIT feedback where the spectral efficiency is greater than open-loop MIMO schemes. However, it may be considered that both modes 3 and 4 implicitly include some elements

---

[4]The term *eigenstructure* is often used in this context even for nonsquare channel matrices, since the right singular values of $\mathbf{H}$ are equal to the eigenvectors of $\mathbf{H}^H\mathbf{H}$.

[5]For each TB.

of space-frequency coding—in rank 3 and 4 transmission, a single code word is mapped across two spatial layers as well as multiple subcarriers, and implicitly is jointly coded across these two dimensions by the scalar Turbo coding applied to the corresponding TB.

9.2.4 Single-Mode Beamforming Modes

Single-mode beamforming is the use of all transmit antennas to form a directional beam to transmit a single stream of data to the UE. In LTE, this concept is applied to both modes 6 and 7.

Closed-loop rank-1 precoding (mode 6) can be considered simply as a special case of the mode 4 closed-loop spatial multiplexing described above, where the rank is set to one. In this case, a single code word is transmitted on a single layer, where the corresponding precoding vector is determined from PMI feedback. In practice, this mode may be preferred when the MIMO channel is significantly spatially correlated; for example, for channels with strong line-of-sight component and/or closely-spaced, co-polarized transmit antennas. Rank-one transmissions may occur particularly in the case of low to moderate SNR, where the UE determines that greater throughput can be achieved by assigning all transmitter power to a single layer and using a higher MCS, rather than sharing it between multiple layers.

Transmission using UE-specific reference signals (mode 7) is a special case where the UE does not report PMI feedback, and instead the precoding vector is determined by the eNB. In practice, the eNB may try to determine the precoding vector based on uplink channel measurements; such as angle-of-arrival or, in the case of TDMA systems, using channel reciprocity to estimate the downlink channel directly. Special UE-specific RS, which are precoded in the same way as the data, are used for demodulation, resulting in higher channel estimate accuracy which can improve coverage at the cell edge. In 3GPP Release 9, dual-layer transmission using UE-specific reference signals is also supported.

9.2.5 Multiuser MIMO Mode

Support for downlink MU-MIMO in mode 5 is also based on the PMI, CQI and RI feedback structure. In principle, this mode is particularly suited to the case of medium to high SNR, where the total transmitter power can be shared between multiple UEs without significantly reducing the per-UE throughput.

In conventional MU-MIMO schemes, the co-scheduled UEs cannot communicate with each other to jointly decode the transmitted signals, and are unaware of the transmission scheme used for each other's signals; thus, energy from signals transmitted to one UE that appears at the receiver of another UE may be considered simply as interference. Therefore, MU-MIMO is better suited to channels with high spatial correlation (where SU-MIMO spatial multiplexing gains are generally low) since low mutual interference between the transmissions to each UE can be more easily obtained.

In LTE, the MU-MIMO implementation is rather basic—the eNB is not precluded from spatial multiplexing of layers to more than one UE, but, if it does, is restricted to transmitting a single layer per UE. Each UE reports its own preferred precoding vector (i.e., PMI) using the same codebook as for rank-1 SU-MIMO transmission.

In general, the PMI is determined at each UE by calculating estimates for the SINR that would result from every possible precoding vector in the codebook. However, these SINR estimates may be inaccurate since the UE is not aware of the precoding vectors assigned to other UEs with which it is co-scheduled. It is left to the eNB to take this into account when determining the MCS used for each TB. In practice, since MU-MIMO gains are generally achieved only if the mutual interference between the UEs is well mitigated, the MU-MIMO implementation in the LTE downlink may be considered somewhat suboptimal. This issue is discussed in more detail in Section 9.3.4.

9.2.6 LTE Reference Signals

Reference signals in LTE support three main functions: channel estimation for MIMO decoding (demodulation), determination of PMI/CQI/RI feedback, and determination of multiuser resource allocation (scheduling).

Downlink

In the downlink, common (cell-specific) reference signals (CRS) are used for all three functions. CRS are transmitted by the eNB on unique REs for each antenna port, sparsely allocated in frequency and time, as shown in Figure 9.1 for the case of two transmit antennas. In the case of four transmit antennas, additional REs are allocated for ports 3 and 4, but with reduced density in the time domain. Since the CRS for each antenna port are mutually orthogonal, conventional channel estimation techniques (such as two-dimensional MMSE interpolation) can be used to determine estimates for the MIMO channel $\mathbf{H}_{(k,l)}$ at all data REs.

The portion of this channel estimate corresponding to the RBs allocated to the UE is used to derive PMI, CQI and RI feedback to adaptively determine the transmission scheme. In addition, further CQI reports may be requested by the eNB over all or part of the system bandwidth (corresponding to a configurable number of subbands) for the purpose of multiuser scheduling.

For demodulation, with reference to Equation 9.1, it is necessary to determine the *effective* channel $\mathbf{H}_{(k,l)}\mathbf{W}_{(k,l)}$ in order to estimate the transmitted data $\hat{\mathbf{x}}_{(k,l)}$. This is possible since the eNB informs the UE of the codebook index of the precoding matrix that is being used, which is generally, but not necessarily, the same as the PMI report.

The exception to this scheme is transmission mode 7, where special UE-specific RS are transmitted for data demodulation. Since these RS are precoded in the same way as the data, the effective channel is estimated directly and it is not necessary to inform the UE of the precoding matrix used.

Uplink

In the uplink, the reference signal scheme is different to that of the downlink since each UE must transmit its own reference signals to the eNB. There are two types of uplink reference signal known as Demodulation Reference Signals (DM-RS) and Sounding Reference Signals (SRS).

DM-RS are used to support data demodulation, and are transmitted by each UE in the fourth SC-FDMA symbol of every slot[6], as shown in Figure 9.3. DM-RS are transmitted only on the RBs to which the UE is allocated. The DM-RS signals are derived from Zadoff-Chu sequences [Chu72] with the Constant Amplitude Zero Autocorrelation (CAZAC) property; therefore, the delay-domain channel impulse response can be estimated by cross-correlation with a copy of the transmitted signal. In addition, a set of such Zadoff-Chu sequences exists where the mutual cyclic cross-correlation between the sequences is approximately constant over all lags. In LTE, a different member of this set is transmitted by UEs associated with different sites; therefore, inter-cell interference is spread evenly over the full length of the impulse response estimate, resulting in improved SINR for the dominant taps of the channel.

In addition, in order to (approximately) orthogonalize DM-RS that are transmitted by multiple UEs on the same RBs and associated with the same site, the *Cyclic Shift* (CS) technique is implemented whereby a unique cyclic time-domain shift (within the SC-FDMA symbol) is applied to the DM-RS sequence transmitted by each UE. The CS technique is used if uplink MU-MIMO is implemented, and also may be applied to multiple cells (sectors) of one site. There are 12 possible cyclic shifts spread evenly over the SC-FDMA symbol length, the spacing of which is greater than the maximum nominal length of the channel impulse response. Therefore, the cross-correlation processing causes the channel impulse responses corresponding to each UE to be separated in the delay domain.

Sounding reference signals (SRS) are used to support multiuser scheduling. Therefore, in principle, SRS should enable the eNB to estimate the channel quality between each UE and the eNB over the entire system bandwidth, not just the RBs currently allocated to the UE. However, a very large resource overhead would be incurred in achieving orthogonality of full-bandwidth RS transmitted in every slot by all UEs in a cell. Therefore, SRS are transmitted only on request by the eNB, and only in the last SC-FDMA symbol of a subframe. The sounding bandwidth is configurable, and SRS from two UEs may be interleaved in the same RB and SC-FDMA symbol using IFDMA, allowing the eNB to manage the trade-off between accuracy of multiuser wideband channel information and RS overhead.

9.3 LTE-ADVANCED MIMO SCHEMES

The standardization of LTE-Advanced has seen significant efforts paid to further enhancing the performance of the MIMO schemes implemented in LTE. The *headline* enhancement in the downlink is support for up to 8×8 MIMO, while in the

[6]For the normal cyclic prefix case.

uplink there is support for up to 4×4 MIMO including the implementation of SU-MIMO for the first time. However, there are other enhancements that are perhaps more important; in particular, extension of the feedback scheme to provide improved support for downlink MU-MIMO.

In this section, in addition to a description of the main enhancements related to MIMO in LTE-A [TR310], the key issues discussed during its development are described in the context of their effect on cellular spectral efficiency and their impact on the LTE system architecture—particularly the feedback scheme, reference signal structure, and backward compatibility with LTE (Release 8) UEs.

9.3.1 LTE-A Transmission Structure

The LTE-A transmission structure has the same basic format as LTE. However, due to the introduction of carrier aggregation, the total system bandwidth may be extended by combining multiple *component carriers*, each of which corresponds to an LTE-compliant bandwidth.

In the downlink, the number of TBs transmitted simultaneously to each UE may be increased with carrier aggregation since the maximum two TBs per subframe is applied per component carrier.

In the uplink, the basic multiple access scheme is enhanced to support noncontiguous allocation of RBs to each UE—so-called *clustered* DFT-spread OFDM. This allows for improved multiuser gains since each UE can be more accurately allocated to RBs where the corresponding channel is strong. However, a side-effect is that the PAPR of the transmitted signal is slightly degraded due to the resulting discontinuities in the mapping of the DFT-spread signal to subcarriers [CMC08].

In addition, due to the introduction of SU-MIMO spatial multiplexing in the uplink, up to two TBs may be simultaneously transmitted per subframe and component carrier. The resulting uplink transmission structure (per component carrier) is shown in Figure 9.5.

9.3.2 Transmit Diversity

Downlink

The downlink transmit diversity scheme in LTE-A is the same as that in LTE since it was found that, in practice, little additional diversity gain is achieved from a scheme

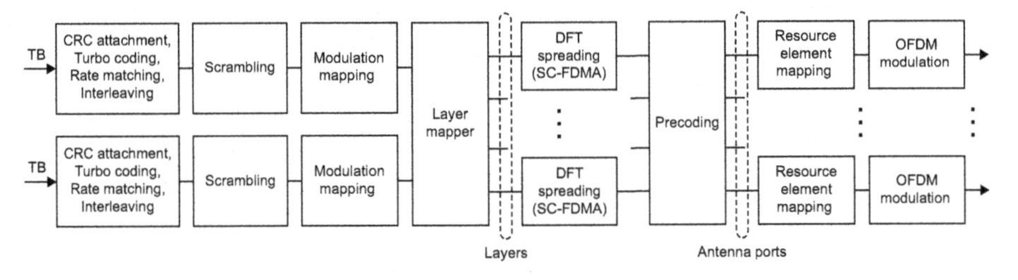

FIGURE 9.5

LTE-A uplink transmission structure.

optimized for eight transmit antennas. If the number of transmit antennas is greater than four, *virtualization* is used to form either two or four virtual transmit antennas to which the LTE scheme is applied. Virtualization is the use of precoding to map each diversity stream to a unique subgroup of physical antennas to form a virtual antenna, and maximizes the total transmit power by allowing the power amplifiers on all transmit antennas to be used. The specification of the virtualization precoding matrix is left to implementation, one suitable scheme is a small-delay version of CDD precoding [DLT09].

Uplink

In the uplink, certain PUCCH transmissions (formats 1/1a/1b) may employ Spatial Orthogonal Resource Transmit Diversity (SORTD) using two transmit antennas, where the same modulation symbol is transmitted from both antennas on orthogonal resources. This simple scheme achieves diversity gain while not degrading the PAPR of the transmitted signal [SOR09]. In the case of four transmit antennas, virtualization is used to form two virtual transmit antennas from which conventional transmit diversity is performed. A Single Antenna Port Mode is also defined, where the transmissions appear equivalent a UE with a single transmit antenna from the eNB perspective.

9.3.3 Spatial Multiplexing Modes

Downlink

The LTE-A downlink supports SU-MIMO spatial multiplexing of up to eight layers, to which up to two code words are mapped. In the case of rank-4 or less transmissions, the mapping of code words to layers is the same as in LTE, as per Table 9.7. For rank-5 to rank-8 transmissions, two code words are transmitted and each code word is mapped to between two and four layers. One exception is the case where a HARQ retransmission is required for a single code word only, in that case the single code word may be mapped across up to four layers.

The baseline implementation for closed-loop spatial multiplexing uses the same basic feedback mechanism as LTE (PMI, CQI and rank), where the PMI codebook is extended to support up to eight transmit antennas. Enhanced feedback schemes were also considered, where the focus in the context of SU-MIMO is reducing the error between the optimal precoding matrix determined at the receiver and the quantized limited-feedback approximation that is fed back to the transmitter. However, it is also preferable for the feedback schemes employed for SU-MIMO and MU-MIMO to be common. The feedback scheme is discussed further in Section 9.3.4.

Uplink

The LTE-A uplink supports closed-loop SU-MIMO spatial multiplexing of up to four layers, to which up to two code words are mapped. Uplink SU-MIMO is a key tool for achieving the spectral efficiency requirements for IMT-A, and the use of multiple transmit signal chains at the UE is considered feasible taking into account the continuing reduction in cost of highly integrated RF components. The mapping of code words to layers is the same as that for the LTE downlink, as shown in Table 9.7. The transmission scheme for each UE (i.e., precoding matrix, MCS and rank) is adaptively determined by the eNB and reported to the UE. For the case of two transmit antennas, a precoder codebook is used for rank-1 transmission, while the identity matrix is always used for rank-2 transmission. For the case of four transmit antennas, a precoder codebook is used for rank-1, 2-, and -3 transmissions, while the identity matrix is always used for rank-4 transmission.

Certain matrices in the rank-1 transmission codebook emulate the transmit antenna switching scheme implemented in the LTE uplink, which allow one of the power amplifiers to be switched off to reduce power consumption. In addition, the codebook for rank-2 and above transmissions is designed to take into account the effect of precoding on the PAPR of the transmitted signal from each antenna. Specifically, note that precoding forms the signal at each transmit antenna by weighted summation of the signals from all layers (where the complex weights are given by each row of the precoding matrix); therefore, in general, the resulting PAPR would be high despite the DFT-spreading applied to each layer since the signals are mutually independent. In order to avoid this, all matrices in the rank-2 codebook for four transmit antennas exclusively map each layer to two antennas; in other words, all rows of the matrices have only a single nonzero element. However, for rank-3 transmission a similar balanced mapping scheme is not possible—both PAPR *friendly* and PAPR preserving codebooks were considered, and the latter was chosen following performance evaluation [Fin10].

9.3.4 Multiuser MIMO Mode

Downlink

A key aim of LTE-A is to improve support for downlink MU-MIMO beyond the basic implementation in LTE. Simulation results provided in the 3GPP Self Evaluation Report [TR309] have demonstrated that enhanced MU-MIMO can provide significant improvements in spectral efficiency.

It is well-known that, for the case where the channel to all receivers is perfectly known at the transmitter, the optimal MU-MIMO scheme in terms of sum-rate capacity is a nonlinear technique known as Dirty Paper Coding (DPC) [WSS04]. However, due to the high computational requirements of DPC, and in order to maintain a common transmission structure for all MIMO modes, the MU-MIMO schemes implemented in LTE-A are linear, and are therefore defined simply in terms of a precoding matrix.

It is useful at this stage to briefly review the well-known Block Diagonalization (BD) linear precoding scheme [SSH04], which can be considered a generalization of zero-forcing precoding for the case of multiple antennas at each receiver. With the assumption of perfect CSIT, BD completely prevents interference between layers transmitted to different receivers, and the MIMO channel to each receiver is spatially decomposed into mutually orthogonal layers. For the case of MU-MIMO transmission from the transmitter to two receivers (A) and (B), the BD precoding matrix for transmission to receiver (A) is defined as:

$$\mathbf{W}^{\mathrm{BD}}_{(k,l)(A)} = \mathbf{V}^{(\mathbf{0})}{}_{(k,l)(B)} \tilde{\mathbf{V}}^{(1)}_{(k,l)(A)} \mathbf{P}_{(k,l)(A)} \tag{9.14}$$

where $\mathbf{V}^{(\mathbf{0})}{}_{(k,l)(B)}$ is obtained from the SVD of $\mathbf{H}_{(k,l)(B)}$ given by

$$\mathbf{H}_{(k,l)(B)} = \mathbf{U}_{(k,l)(B)} \mathbf{\Sigma}_{(k,l)(B)} \left[\mathbf{V}^{(\mathbf{1})}{}_{(k,l)(B)} \quad \mathbf{V}^{(\mathbf{0})}{}_{(k,l)(B)} \right]^H \tag{9.15}$$

and where $\tilde{\mathbf{V}}^{(1)}_{(k,l)(A)}$ is obtained from the SVD of $\mathbf{H}_{(k,l)(A)} \mathbf{V}^{(\mathbf{0})}{}_{(k,l)(B)}$ given by

$$\mathbf{H}_{(k,l)(A)} \mathbf{V}^{(\mathbf{0})}{}_{(k,l)(B)} = \tilde{\mathbf{U}}_{(k,l)(A)} \tilde{\mathbf{\Sigma}}_{(k,l)(A)} \left[\tilde{\mathbf{V}}^{(1)}_{(k,l)(A)} \quad \tilde{\mathbf{V}}^{(0)}_{(k,l)(A)} \right]^H \tag{9.16}$$

and where $\mathbf{H}_{(k,l)(A)}$ and $\mathbf{H}_{(k,l)(B)}$ are the MIMO channels between the transmitter and receivers (A) and (B), respectively, and $\mathbf{P}_{(k,l)(A)}$ is a diagonal matrix representing the power allocation for each layer. The BD precoding matrix $\mathbf{W}^{\mathrm{BD}}_{(k,l)(B)}$ for receiver (B) is calculated from the above by simply swapping A and B.

It can be seen that calculation of the BD precoding matrix for one receiver requires knowledge of the basis for the *null space* $\mathbf{V}^{(\mathbf{0})}{}_{(k,l)}$ of the co-scheduled receiver(s). On the other hand, the PMI feedback in LTE (as described in Section 9.2.3) provides information about only the *signal space* of the channel to each UE; that is, an approximation of the dominant singular vectors from each $\mathbf{V}^{(\mathbf{1})}{}_{(k,l)}$. With knowledge of only this partial subspace information, the ability of the transmitter to mitigate interference is limited to spatial separation of the signals to each UE at the transmitter side with judicious selection of co-scheduled UEs. On the other hand, if full subspace information were available, the transmitter would be able to *jointly* determine the precoding matrices for all UEs in order to minimize (or even completely avoid) mutual interference[7].

Given the above discussion, there are three main issues in specifying an enhanced feedback scheme for downlink MU-MIMO in LTE-A:

- It may, depending on the impact on overhead, be preferable for the feedback scheme to include information on both the signal and null subspaces and their relative significance (i.e., the corresponding singular values).
- Improvement of the accuracy of feedback information; that is, minimizing the effect of error due to limited feedback.
- Commonality of the SU-MIMO and MU-MIMO feedback schemes, and testability to ensure the expected performance is achieved in all scenarios.

Various proposals have been made for enhanced feedback schemes to meet these requirements. Regarding the first issue, the *companion PMI* scheme [Com09] involves feedback of both the preferred and least preferred precoding matrices from a codebook, which can be considered as (low rank) approximations to the bases for the signal and null subspaces. As a basic implementation, UEs might be co-scheduled (*paired*) where the preferred PMI for one is the least preferred for the other, and vice versa.

[7]Note that, in general, the jointly determined precoding matrices will not be equal to a codebook entry, even if the feedback scheme is codebook-based.

An alternative proposed scheme involves quantized feedback of the *short-term* (sample) spatial covariance matrix of the channel to each UE $\mathbf{R}_{(l)} = \mathcal{E}[\mathbf{H}^H_{(k,l)}\mathbf{H}_{(k,l)}]$ (averaged over a sub-band), which can be written as

$$\mathbf{R}_{(l)} = \mathcal{E}\left[\mathbf{V}_{(k,l)}\mathbf{\Sigma}^H_{(k,l)}\mathbf{\Sigma}_{(k,l)}\mathbf{V}^H_{(k,l)}\right] \quad (9.17)$$

It can be seen that $\mathbf{R}_{(l)}$ contains information on both the full subspace (i.e. all right singular vectors) and the corresponding singular values [Eff10]. These covariance matrices might then be used with an adaptation of the BF scheme, or some other criterion such as maximum signal to leakage plus noise ratio (SLNR), in order to jointly determine the corresponding precoding matrices. The covariance matrices may also be used to select co-scheduled UEs, taking into account both the strength and degree of mutual orthogonality of the corresponding channels. This latter proposed scheme is a type of *explicit* feedback, whereas feedback schemes that are conditional on the transmission mode (e.g., the LTE feedback scheme, where the PMI assumes a certain RI and the CQI assumes a certain PMI) are described as *implicit* feedback.

Regarding the second issue, improving the accuracy of channel information feedback is relevant to the performance of both the MU-MIMO and SU-MIMO schemes. In principle, this might be achieved by increasing the codebook size, however this comes at the expense of greater feedback overhead. An alternative approach is to support a dynamic codebook, where the definition of its entries can be changed in response to slowly varying channel characteristics that are known at both the transmitter and receiver. One generic framework for this approach defines the precoding matrix as [Fle10]:

$$\mathbf{W} = \mathsf{G}(\mathbf{W}_{\text{PMI1}}, \mathbf{W}_{\text{PMI2}}) \quad (9.18)$$

where $\mathsf{G}()$ represents an arbitrary matrix operator, $\mathbf{W}_{\text{PMI1}}$ is a matrix that represents the slowly varying channel characteristics defined by occasional feedback of PMI1, and $\mathbf{W}_{\text{PMI2}}$ is a matrix that represents the instantaneous channel defined by frequent feedback of PMI2.

For example, if $\mathsf{G}()$ is the matrix product operator and $\mathbf{W}_{\text{PMI1}}$ represents the square root of the *long-term* spatial covariance matrix of the channel $\mathbf{R}^{\frac{1}{2}} = \mathcal{E}[\mathbf{H}^H_{(k,l)}\mathbf{H}_{(k,l)}]^{\frac{1}{2}}$ (averaged over both a sub-band and multiple slots), then this scheme resolves to conventional covariance-based adaptive codebook feedback [Ada09] where PMI2 is determined from a base codebook. The accuracy of feedback is improved without significantly increasing the feedback overhead since the base codebook is effectively conditioned by the spatial statistics of the channel.

Regarding the third issue, a common feedback structure for both SU-MIMO and MU-MIMO would be attractive since it would allow for dynamic switching between the two modes depending on the instantaneous channel conditions. On the other hand, while in principle explicit feedback including null space information is optimal for MU-MIMO, implicit feedback of the preferred precoding matrix and corresponding CQI is sufficient to optimize for SU-MIMO. Since in practice the feedback overhead (e.g., codebook size) must be quite small, determination of the overall best scheme is non-trivial, and requires exhaustive system-level simulations of realistic deployment scenarios.

Finally, it was decided that the basic feedback structure will remain similar to that of LTE. For the case of eight transmit antennas, a dual codebook scheme will be used as per Equation 9.18, where the recommended precoder is equal to the product of the two matrices. Codebook entries for the left-hand matrix have a block-diagonal structure in order to optimize for the spatial channel correlation that occurs with a transmit array formed from dual-polarized antennas. For the case of four transmit antennas, due to the reduced gains found with the proposed techniques, it was decided to defer any codebook enhancements to future releases and reuse the existing LTE scheme. Note that this does not imply that there will be no enhancement of MU-MIMO gains compared to LTE (Release 8). For example, the specification of DM-RS (which removes the restriction that the precoder matrix used by the eNB must be an entry in the codebook), improvements to dynamic SU/MU-MIMO switching, and CQI feedback accuracy enhancements will in themselves lead to significant performance gains.

Uplink

In the uplink, MU-MIMO is supported in a similar way to LTE, where the eNB may co-schedule multiple UEs, and multiuser detection at the receiver is implementation specific. The precoding matrix for each UE is determined by the eNB—the scheme is much simpler than for the downlink since the eNB can directly estimate the MIMO channel for all eNBs from uplink reference signals. The precoding matrix is indicated to each UE using the same codebook-based scheme as for uplink SU-MIMO. In principle, the eNB may additionally take into account the separability at the receiver of signals from each UE when determining the corresponding precoding matrices.

9.3.5 Coordinated MultiPoint Processing

Coordinated MultiPoint (CoMP) processing is a new technique studied for LTE-A, which is intended particularly for extending coverage of high data rates into cell-edge regions. CoMP involves the coordination of multiple *points* to assist in communications with each UE, and incorporates techniques sometimes categorized in the literature as *network MIMO*. In general, a point is a transceiver at an eNB associated with one cell, in principle it may also be a relay node.

Two types of CoMP are possible: *intra-eNB* CoMP involves multiple points within a single eNB (i.e., multiple cells of the same site), while *inter-eNB* CoMP involves points associated with different eNBs [3GP10a]. The multiple points that coordinate for transmission to a given UE form a *cooperating set.*

Downlink

For the purpose of studying CoMP for the LTE-A downlink, two categories of CoMP have been defined: *Joint Processing* (JP) requires the availability of data to be transmitted to a given UE at multiple points, while *Coordinated Scheduling/Beamforming* (CS/CB) requires the availability of data at just a single point, but multiple points cooperate in order to reduce interference with that transmission.

Within the JP category there are two subcategories defined: *Joint Transmission* (JT) involves the transmission—either mutually coherent or incoherent—of data signals from multiple points simultaneously, while *Dynamic Cell Selection* involves transmission of data signals from just a single point, where the transmission point may rapidly and dynamically change over time.

Coherent JT essentially combines all transmit antennas within the cooperating set into a single distributed array, with which in principle the greatest MIMO gains can be realized. However, it implies very stringent requirements on the synchronization and mutual phase coherency of the cooperating points. On the other hand, incoherent JT and Dynamic Cell Selection can be considered as (macro)diversity combining and switching techniques, respectively, which can also take advantage of the MIMO framework. They do not require RF phase coherency, but still require time-domain synchronization with accuracy much better than the cyclic prefix length.

Techniques in the CS/CB category do not necessarily imply additional synchronization requirements, but still require cooperative resource scheduling and joint determination of precoding matrices among all points in each cooperating set.

In fact, it has been decided that, for LTE-A, only intra-eNB CoMP techniques will be possible, in order to avoid impacting the backhaul load and synchronization requirements (e.g., the X2 interface between eNBs), and because the performance gains using techniques developed so far are not yet well defined. Note that this does not imply that the points must be co-located—the physical antennas for each cell may be implemented in Remote Radio Heads (RRH) that are physically separated from the base-station processor and connected by optical fibre. However, since most of the complexity is limited to within a single eNB, most aspects can be left to implementation. Nevertheless, support for CoMP does imply specific support in the reference signal structure in order to obtain the necessary information on the channels between all cooperating points and the UE: this issue is described in Section 9.3.6.

Uplink

Uplink CoMP involves the capability for signals transmitted from each UE to be received at multiple points. In principle, this does not require explicit support in the standard, since the way that the points may cooperate to jointly decode the received signals can be implementation specific. However, it does imply support in the reference signal structure so that the RS transmitted from each UE can be received at all cooperating points without significant interference from other UEs in the cooperating cells that are allocated the same resources, as discussed in Section 9.3.6.

9.3.6 LTE-A Reference Signals

Downlink

There are three types of reference signal defined for the LTE-A downlink:

1. Common Reference Signals (CRS),
2. Demodulation Reference Signals (DM-RS),
3. CSI Reference Signals (CSI-RS).

CRS are cell-specific reference signals that are sparsely allocated in frequency and time as per the LTE common RS—they are to be used primarily for back compatibility with LTE UEs.

DM-RS are UE-specific reference signals that are also sparsely allocated in frequency and time (on different resources to CRS), they are used by LTE-A UEs for data demodulation. DM-RS are precoded in the same way as the data, and so allow demodulation even if the actual precoding matrix is unknown at the receiver. Therefore, the DM-RS overhead depends on the transmission rank, and a CDM-based technique Orthogonal Cover Coding (OCC) is used to reduce the number of REs dedicated to DM-RS [Way10].

CSI-RS are cell-specific reference signals that are used by LTE-A UEs for determining channel state information feedback. CSI-RS are sparsely *punctured* into the data region of selected (configurable) subframes; that is, the data modulation symbols on certain REs in these subframes are replaced by CSI-RS symbols. Since LTE (Release 8) UEs are unaware of CSI-RS, they will act as interference for data decoding. However, since the puncturing is relatively sparse (the design requirements for CSI-RS are less severe than for DM-RS since they are not used for channel equalization), the degradation in effective SINR is minimal.

In a CoMP configuration, each UE may be required to estimate the MIMO channel between multiple points and itself using the CSI-RS transmitted from each point. Therefore, CSI-RS from each point in a cooperating set can be scheduled in different subframes, and schemes have been proposed so that certain data REs transmitted from other points can be muted in order to minimize the interference caused to CSI-RS [CSI09].

Uplink

In the LTE-A uplink, the two reference signal types DM-RS and SRS from LTE are retained.

Since the LTE-A uplink supports multiple transmit antennas per UE, the DM-RS are precoded in the same way as the data. Therefore, the number of Cyclic Shift (CS) positions required per UE is equal to the transmission rank. However, while 12 cyclic shift positions are available over the SC-FDMA symbol dedicated to DM-RS, in practice it is preferred that not all are used in order to reduce residual interference that exists between the corresponding channel impulse responses. In addition, since CS are used to orthogonalize DM-RS from co-scheduled UEs in MU-MIMO and also associated UEs in uplink CoMP, it is considered that the number of cyclic shift positions is inadequate. Therefore, it has been proposed that the two SC-FDMA symbols dedicated to DM-RS in one subframe are combined using an Orthogonal Cover Code (OCC), which can be considered as a simple CDM resulting in 24 effective cyclic shifts. OCC is suitable only for the low/medium mobility case where the channel is approximately constant over the two adjacent slots, however in practice this is also the case where high-rank transmission with MU-MIMO and/or CoMP can be expected.

CHAPTER

Multiple Antenna Terminals 10

Buon Kiong Lau

Implementation of (co-band[1]) multiple antennas in terminal devices is a significant challenge. On one hand, it is desirable to include many antennas, in order to take full advantage of MIMO technology; for example, to increase spectral efficiency. On the other hand, the number of antennas that can be effectively employed is fundamentally limited by the compact sizes of today's terminals. In this context, this chapter is organized as follows:

- The problem at hand is presented as a size-performance trade off.
- An example compact prototype quantifies performance degradation due to electromagnetic coupling of closely spaced antenna elements.
- Two classes of techniques that enable the use of more antennas per unit volume are summarized, namely antenna decoupling and antenna-channel matching.
- The chapter concludes with a discussion on some related issues and future outlook.

Due to the complexity of the problem, and the myriad of antenna types that one could employ, it is very difficult to make definitive and general conclusions. Instead, the goal of this chapter is to provide readers with a framework to correctly model and understand compact antenna arrays, together with the resulting insights on good engineering practices. Moreover, since a complete general theory for multiple antenna transmission limits between confined volumes that includes mutual coupling is still not available, the approach here is to apply the knowledge and evidence at hand to gain intuition into how compact arrays behave, and how to reasonably design them and improve their performance.

10.1 SIZE-PERFORMANCE TRADE OFF

Even though the implementation of multiple antennas in mobile terminals has attracted growing interests in recent years, research in the area can be traced back to

[1]The co-band case is treated here, since the interest is to use multiple antennas of the same band to exploit the spatial dimension for performance gains. Moreover, co-band antennas represent in the worst case of mutual coupling.

MIMO. DOI: 10.1016/B978-0-12-382194-2.00010-1

the classic work on spatial diversity reception by Beverages and Peterson in the 1920s and 1930s [Dun44, BP31, PBM31]. The pioneers found by experiments that signals received at physically separated locations fluctuated independently of one another. The diversity system in those days operated in the short wave range (3–30 MHz) and the antennas are distributed across acres of land. In the 1960s and 1970s, important new results were published in the area of diversity reception [Cla68, Jak71]. However, it was not until the major deployment of diversity reception technique in GSM base stations that multiple antenna systems found recognition as a mainstream wireless technology.

On the other hand, diversity-based multiple antenna systems remained out-of-bounds for mobile terminals, due to cost and complexity considerations, which were invariably tied to the power consumption and battery life of such devices. In addition, the introduction of transmit diversity [Ala98] further suggests that system complexity can be largely confined to the base station for both uplink and downlink transmissions. Therefore, implementing multiple antennas in terminals remained a secondary issue in the standardization of third generation WCDMA and CDMA2000 systems.

Nevertheless, with the ever growing demand in the data rate of wireless systems, employing multiple antennas in terminals has become a key technology in achieving the required performance, as can be seen in IEEE802.11n [Per08], LTE [DPS08], and WiMAX [LLZR09] standards. For example, LTE requires mobile terminals to be equipped with at least two antennas [DPS08]. As an intermediate step, some companies have started to offer mobile devices (for example, USB dongle and PC card) with HSPA diversity reception capability [PL09].

In reality, even though recent advances in technology have allowed multiple antenna terminals to be implemented at acceptable costs, the number of antennas that can be implemented in today's terminal devices is severely limited by their compact sizes. In contrast, the space restriction is far less severe at the base station. And even though the angular spread of incoming electromagnetic waves as seen at the base station is typically smaller than that at the mobile terminal, techniques such as macro diversity may be applied to alleviate the problem of higher spatial correlation. Therefore, terminal size is the primary factor limiting theoretical MIMO capacity gains in wireless systems.

The two fundamental limitations on the number of antennas that can be implemented on a mobile terminal are:

1. Physical size – Since practical antenna elements occupy finite volumes, the number of terminal antennas is upper bounded by the volume of the terminal (or more strictly, the available volume in the terminal for antenna implementation).
2. Spatial correlation and electromagnetic coupling – Closely spaced antennas suffer from high spatial correlation, which reduces the degrees of freedom (DoFs) of the antenna array. In addition, strong mutual coupling exists between closely spaced antennas, and coupling directly reduces the radiated and received power[2].

[2]It will be shown later in this chapter that the problems of correlation and coupling may be compensated with multiport impedance matching networks, at the cost of bandwidth.

The second aspect can be intuitively understood from the point of view of spatial sampling: MIMO relies on sampling in the spatial domain, which is analogous to a conventional single antenna system's sampling in the frequency domain. Therefore, confining the MIMO transmit and receive antennas (or spatial samples) within a fixed volume has a similar effect as limiting the spectral resource available to the communication system, that is, the performance saturates when the available DoFs are exhausted.

Although the near term goal is to implement a modest number of elements, such as the two-antenna prototype in [PLDY09], MIMO systems with significantly more antenna elements have been tested. As an example, NTT DoCoMo in Japan had conducted a field test for 12×12 MIMO at 4.6 GHz, where the mobile station (i.e., a large van) had a roof mounted 12-element dipole array with adjacent element spacing of 20 cm (or 3.1 wavelengths) [TDHS07, TH07]. Therefore, it is both relevant and important to consider the fundamental limit which governs the size-performance trade off.

Unfortunately, there is no rigorous extension of Chu's limit [Chu48] to multiple antennas. For single antenna, Chu's limit tells us the upper bound of the gain-bandwidth performance as a function of a, when the antenna structure is completely enclosed in a sphere of radius a. However, Chu's limit is a loose bound, since most antennas do not make full use of the available spherical volume. Recently, a new performance bound has been derived for an antenna enclosed in a volume of arbitrary shapes [GSK09]. As a result, it provides a tighter performance bound than the Chu's bound for commonly considered antenna structures.

Even though a rigorous performance bound for multiple antennas is an open problem, there is no question that the bound exists. To illustrate this, one can apply Chu's limit to the context of multiple antennas, if we assume that each of the antennas is ideally coupled to one of the orthogonal electromagnetic radiation modes; that is, the so-called mode-coupled antennas [GN07]. The Bode-Fano criterion [Poz05] can then be applied to optimally match the impedance at the output ports of each mode-coupled antenna. However, this approach assumes ideal antennas that are not mutually coupled, which do not exist in reality. Hence, the derived bound is optimistic.

From a more practical standpoint, it has been found in the context of circuit theory that multiport impedance matching network can be used to maximize power transfer from source to load in a multiport electrical circuit [HA59], in the same way as the complex conjugate match maximizes power transfer for the single-port case. This idea has since been carried over to antenna theory to match multiple antennas [AR76, VBA03, WJ04a, WJ04b].

As it turns out, the so-called multiport conjugate match [WJ04a] is not only optimal from the point of view of power transfer irrespective of antenna separation, but it also brings about zero signal correlation in propagation environments with uniform 3D Angular Power Spectrum[3] (APS) (i.e., in the limit of rich scattering). In fact,

[3] In general, this implies that incoming waves are distributed uniformly in Direction-of-Arrival (DoA) and polarization.

signal correlation can also be expressed in terms of the scattering (or S-) parameters of reflection and coupling coefficients [BRC03], a result that can be traced to the earlier work of Stein [Ste62]. For two antennas[4], the envelope correlation ρ_e (or correlation between the envelopes of the signals), which should not be confused with complex correlation ρ_c [VBA03], can be approximated by their reflection (S_{11}, S_{22}) and coupling (S_{21}, S_{12}) coefficients:

$$\rho_e \approx |\rho_c|^2 = \frac{\left|S_{11}^* S_{12} + S_{21}^* S_{22}\right|^2}{(1 - |S_{11}|^2 - |S_{21}|^2)(1 - |S_{22}|^2 - |S_{12}|^2)}. \tag{10.1}$$

Since the multiport conjugate match ensures that $S_{11} = S_{22} = S_{21} = S_{12} = 0$ at the center frequency, zero correlation is achieved at that frequency. In non-uniform 3D APS environments, the multiport conjugate match can be tweaked to achieve zero correlation, though this additional procedure does not fundamentally improve the performance of the MIMO system [WJ04b].

Since MIMO capacity depends on both correlation and received power, and the multiport conjugate match can offer zero correlation and maximum power transfer at any antenna separation, the matching condition appears to facilitate linear capacity gain with the number of antennas, even if the antennas are packed arbitrarily close together. This implies that the capacity is only limited by the number of antennas that can be physically accommodated within the terminal. Not surprisingly, this simple logic is flawed, due to the requirement of nonzero bandwidth for any communication system.

A simple example based on two identical, parallel half-wave ($\lambda/2$) dipoles can be used to illustrate the impact of antenna separation on bandwidth [LAKM06b]. The separation between the dipoles is varied in steps, and at each step, the frequency responses of the dipoles with multiport conjugate match are obtained (see Figure 10.1). In Figures 10.1(a) and 10.1(b), the frequency response of S-parameters for reflection (S_{11}) and coupling (S_{21}) coefficients are shown for multiport conjugate match realized using the approach in [DBR04] (realization 1). As can be observed, the bandwidths of both S_{11} and S_{21} for realization 1 becomes smaller with decreasing separation. The uneven frequency response over antenna separation is the result of the existence of multiple local solutions and the design procedure in [DBR04] not giving the global solution [LAKM06b].

Figures 10.1(c) and 10.1(d) give the corresponding results for the odd and even modes produced by an alternative realization of multiport conjugate match obtained using hybrid 180° couplers [DBR05] (realization 2). The coupling between the odd and even modes is not shown in Figure 10.1, since it has a wider bandwidth and thus not the limiting factor. In this case, the even mode stays relatively wideband at small antenna separations, whereas the bandwidth of the odd mode vanishes with decreasing separation. This can be understood by that the two antennas effectively become only one (i.e., the even mode) as the separation becomes arbitrarily small.

[4] It is trivial to extend the result for an arbitrary number of antennas (see [Stj05]).

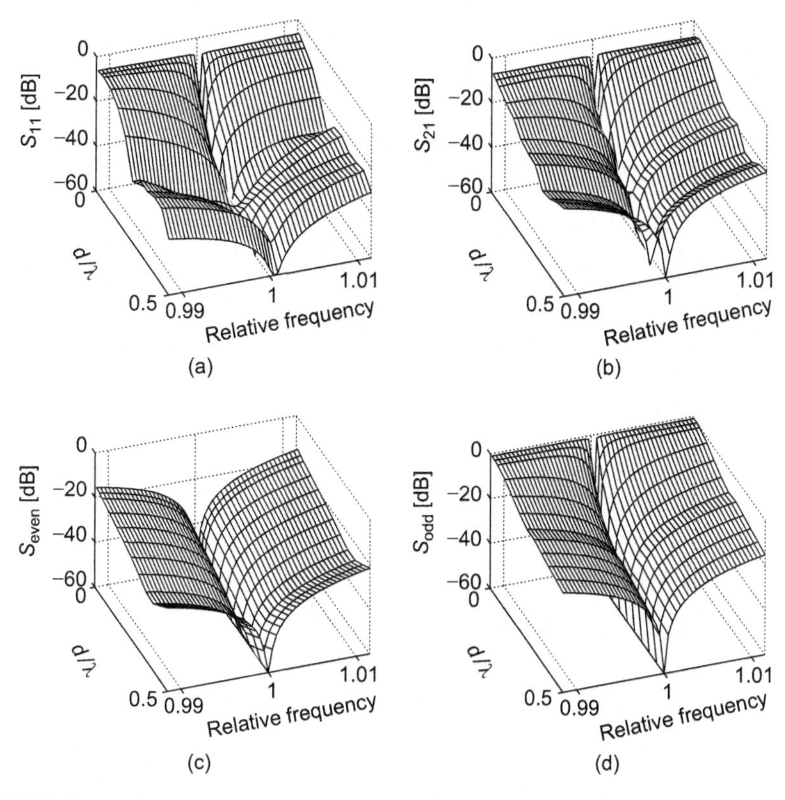

FIGURE 10.1

(a) Reflection coefficient and; (b) coupling coefficient for realization 1 [LAKM06b]; (c) Even mode and; (d) odd mode reflection coefficients for realization 2.

It is clear from Figure 10.1 that bandwidth is the price to pay for achieving very small antenna separations for both circuit realizations of the multiport conjugate match. Consequently, for MIMO systems, and particularly for compact implementations, it would be more appropriate to characterize performance in terms of spatial-spectral efficiency, instead of spectral efficiency in bits/s/Hz.

A step in this direction had been taken by Poon et al., who considered spatial DoFs using a signal sampling approach, which conventionally only takes into account spectral-temporal characteristics of the signal [PBT05]. The "ultimate spatial DoFs" is given by [PBT05]:

$$\min\{2\mathcal{A}_t|\Omega_t|, 2\mathcal{A}_r|\Omega_r|\}, \tag{10.2}$$

for $\mathcal{A}_t|\Omega_t|, \mathcal{A}_r|\Omega_r| \gg 1$, where $\mathcal{A}_t$ and $\mathcal{A}_r$ denote the effective apertures of the transmit and receive arrays, respectively. $|\Omega_t|$ and $|\Omega_r|$ are solid angles spanned by scattering clusters in the environment of interest as viewed from the transmitter and receiver, respectively. However, it should be emphasized that since the focus of

[PBT05] is on the signal space, the gain-bandwidth issue of physical antennas does not appear in the derivation of (10.2).

10.2 PERFORMANCE OF COMPACT DESIGN

As discussed in Section 10.1, implementing multiple antennas in a compact terminal necessarily involves a performance trade off, when the number of antennas exceeds a certain threshold. In this section, we provide an implementation example of a representative compact multiband dual-antenna prototype to illustrate and to quantify the impact of mutual coupling on diversity and capacity performance [PLDY09, PSLDY09].

It should be noted that several other compact multiband dual-antenna prototypes had been designed and evaluated in simulations [Ave07]. However, it suffices for our present purpose to only focus on one representative design. Moreover, other compact designs in [Ave07] have been observed to exhibit similar behavior.

10.2.1 Dual-Antenna Prototype

The dual-antenna prototype is enclosed in a volume of $100 \times 43 \times 9\ \mathrm{mm}^3$, which corresponds to the typical size of today's mobile handsets. As shown in Figure 10.2(a), the two multiband antennas are mounted on a ground plane with the dimensions of $85 \times 40 \times 2\ \mathrm{mm}^3$ [PLDY09].

The main antenna (port 1) is a multiband monopole, with one of the branches forming a patch with dense meandering end for the WCDMA850 band. The antenna is placed at the bottom end of the prototype (i.e., the left side in Figure 10.2(a)). The diversity antenna (port 2) is a multiband Planar Inverted F Antenna (PIFA) with a shorted parasitic branch for the UMTS band. Each of the antennas cover the entire receive bands of 869–894 MHz (WCDMA850), 1805–1880 MHz (WCDMA1800)

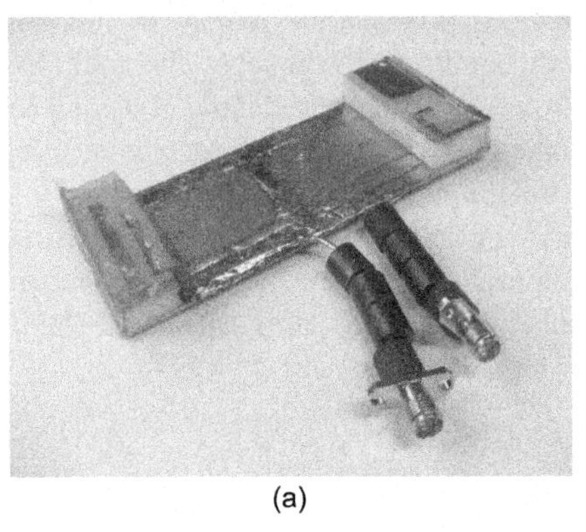

(a)

(b)

FIGURE 10.2

(a) Diversity prototype, and; (b) reference prototype [PLDY09].

and 2110–2170 MHz (UMTS) at 6 dB impedance bandwidth. The monopole antenna also covers the corresponding transmit frequency bands. In addition, since the motivation of implementing multiple antennas in terminals depends on their performance gain with respect to single antenna terminals, a reference multiband single-antenna prototype is also constructed based on only the main antenna of the diversity prototype, as shown in Figure 10.2(b).

One benefit of studying the multiband antenna structure is that the effect of coupling at different frequency bands may be compared. This is particularly the case when each of the two multiband antennas has only one feed point and approximately co-located phase centers. The separation distance between the feeds is 85 mm, i.e., the feeds are nearly maximally separated along the $85 \times 40 \times 2\ mm^3$ ground plane[5]. In terms of electrical wavelength λ, it corresponds to 0.24λ, 0.52λ and 0.61λ for the WCDMA850, WCDMA1800 and UMTS bands, respectively [PLDY09]. Since the higher bands of WCDMA1800 and UMTS are close to each other in frequency, there are effectively only two cases to consider - the "low band" (WCDMA850) and the "high band" (WCDMA1800 or UMTS).

10.2.2 S-parameters and Envelope Correlation

The S-parameters of the diversity and reference prototypes are given in Figure 10.3. For those who are familiar with common array antennas such as quarter-wave ($\lambda/4$) monopoles and half-wave ($\lambda/2$) dipoles, coupling coefficient of -4 dB at the low band in Figure 10.3(c) may seem excessively high for the 0.24λ feed separation distance. In fact, it has been found in [WJ04a, WJ04b] that when there is rich scattering, capacity and diversity gain only start to deteriorate when spacing between two dipoles is reduced to below 0.2λ[6]. Moreover, it was shown in [OC08] that severe degradation in antenna performance is observed only when the distance between adjacent monopoles implemented on a fixed ground plane (for PC card application) is reduced from 0.15λ to 0.1λ.

However, one important distinction between the low and high bands is the extent to which the ground plane contributes to the radiation of the antenna elements. At the low band, the largest dimension of the prototype is only 0.27λ and the entire structure (including its ground plane) is integral to each element's characteristics[7].

[5] In general, larger feed separation implies lower coupling. However, as will be explained in Section 10.2.2, other mechanisms can dominate the coupling performance. The ability of the prototype to excite orthogonal polarizations will also be discussed.

[6] The main reason that the full diversity (or capacity) gain is maintained down to 0.2λ is that the drop in antenna efficiency due to an increase in coupling is partially offset by the increase in angle diversity of the radiation patterns. At even smaller spacing, the loss of efficiency is even more severe *and* the angle diversity reduces, thus the MIMO performance degrades. Incidentally, this is the reason that earlier studies that neglect the loss of efficiency, such as [SR01], mistakenly concluded that coupling can actually *increase* capacity as the antenna spacing is reduced.

[7] Incidentally, this is also the reason behind the typical 40–50% efficiency value for terminal antennas at the 850 MHz band, as opposed to 60–70% for the 1800 MHz and 2100 MHz bands.

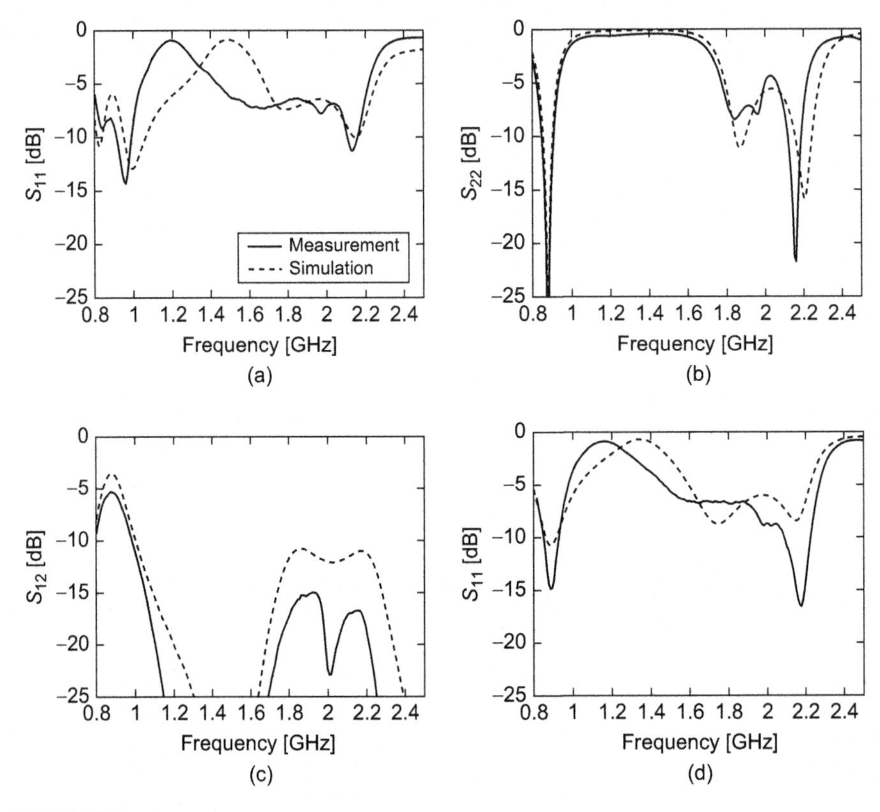

FIGURE 10.3

For the diversity prototype, reflection coefficients of the (a) main antenna and; (b) diversity antenna, and; (c) coupling coefficient; and (d) reflection coefficient of the single antenna reference prototype [PLDY09].

The antennas will be severely detuned if the size of the ground plane is modified. Therefore, each "antenna element" acts primarily as a coupling element for radiation in the ground plane [VOKK02, VOKV06]. In this context, it is easy to understand that the mutual coupling observed between the two antennas at the low band can be significantly higher than the coupling between two $\lambda/2$ linear dipoles or two $\lambda/4$ monopoles for the same feed separation distance. In other words, the significant current distribution on the ground plane at the low band has a more dominant influence on coupling than the feed separation distance.

Another direct consequence of the ground plane being an integral part of the low-band antenna radiation is that the dimensions of the ground plane determine the ability of the antenna structure to excite different polarization. In this case, since the length of the shorter side of the ground plane is merely 0.12λ, which is approximately half the length of its longer side, the structure is expected to support only one dominant polarization.

Since the ground plane plays a lesser role in the antennas' radiation for the high bands, the antenna separation (in wavelength unit) could potentially be smaller than that of the low band for a given level of mutual coupling, as in the case of dipole arrays. One such example can be found in [GCYP07], where a conventional dual-element PIFA array that operates in the 2.5 GHz band and has an antenna separation of 0.2λ gives a coupling coefficient of between –7 and –8 dB. This is in stark contrast to the -4 dB coupling coefficient obtained for the dual-antenna prototype in the low band with an antenna separation of 0.24λ.

As can be expected from (10.1), the severe coupling between the prototype's antennas at the low band causes an envelope correlation of around 0.6 [PLDY09], which is higher than the rule of thumb threshold of 0.5 for good diversity gain [VA87]. On the other hand, the correlation at the high band is negligibly small [PLDY09].

10.2.3 Diversity and Capacity Performance

In order to relate the antenna and correlation parameters to system level performance, diversity and capacity performance metrics are evaluated. For simplicity, the propagation channel is modeled by uniform 3D APS, which is a fairly good match to indoor environments with rich scattering. 3D radiation patterns are obtained for three frequencies (upper, middle and lower frequencies) in each of the three operating bands.

For dual-antenna diversity with selection combining, one can obtain the Cumulative Distribution Function (CDF) for the received power of the combined signal from the average SNR of each antenna branch and the correlation coefficient [PLDY09]. Mean effective gain (MEG) can be used in place of average SNR, if the noise is assumed to be the same across the branches. The correlation coefficient and MEGs are calculated from the radiation patterns and the assumed APS. The diversity gain is then obtained as a ratio of the SNR of the combined signal relative to that of a reference signal at a given probability level. As a result, the choice of the reference case can have a dramatic impact on the diversity gain. Specifically, *apparent* diversity gain (or simply diversity gain) is the SNR gain (in dB) of using two coupled elements compared to using the best coupled element alone (in other words, the element with the stronger average received signal). *Effective* diversity gain utilizes a 100% efficient antenna in free space as the reference, and *actual* diversity gain is referred to a standalone single antenna with no coupling. Since coupling reduces the efficiency of the coupled antennas, including the one used as the reference antenna, it is likely that apparent diversity gain will be artificially high. On the other extreme, with its conservative 100% efficient reference antenna, effective diversity gain provides the lower bound for diversity gain performance. Therefore, the focus here is on actual diversity gain, which compares the dual-antenna prototype to a realistic single-antenna prototype [PLDY09]. In the event of user interaction (as in the following section), the reference antennas for the apparent and actual diversity gains are both subjected to the same user interaction as the dual-antenna prototype, to ensure fair comparisons.

Table 10.1 Simulated Actual Diversity Gain (dB) with Selection Combining at the 1% Probability Level for Different User Interaction Cases

Frequency Band	Free Space	Hand	Head	Hand & Head
850 MHz	5	6	9.5	8
2100 MHz	9	9	10	9

It is found in [PLDY09] that the (apparent) diversity gain with selection combining at the 1% probability level for the prototype in free space is 8 dB for the low band, which is 2 dB lower than the i.i.d. Rayleigh case. On the other hand, the actual diversity gain is only 5 dB (see Table 10.1). The 3 dB difference is due to the reference antenna for diversity gain being the best (coupled) antenna of the dual-antenna prototype (i.e., a degraded reference due to coupling), whereas the reference antenna for actual diversity gain is the single-antenna prototype shown in Figure 10.2(b). Since the efficiency of either antenna in the dual-antenna prototype is significantly lower than the that of the antenna in the single-antenna prototype due to coupling loss, the latter reference offers a lower diversity gain value. Nevertheless, actual diversity gain as defined in this manner is expected to be more practical for antenna designers who must decide whether it is worthwhile to replace conventional single (co-band) antennas with two or more (co-band) antennas [PLDY09]. For the high band which does not suffer from severe coupling, the diversity gain and actual diversity gain stay close to the i.i.d. Rayleigh case of 10 dB [PLDY09]. A comparison of the actual diversity gain of the prototype between the low and high bands (in free space) in Table 10.1 reveals that coupling is mainly responsible for the 4 dB loss in performance in the low band, as compared to the high band.

The ergodic capacity is evaluated in a simulation environment by combining the uniform 3D APS channel with the antenna radiation patterns (see e.g., [WJ04a]). For the MIMO case, the 2×2 channel assumes ideal uncoupled and uncorrelated antennas at the transmit end and the dual-antenna prototype at the receive end. Full channel information is assumed at the transmitter, so that the waterfilling procedure is performed on the synthesized channel.

SISO and MIMO ergodic capacities for the single- and dual-antenna prototypes, respectively, are presented in Figure 10.4 for two frequencies: 850 MHz (low band) and 2100 MHz (high band) [PSLDY09]. The capacity calculations assume a reference SNR of 20 dB. For comparison, the capacity of the 2×2 i.i.d. Rayleigh case with waterfilling is calculated to be 11.3 bits/s/Hz. It is observed in Figure 10.4(b) that the free space ergodic capacity of 9.5 bits/s/Hz at the high band is lower than the i.i.d Rayleigh case, and this is mainly due to the nonideal efficiency of the terminal antennas at the high band. A quick comparison of the free space case between the low and high bands reveals a difference of 2 bits/s/Hz, which highlights the degradation at the low band due to coupling and correlation effects.

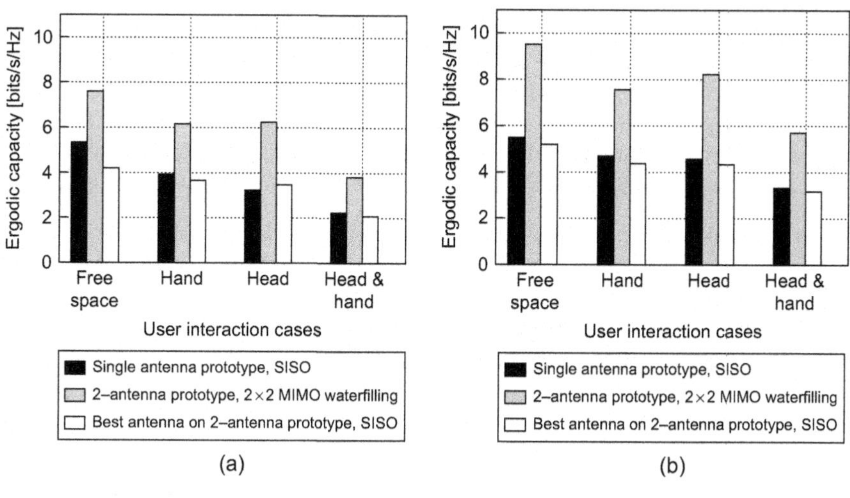

FIGURE 10.4

Ergodic capacity for single- and dual-antenna prototypes at (a) 850 MHz and; (b) 2100 MHz.

10.2.4 Effect of User Interaction

It is important to consider the effect of user on the performance of compact multiple antenna terminals, since the use of such a device almost invariably involves the presence of hand and/or head of the user. The diversity and capacity performances of the aforementioned dual-antenna and reference single-antenna compact prototypes have been studied for such user interactions in [PSLDY09, PLDY09].

In addition to the reference case of no user interaction (or free space), three user interaction cases are investigated: Data mode (hand held position), talk position without the hand (head only position), and talk position with hand (head and hand position). The phantom hand and head used in the study are obtained from IndexSAR [PLDY09].

The actual diversity gain for different user interaction cases are summarized in Table 10.1. As can be seen, the actual diversity gain at the high band (2100 MHz) are 9–10 dB for all four cases, which is nearly identical to the case of i.i.d. Rayleigh. This indicates that the introduction of user hand and/or head does not significantly impact on the diversity gain, when comparing the dual-antenna prototype to the single-antenna prototype for a *given* user interaction case. Moreover, the correlation coefficients are very low for all user interaction cases (see Table III in [PLDY09]). However, the effective diversity gain for the dual-antenna prototype in the presence of hand and/or head is significantly lower than that of the free space case. This is because effective diversity gain always assumes an *ideal 100% efficient reference antenna in free space*, regardless of user interaction. On the contrary, the reference single-antenna prototype used to obtain actual diversity gain experiences body absorption and detuning (or mismatch) in the presence of hand and/or head (see Figure 8 in [PLDY09]).

More interestingly, the data from the low band reveals that when the reference single-antenna prototype is also subjected to the same user interaction as the dual-antenna prototype, the actual diversity gain can actually improve relative to the free space case. This phenomenon is primarily the result of two related mechanisms:

1. The body absorption and detuning experienced (by the monopole) in both the single- and dual-antenna prototypes are comparable. However, the presence of the user reduces the coupling that is experienced by the dual-antenna prototype compared to the free space case.
2. The correlation performance improves in the presence of a user, which results from the interaction between the user and the antennas giving higher pattern diversity than the free space case.

The ergodic capacity for both the single- and dual-antenna prototypes under different user interactions is shown in Figure 10.4. In the low band, the introduction of the user hand or head causes the capacity to drop by 1.5 bits/s/Hz. This is primarily due to a drop of up to 5 dB in the MEG of the antennas, which is only marginally offset by a corresponding drop in envelope correlation [PSLDY09]. On the other hand, if the comparison is made in the framework of capacity gain of the dual-antenna prototype with respect to the single-antenna prototype for a given user interaction case, the introduction of the user hand or head is actually favorable. In particular, the percentage increase in capacity for the hand only and head only cases are 50% and 90%, respectively, as compared to the free space case of 40%. A further 5 dB drop in the MEG when both hand and head are present causes a further 1 bit/s/Hz drop in the dual-antenna capacity, although the improvement is 71% relative to the single-antenna prototype in the same user interaction case.

As can be seen in Figure 10.4, the high band gives consistently better capacity performance than the low band. This is due to the antennas having higher efficiency due to improved isolation and the prototype's larger electrical size. In addition, it exhibits greater robustness to user interactions, which is a consequence of the antenna element being the dominant source of radiation, instead of the entire structure including the ground plane. The capacity increase from the SISO setup to the 2×2 MIMO setup for any given user interaction case is between 60–80%.

Moreover, the SISO capacity performance of the best antenna on the dual-antenna prototype is found to be slightly lower than that of the single-antenna prototype, with only one exception (i.e., user head interaction at the low band). This is mainly the result of the degradation in MEG due to coupling, and it suggests that comparisons with the best antenna on the multiple antenna prototype is often optimistic.

For all cases, the gain imbalance for the dual-antenna prototype is modest (at 0–3 dB), and is maximum at 3 dB for the user hand interaction [PSLDY09]. Therefore, the imbalance is only a minor contribution to any degradation in diversity and capacity performance.

10.3 COMPACT DESIGN TECHNIQUES – ANTENNA DECOUPLING

In the preceding section, we have shown that the performance of compact multiple antenna terminals can degrade significantly due to coupling and correlation effects, which illustrates the size-performance trade off given in Section 10.1. Therefore, the proper question to pose is the following, "Given the fundamental limitation and trade off, how can we make the best of what is available?" In this and the following sections, we focus on two classes of techniques that are suitable for improving the performance of compact multiple antenna terminals:

1. Antenna decoupling – The goal is to mitigate coupling of multiple antennas, without any regard to the propagation environment.
2. Antenna-channel matching – The goal is to achieve a good match between the antenna and the corresponding propagation channel (see Section 10.4).

Complete decoupling of N antennas at the transmit and/or receive ends implies that the mutual impedances of the N-antenna array are reduced to zero, i.e., $Z_{ij} = 0, i \neq j, \{i,j\} = 1,\ldots,N$. This is identical to achieving zero in all transmission (or coupling) coefficients $S_{ij}, i \neq j$ in the S-parameter representations for the given array. In practice, some decoupling techniques aim to reduce coupling, rather than completely cancel coupling (e.g., the use of modified ground plane, as summarized in Section 10.3.2). In any case, since the impedance mismatch of an array is a function of both self-matching and coupling characteristics [LAKM06b], the goal of decoupling antennas implicitly requires self-matching, i.e., S_{ii} should be small.

Apart from improving efficiency, another motivation for decoupling is that a reduction in mutual coupling is often linked to a reduction in correlation. In an environment with uniform 3D APS and when certain conditions are satisfied, the relationship between the two quantities is given by (10.1). As can be seen, a reduction in the magnitude of S_{ij} at a given phase necessarily reduces the correlation. From another perspective, one can utilize decoupling to maintain a given level of array efficiency and correlation (and thus capacity), despite adding more antennas into a compact terminal.

In order to achieve decoupling between closely spaced antennas, one can employ external matching circuits for a *given* antenna array. Alternatively, one can also modify the antenna/array structure to achieve decoupling (*and* good self-matching). The latter optimization approach is optimal for a given form factor from a decoupling point of view, since decoupling is achieved within the existing antenna/array structure and no additional matching circuit is needed. Both approaches are presented below.

To highlight the potential benefits and limitations of the RF (including antenna) based design techniques presented in this section and Section 10.4, comparisons can be made against baseband techniques. Unlike spatial filtering in baseband with signal processing algorithms, spatial filtering on the RF level with antennas and matching circuits is able to increase the information capacity of the transmission channel. However, for the *receive* end, this requires that the signal can be processed independent of the noise, meaning that the dominant noise should only be introduced to the signal path after the antenna and the matching circuit. This is often referred to as a noise-limited environment.

In an interference-limited environment, if the dominant noise is spatially unstructured interference, there is no advantage in performing receive spatial filtering in RF rather than in baseband. On the other hand, if the interference has a spatial structure (or signature) which is distinct to that of the desired signal, (e.g., in a cellular network with only a few dominant interfering users), then signal processing algorithms can be used to strongly suppress interference, such that signal-to-interference (SIR) ratio is no longer the limiting factor in system performance. As a result, the processed channel effectively becomes noise-limited again. In this situation, it pays to improve SNR via RF-level processing.

10.3.1 Circuit Level Decoupling

The multiport conjugate match introduced in Section 10.1 in the context of size-performance trade off gives the necessary and sufficient condition for the impedance matching network to achieve complete decoupling at the center frequency. However, the realization of the multiport matching network on actual circuit elements is nonunique and the study of different circuit realizations has been an active area of research in recent years. Broadly, the realizations differ from one another in several ways:

- Use of lumped elements (e.g., capacitors) [CW04, WVB$^+$06] or distributed elements (e.g., transmission lines, open/short-circuited stubs) [DBR04].
- Choice of circuit elements, e.g., (rat race) hybrid 180° coupler [DBR05, VWS$^+$08, CY08c, LW08] versus decoupler line [CW04, DBR04].
- Design approach – closed form or exact approach [DBR05, CY08a], as opposed to optimization based approach [DBR04], which can have the problem of local minima [LAKM06b].

As a result of the different matching circuit architectures, the S-parameter performance over frequency differs among them, except at the center (or design) frequency where complete decoupling is achieved. Therefore, different overall bandwidth performance [LAKM06b] can be expected in general[8]. Nevertheless, the bandwidth performance due to any passive circuit realization of the multiport conjugate match is fundamentally limited by the size-performance trade off illustrated in Figure 10.1.

Figure 10.5 illustrates a typical setup of a decoupling circuit (in a black box) for two vertically oriented receive dipoles that are separated by a distance d. The input port to the multiport matching circuit is connected to the antenna ports, whereas the

[8]There are exceptions. The bandwidth performance of the two realizations in Figure 10.1 is nearly the same, despite the use of different (lossless and distributed) circuit elements.

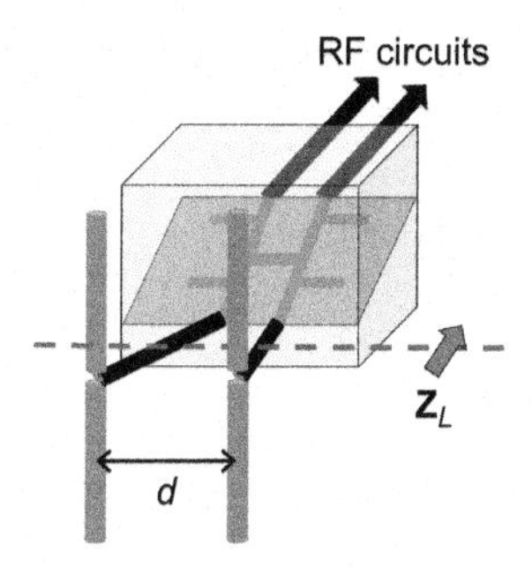

FIGURE 10.5

Multiport matching for two dipoles.

output ports are connected to the rest of the RF circuits. In the diagram, the coupling circuit is a Printed Circuits Board (PCB) with transmission lines, a decoupler line and open-circuited stubs, following the architecture proposed in [DBR04]. The condition for multiport conjugate match is that the load impedance as seen by the antenna ports $\mathbf{Z}_L$, which is the input impedance of the matching network plus the RF circuits, must satisfy[9] $\mathbf{Z}_L = \mathbf{Z}_A^H$, where $\mathbf{Z}_A$ is the antenna impedance matrix.

Recall that the S-parameter performance of two realizations of the multiport conjugate match had been provided in Figure 10.1. Apart from illustrating the trade off between antenna separation against antenna system bandwidth, Figure 10.1 can be used to determine the smallest antenna separation that satisfies a given bandwidth requirement. In order to relate bandwidth requirement to the more system related metric of ergodic capacity, the identical dipoles and multiport conjugate match (realized with a hybrid 180° coupler [DBR05]) of Figure 10.1 are simulated on the receive end of a MIMO system. The transmit end consists of two self-impedance matched dipoles with a fixed antenna separation of 1λ, to ensure low mutual coupling. A propagation environment with uniform 2D APS is assumed. The ergodic capacity with water-filling averaged over fractional bandwidth is presented in Figure 10.6 for several receive antenna separations within the range $d = [0.01\lambda, 0.2\lambda]$. Note that normalization of the channel matrix is performed against a reference uncoupled dual-dipole setup, in order to take into account the difference in antenna gain due to matching. The SNR for the reference setup is 10 dB. As an example, if a bandwidth of 2% is sufficient for a given application, then it can be seen in Figure 10.6 that one can reduce the antenna separation down to 0.1λ with only a small loss in ergodic capacity.

So far it has been explained that the multiport conjugate match facilitates good matching characteristics for compact multiple antenna systems, albeit with a penalty in bandwidth performance. However, the above presented simulation results assume

[9]For reciprocal antennas, $\mathbf{Z}_L = \mathbf{Z}_A^*$ [VBA03].

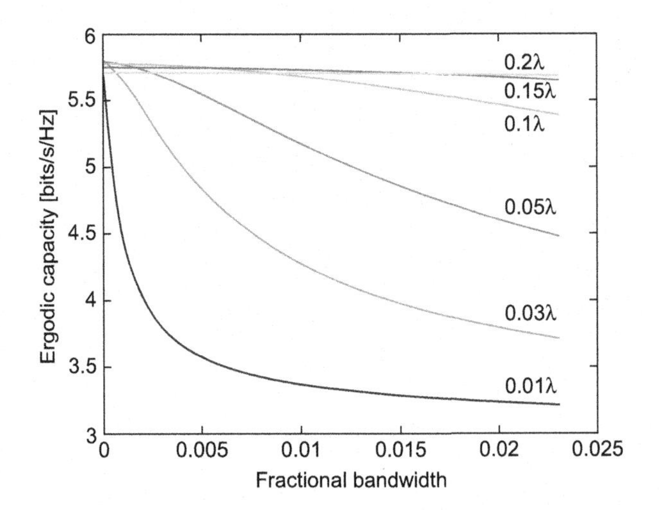

FIGURE 10.6

Ergodic capacity averaged over fractional bandwidth for different antenna separations.

lossless antennas and matching networks, as the intention has been to demonstrate some basic principles. For the purpose of real implementation, practical issues should also be examined:

- Impact of losses and countermeasures [WVB+06, VWS+08, LW08] – Severe losses can be expected for closely spaced antennas due to the compact array's inherent superdirectivity (or supergain) characteristics [MJW05].
- Broadband matching – Since bandwidth is a serious drawback for decoupling networks at small antenna separations, one may attempt to apply broadband matching techniques to alleviate the problem. For example, the addition of resonance(s) is a standard broadband matching technique. Adding more decoupler lines in the realization of multiport conjugate match in [DBR04] will allow more than one resonance to be created in the coupling coefficients $S_{ij}, i \neq j$, providing that a good optimization procedure is formulated for the network. In [VSW+08], broadband matching is applied to the multiport conjugate match that is realized with a hybrid 180° coupler. Since there is good isolation between the odd and even modes over a broad frequency range, common single-antenna broadband matching techniques can be used on the odd mode of the compact antenna array, with the odd mode being the limiting factor for the antenna system bandwidth. However, the higher excitation current associated with the odd mode (with supergain characteristics) results in significantly lower radiation efficiency in the decoupling network, such that the overall odd mode efficiency for the 4% fractional bandwidth case (i.e., 100 MHz bandwidth at 2.45 GHz) is measured to be between 10% and 25%, even though the reflection coefficient is lower than −5 dB. On the contrary, the overall efficiency of the even mode stays above 70%.

- Antenna types – Although the study of multiport conjugate match is mainly based on monopoles and dipoles, in simulation and/or measurements, patch arrays [DRJ+06] and PIFA arrays [VWS+09] have also been considered.
- Size of matching circuits – The reduction in physical size of the multiport matching network is important for compact implementations, especially for distributed elements with lengths of the order of one wavelength [CY08b]. In addition, many existing microwave circuit techniques may be applied to reduce the size of the hybrid 180° coupler (see e.g., [SSWT00]). Realization of the match with small lumped circuit elements is favorable in this respect, provided that the operating frequency is not too high. Its use has been successfully demonstrated at 2.45 GHz [CW04].

10.3.2 Antenna Level Decoupling

The circuit-level multiport conjugate match presented above provides complete decoupling for a *given* antenna array, in the sense that it effectively reverses or cancels the strong mutual coupling as seen at the antenna ports. However, if the structure of the antenna array can be modified, several techniques may be used to realize decoupling among closely spaced antennas. In particular, three groups of techniques are considered here:

1. Modified ground plane,
2. Neutralization line, and
3. Parasitic scatterer.

Modified Ground Plane

Changes introduced in the ground plane of a compact mobile terminal can alter its current distribution and surface waves, and hence the extent of coupling among the terminal antennas. There are many different approaches to modifying the compact ground plane for the purpose of decoupling, and some examples are listed here:

- The introduction of multiple slits or "slitted pattern" on the ground plane has been found to have a bandstop filter effect, which reduces coupling [CCMR07]. The technique has also been shown to work better than having completely disjoint ground planes for the antennas. Implementation examples are provided for monopoles, patches and PIFAs, and the technique has been applied to arrays of up to four elements. A closely related approach is given in [KLF08], where two quarter-wavelength slits in the ground plane between two antenna elements provide the decoupling effect via the magnetic resonance generated in the band of interest.
- A ring shaped defect in the ground plane can be used to reduce coupling for two cylindrical dielectric resonator antennas (DRAs) [GBJS08].
- The use of a small local ground plane for each of the multiple PIFAs that is separated from the main ground plane forces the currents induced by each antenna element to be mainly confined to its local ground plane [GCYP07]. Hence, a degree of decoupling is achieved.

One possible drawback with ground plane modification is that, in practice, the ground plane is shared for many other purposes. Thus, it may not be feasible to introduce a significant change in the ground plane. Another point to note is that, in general, these techniques reduce mutual coupling at the center frequency, but do not induce a deep null in $S_{ij}, i \neq j$, as can be achieved by the two other methods of neutralization line and parasitic scatterer.

Neutralization Line

Instead of remodeling the ground plane, one can simply add a conducting wire of an optimized length to connect between two closely coupled PIFAs on a common ground plane, in order to enhance isolation between them [DLLT$^+$08, DLTK08]. The technique was originally conceived to improve isolation between two single-band PIFAs operating in the adjacent DCS1800 and UMTS bands [DLLT$^+$06].

The technique may seem counterintuitive at first sight, since adding a direct path for current does not seem to be a good solution for reducing coupling. Recall that in our previous discussion on ground plane modification, the reduction of coupling is typically achieved by physically altering the direct current path between the coupled antennas. However, the dilemma can be resolved by considering that an appropriate length of wire also changes the current phase of the additional current path, which can actually help to reverse the effect of coupling. Another way to understand the phenomenon is that the entire compact terminal can be considered a multiport antenna structure. The addition of the line changes the current distributions in such a way that the two ports are now able to excite the orthogonal modes of the structure more effectively.

Measured results in a reverberation chamber, which closely resembles an environment with uniform 3D APS, show that the technique can provide up to 2 dB improvement in diversity gain for two PIFAs working in the UMTS band [DLTK08]. The neutralization line technique has been shown to work well, even when the closest sides of the two PIFAs are only 0.12λ apart [CLD$^+$08]. Moreover, a slight variant of the method has also been introduced in [CLD$^+$08], where a wire connects each PIFA feed to optimized locations on the ground plane, instead of directly connecting the PIFA feeds. In this case, the spacing between the closest sides can be as small as 0.027λ.

Parasitic Scatterer

Another somewhat counterintuitive candidate technique for the decoupling of two closely spaced antennas is to introduce a parasitic scatterer between them. The idea was first proposed as a patent application in [LA07b]. The decoupling scatterer, which can be a replica of one of the coupled antennas, can be designed as a reflector or shield that completely removes mutual coupling [LA07b, LA09]. The use of parasitic elements has been a subject of investigation for many years, although the focus had been to enhance single antenna structures, such as enabling pattern reconfigurability [KKP08, LL07] and creating additional resonances to either enlarge bandwidth [dCDNL07] or enable multiband operation [YD02]. In the context of multiple antennas, an array of parasitic elements has also been applied to lower the sidelobes and improve the directivity of an (active) antenna array [AFRGAP09].

The basic principle of the decoupling parasitic scatterer applies to all single-mode antennas, as well as multimode antennas with one dominant mode. Examples include dipoles, monopoles, PIFAs and variants of the latter two on a compact ground plane. For simplicity, an array of two closely spaced dipoles is used in the following example to illustrate the design procedure [LA09]. In Figure 10.7, the two closely coupled active antennas are labeled antennas "1" and "3", and the parasitic scatterer (in this case another dipole) that is inserted between them is antenna "2". The parasitic scatterer is terminated by a load with impedance Z_L, whereas each of the active antennas is connected to an uncoupled impedance matching network, for reasons that will be explained later. The diameter of the chosen dipoles is $\lambda/400$ and the lengths are identical $L = L_1 = L_2 = L_3$, even though in general this restriction is not necessary. The antenna impedances and radiation patterns are obtained from a method-of-moments (MoM) Matlab script provided by [Mak02]. The self and mutual impedances of the three-element array are denoted by Z_{ii} and $Z_{ij}, i \neq j, \{i,j\} = 1,2,3$. The separation between the two active antennas is set at $2d = 0.1\lambda$.

We begin with the voltage and current relationship for the above setup:

$$\begin{bmatrix} V_1 \\ V_2 \\ V_3 \end{bmatrix} = \begin{bmatrix} Z_{11} & Z_{12} & Z_{13} \\ Z_{21} & Z_{22} & Z_{23} \\ Z_{31} & Z_{32} & Z_{33} \end{bmatrix} \begin{bmatrix} I_1 \\ I_2 \\ I_3 \end{bmatrix} \tag{10.3}$$

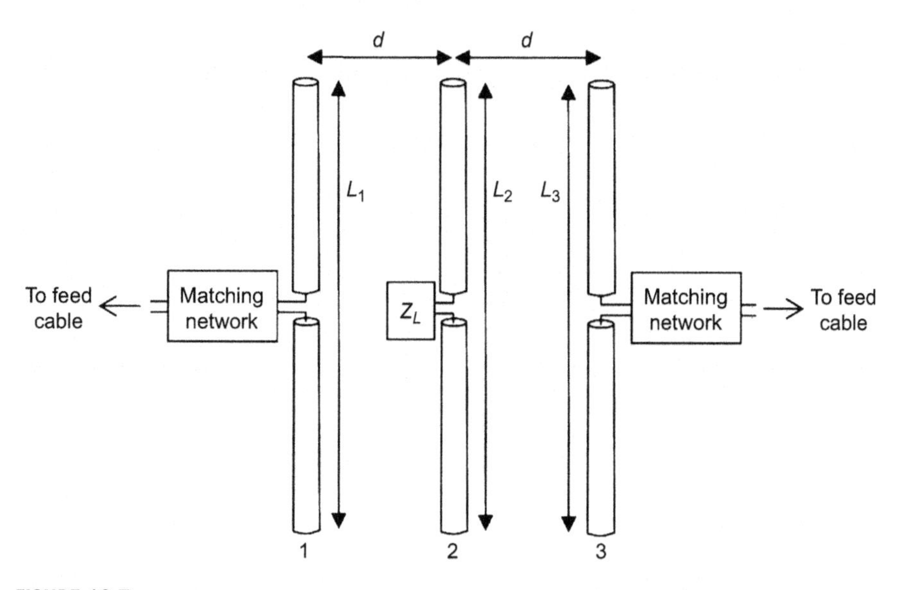

FIGURE 10.7

A decoupling setup with the middle dipole acting as a parasitic scatterer.

or in matrix notation $\mathbf{V} = \mathbf{Z}_A\mathbf{I}$, where V_i and I_i are the voltage and current across the ith antenna port. Given that the antennas are identical and reciprocal for this example, $Z_{11} = Z_{22} = Z_{33}$, $Z_{12} = Z_{21} = Z_{23} = Z_{32}$ and $Z_{13} = Z_{31}$. Moreover, the termination condition for the parasitic scatterer implies that $V_2 = -Z_L I_2$, which upon substitution into (10.3) and rearrangement gives the reduced voltage and current relationships across the ports of the active antennas:

$$\begin{bmatrix} V_1 \\ V_3 \end{bmatrix} = \begin{bmatrix} Z'_{11} & Z'_{13} \\ Z'_{13} & Z'_{11} \end{bmatrix} \begin{bmatrix} I_1 \\ I_3 \end{bmatrix}, \tag{10.4}$$

where $Z'_{11} = Z_{11} - Z_{12}^2/(Z_{11} + Z_L)$, $Z'_{13} = Z_{13} - Z_{12}^2/(Z_{11} + Z_L)$. In order to perfectly decouple the active antennas, we require $Z'_{13} = 0$, i.e.,

$$Z'_{13} = Z_{13} - Z_{12}^2/(Z_{11} + Z_L) = 0 \Rightarrow Z_L = \left\{ \frac{Z_{12}^2}{Z_{13}} \right\} - Z_{22}. \tag{10.5}$$

By separating the real and imaginary parts, and noting that the load resistance should be set to zero which will ideally prevent any ohmic loss in the scatterer

$$R_L = \Re\{Z_L\} = \Re\left\{ \frac{Z_{12}^2}{Z_{13}} \right\} - R_{22} = 0, \tag{10.6}$$

$$X_L = \Im\{Z_L\} = \Im\left\{ \frac{Z_{12}^2}{Z_{13}} \right\} - X_{22}, \tag{10.7}$$

where $R_{22} = \Re\{Z_{22}\}$ and $X_{22} = \Im\{Z_{22}\}$.

The condition (10.6) can be achieved by tuning the identical length of the dipoles L. As illustrated in Figure 10.8, two solutions satisfy this condition, i.e., $L = \{0.377\lambda, 0.482\lambda\}$ and the corresponding load reactances are $X_L = \{137.2\Omega, 21.2\Omega\}$. In general, the input impedances of antennas 1 and 3 are not equal to the reference impedance of 50Ω when the load reactance of antenna 2 is set to one of these two values. Therefore, an impedance matching circuit is needed to transform the impedance of each of the two active antennas (i.e., Z'_{11} and Z'_{33}) to 50Ω, which can be conveniently realized with transmission lines and open-circuited stubs[10] [Poz05]. Note that similar uncoupled impedance matching networks are required for a given realization of multiport conjugate match, except in the present case the decoupling function of the decoupler line [DBR04] or the rat race hybrid 180° coupler [DBR05] is provided by the parasitic element.

The S-parameters versus normalized frequency of the decoupled active antennas using either of the two solutions are shown in Figure 10.9, where S'_{11} and S'_{13} are the S-parameters of the active antennas after the decoupling *and* 50Ω matching procedure. As a reference case, the S-parameters of the two dipoles without the

[10]Lumped elements may also be used for circuit miniaturization.

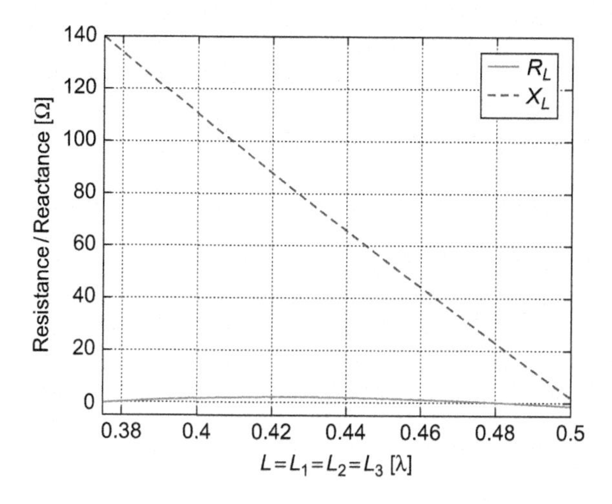

FIGURE 10.8

Load resistance and reactance of the parasitic scatterer for complete decoupling between active antennas versus antenna length.

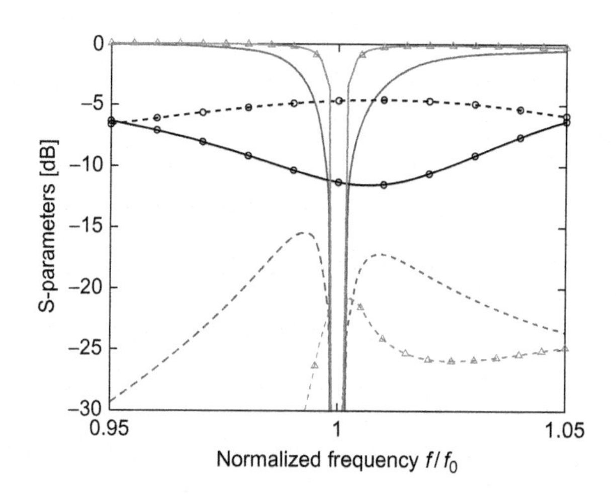

FIGURE 10.9

S-parameters of the two dipole arrays with and without parasitic scatterer. Solid and dashed lines are S'_{11} and S'_{13} for $L = 0.482\lambda$, solid and dashed lines with triangle markers are S'_{11} and S'_{13} for $L = 0.377\lambda$, and solid and dashed lines with circle markers are S_{11} and S_{13} without the parasitic scatterer.

parasitic scatterer are also shown. As expected, complete decoupling is achieved at the center frequency for either of the two solutions. However, the solution with the shorter dipoles gives a more narrowband behavior in S'_{11}, as can be expected from the higher reactance load needed. Comparing with the reference case, it is clear that the decoupling approach gives very good matching performance, albeit for a relatively small bandwidth. It should be noted that the odd mode of the hybrid 180° coupler realization of multiport conjugate match [DBR05], which has a smaller bandwidth than the even mode, has been found to yield similar bandwidth performance.

It is noted that several other papers have recently proposed the use of parasitic scatterer between closely spaced antennas for the purpose of decoupling [MRM08, FGAR08, AMY+09, HCL+08]. However, [LA07b] is unique in recognizing that complete decoupling $S_{ij} = 0, i \neq j$ can be exactly achieved by satisfying the conditions (10.6) and (10.7) using the simple and rigorous procedure described above.

Even though decoupling via antenna-level modifications removes the need for decoupling networks, it is no silver bullet for the problem of bandwidth reduction when antenna separation is reduced, as was demonstrated for the case of decoupling networks. In order words, the size-performance trade off discussed in Section 10.1 is fundamental and necessarily applies to this case as well. The bandwidth problem illustrated for the parasitic scatterer is likewise observed for the techniques of ground plane modification and neutralization line, i.e., the reduced coupling leads to a degradation in the bandwidth of S_{ii}.

10.4 COMPACT DESIGN TECHNIQUES – ANTENNA/CHANNEL MATCHING

Although the decoupling of compact antenna arrays is important for improving system performance, the whole approach implicitly assumes that the antennas are immersed in an environment with uniform 3D APS. Recall that capacity is maximized in a uniform 3D APS environment when the antennas are perfectly matched to cancel coupling, since the signal correlation between any antenna pair is also forced to be zero, as can be seen from (10.1). However, real mobile environments do not often resemble the ideal uniform 3D APS, and is non-stationary. Therefore, a more comprehensive design approach should also take into account both the antennas *and* the propagation environment.

A further motivation of enabling the antennas to be "tuned" to the propagation environment is the significant impact of user interactions on the characteristics of multiple antenna terminals, as can be seen in the example dual-antenna prototype presented in Section 10.2. Therefore, it is important for the antenna-channel tuning to also account for time-varying nearfield user interactions.

In this section, we begin by considering the simpler case where the interest is to optimize the MIMO performance of a *given* compact array in an environment characterized by the second-order statistics of APS [AL06], [LA07a] (see Section 10.4.1). This is followed by a short introduction into one generalization of the approach of Section 10.4.1, which enables the matching of arbitrary radiation/reception characteristics to the APS for optimal diversity performance [JQB08] (see Section 10.4.2).

10.4.1 Circuit Level Matching

Here, we will only treat uncoupled matching networks, i.e., there is one matching network per antenna, and there exists no interconnection between the matching networks. This is because for coupled matching networks, perfect decoupling can be achieved for the antennas, regardless of the environment[11]. Hence, what is left is to combine the *decoupled ports* differently, i.e., in a beamforming setup. Such a beamforming operation may be more conveniently performed at the baseband level, since beamforming in RF using real circuits are lossy and is less flexible than baseband beamforming with today's technology.

Despite the inability of uncoupled matching networks to perform complete decoupling, they may be preferable for their simplicity and physical size, as they do not require circuit interconnections like the hybrid 180° coupler, which also implies additional lossy circuit elements. In general, uncoupled matching networks have also found to give more well behaved solutions than the supergain solutions of multiport conjugate match at very small antenna separations, and consequently they provide a larger bandwidth.

In order to illustrate the dependence of different MIMO performance metrics on uncoupled matching networks, we present in Figure 10.10(a) a MIMO system model consisting of two uncoupled and uncorrelated transmit antennas, a MIMO channel, and two closely coupled receive antennas with uncoupled matching circuits. The assumption of uncoupled and uncorrelated transmit antennas is made to simplify the presentation, without any loss in generality. The equivalent circuit for the two receive antennas of Figure 10.10(b), where the environment is represented as open-circuit voltages V_{oc1} and V_{oc2} for antennas 1 and 2, respectively. The voltages at the load are derived as:

$$\begin{bmatrix} V_1 \\ V_2 \end{bmatrix} = \begin{bmatrix} Z_{L1} & 0 \\ 0 & Z_{L2} \end{bmatrix} \underbrace{\begin{bmatrix} Z_{11}+Z_{L1} & Z_{12} \\ Z_{21} & Z_{22}+Z_{L2} \end{bmatrix}^{-1} \begin{bmatrix} V_{oc1} \\ V_{oc2} \end{bmatrix}}_{\begin{bmatrix} I_1 \\ I_2 \end{bmatrix}}. \quad (10.8)$$

Based on (10.8), closed form expressions for average received power (relative to that of a conjugate matched single antenna) and output correlation at the load can be derived as functions of antenna impedances, load impedance and open-circuit correlation $\alpha = \mathcal{E}\left\{V_{oc1} V_{oc2}^{*}\right\}$ [AL06]. α can be calculated from the APS of the environment and the radiation pattern obtained from each antenna port, with all other ports open-circuited. Similarly, the 2×2 capacity expression for a given channel realization, which assumes two uncoupled and uncorrelated transmit antennas and Kronecker channel model, is derived in [LA06, FFLT08]. For convenience, a closed

[11] One should be aware that, in general, perfect decoupling does not imply zero correlation. In fact, [WJ04b] shows that forcing correlation to zero with multiport conjugate match does not improve the diversity performance.

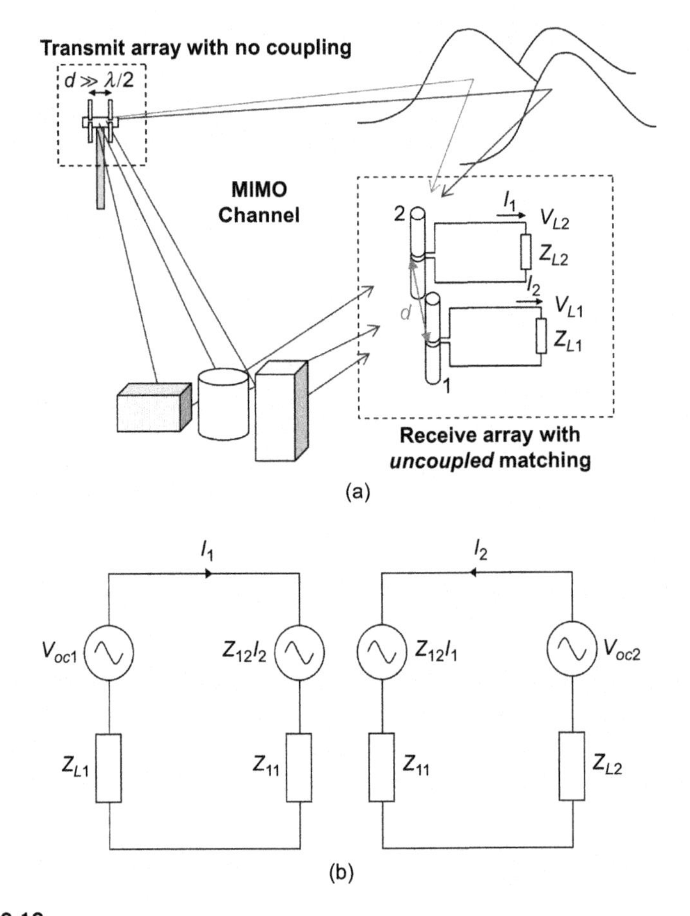

FIGURE 10.10

(a) MIMO system with two uncoupled and uncorrelated transmit antennas, propagation channel, and two closely coupled receive antennas with uncoupled matching; (b) Equivalent circuit for the two receive antennas with uncoupled matching.

form expression for approximate ergodic capacity can also be derived, as was done in [LAKM06a]. Furthermore, these derivations can be extended for more than two antennas, i.e., $N > 2$ [TL08, FFLT08].

For the first example, we impose the constraint $Z_{Li} = Z_L \in \mathbb{C}, i = 1, \ldots, N$, i.e., the load impedances are the same across N antennas. This restriction gives a simpler implementation, in the sense that the performance optimization is performed over the the real and imaginary parts of Z_L (i.e., two parameters), as opposed to the general case of optimizing over the real and imaginary parts of $Z_{Li} \in \mathbb{C}, i = 1, \ldots, N$ (i.e., $2N$ parameters). It is also easier to visualize the two-parameter case, since the performance metric can be plotted over the $\Re\{Z_L\}$-$\Im\{Z_L\}$ (or R_L-X_L) plane. The two identical and parallel dipoles in this example has a diameter of $\lambda/400$ and $d = 0.1\lambda$. As in Section 10.3.2, the antenna impedances and radiation patterns are obtained from a MoM implementation in Matlab [Mak02]. A uniform 2D APS environment is assumed.

Figure 10.11(a) plots the total average received power of the two closely coupled antennas versus load impedance at the center frequency, relative to that of a complex conjugate matched single antenna. In other words, 0 dB implies that the coupled array has the same average received power as the single antenna case. The conjugate match

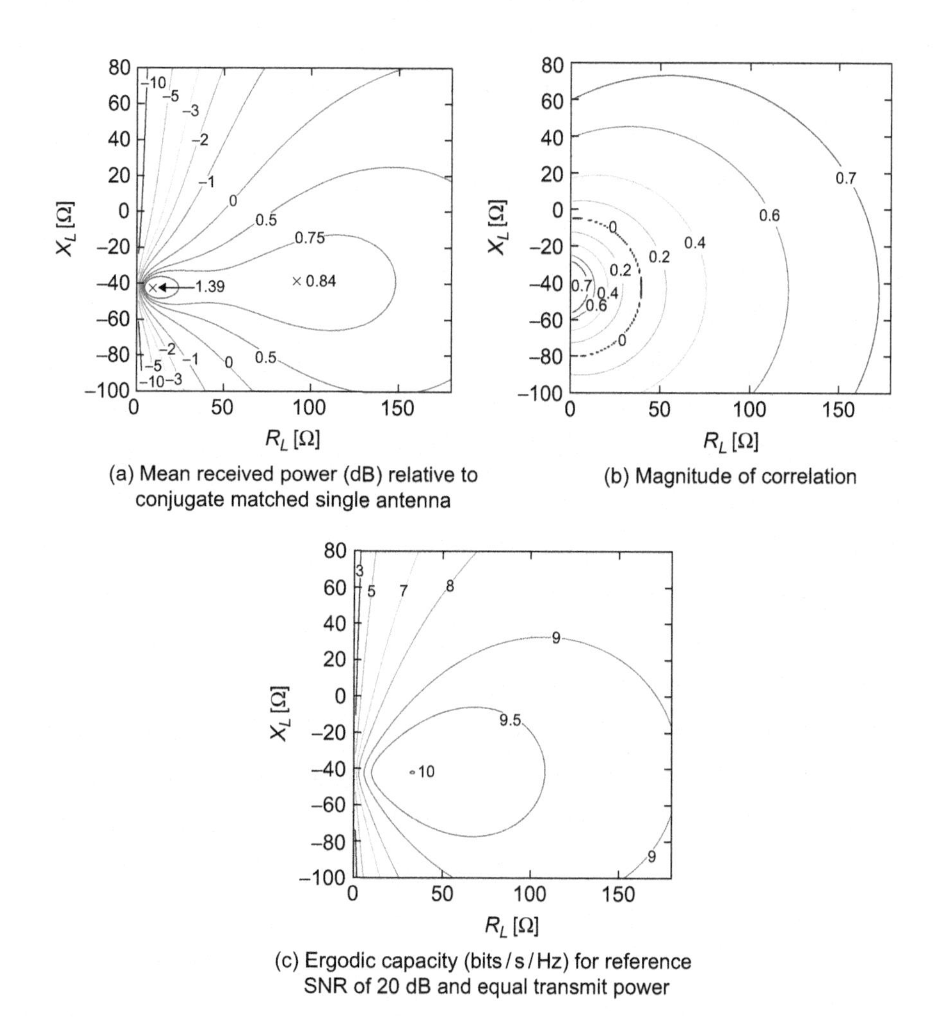

FIGURE 10.11

MIMO Performance for two dipoles with $d = 0.1\lambda$ and uncoupled impedance match in uniform 2D APS.

is used because it maximizes the power received by the single antenna. As can be seen in Figure 10.11(a), there exist two maxima for the received power [AL06]. Apart from the global maximum of 1.39 dB, a local maximum of 0.84 dB is also observed. It turns out that the global maximum is a mild supergain solution, which will gradually disappear as the bandwidth increases. On the other hand, the relatively flat contour around the local maximum of 0.83 dB indicates that it is far less sensitive to changes in different antenna parameters, including bandwidth.

Turning to the correlation performance in Figure 10.11(b), it can be observed that in this case, the solution for zero correlation is nonunique and is achieved along a circular curve [AL06]. This means that one can optimize for received power along the curve of zero correlation. However, in general, zero correlation may not exist when $Z_{Li} = Z_L \in \mathbb{C}, i = 1,2$, as is the case for the nonuniform APS investigated in [LA06].

The ergodic capacity is found for the case of equal transmit power and a reference SNR of 20 dB. The capacity contours in Figure 10.11(c) shows a well behaved capacity surface, with a broad capacity peak of 10 bits/s/Hz. Since the corresponding 2×2 Rayleigh i.i.d. case has a capacity of 11 bits/s/Hz, the close spacing of the dipoles incur a relatively small penalty in capacity of 1 bits/s/Hz. Comparing the optimum load for capacity to the optimum load for maximum received power and that for zero correlation (i.e., the point along the zero correlation curve with the maximum received power), it is observed that the optimum load for capacity is closer to the optimum load for correlation. On the other hand, in the extreme case of small antenna separation (e.g., $d = 0.01\lambda$) or very small angular spread, uncoupled matching is unable to reduce the correlation sufficiently for good capacity. Consequently, the optimized load for capacity is closer to the optimum load for received power [LAKM06a]. In general, the optimum matching condition for capacity is a compromise between the correlation and received power performance, as can be concluded from a comprehensive parametric study reported in [LAKM06a]. Similarly, the merits of adaptive matching over a given fixed matching network also depend on the propagation environment and the antenna separation [LAKM06a].

Simple experiments based on monopoles and uncoupled matching circuits on PCBs have been performed to verify the key principles of optimizing uncoupled matching for MIMO performance and good agreements have been found between measured and simulated results [FLS+06, FFLT08]. The work in [FFLT08] also obtained approximate closed form expressions for optimum capacity match by optimizing for the upper bound of ergodic capacity with Jensen's inequality. The approximate solution has been found to agree well with numerical calculations based on Monte Carlo simulations.

The array configuration can also play a major role in determining MIMO performance, as shown in Figure 10.12 for the cases of a three-element uniform linear array (ULA) and a three-element uniform triangular array (UTA), for adjacent element spacing $d = 0.1\lambda$ and a reference SNR of 20 dB [TL08]. Moreover, comparisons between the use of identical matching loads $Z_{Li} = Z_L \in \mathbb{C}, i = 1,2,3$ and arbitrary loads $Z_{Li} \in \mathbb{C}, i = 1,2,3$ with Laplacian 2D APS of different mean angles (i.e., $0°, 45°, 90°$, with $0°$ being the broadside direction for the ULA) reveal that the

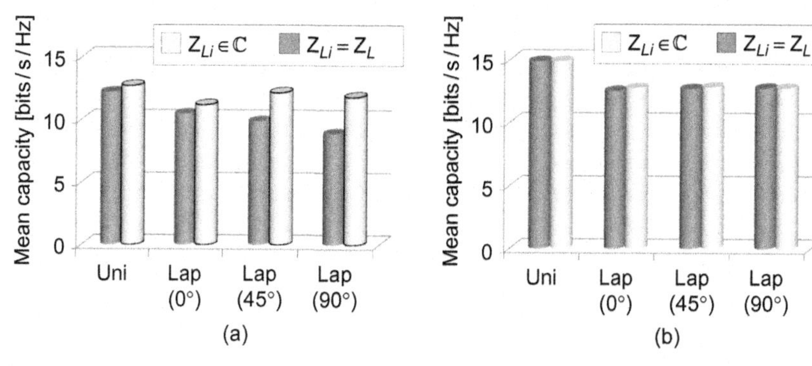

FIGURE 10.12

Ergodic capacity of (Uni)form 2D and (Lap)lacian 2D APS for (a) uniform linear array, and; (b) uniform triangular array, for a reference SNR of 20 dB. The adjacent element spacing is 0.1λ and the standard deviation for the Laplacian 2D APS is 30°.

capacity obtained from identical loads is only slightly lower, if the symmetry between the propagation environment and the array geometry is maintained. As can be seen in Figure 10.12, this condition is approximately satisfied for the UTA, whereas asymmetry is severe for the ULA when the Laplacian cluster is centered at its endfire direction of 90°. To understand this phenomenon, an analogy can be drawn between matching load impedances and array weights in a beamformer, where the beamformer steers its beam towards nonbroadside directions of the signal cluster through the use of nonidentical weights.

Finally, it can be shown that the same decoupling condition $\mathbf{Z}_L = \mathbf{Z}_A^H$ for the coupled matching networks as presented in Section 10.3.1 can be achieved with uncoupled matching networks for a given set of open-circuit voltages [JL10]. This implies that if the uncoupled matching network can adapt to the decoupling condition for an instantaneous snapshot of the channel, then the resulting MIMO performance is equivalent to using the multiport conjugate match. Such an approach would circumvent the need for decoupling circuits. Even though such rapid instantaneous adaptation may be impractical for today's RF circuit technology, it may be an interesting approach for the future.

10.4.2 Antenna Level Matching

Although impedance matching networks can improve the matching between the antennas and the propagation environment, the approach is fundamentally limited by the excitation (and sampling) characteristics of the given transmit (and receive) antenna array. For the receive case, this can be understood by the Thevenin equivalent representation of the antennas, where the open-circuit voltages are sources with fixed values for a given snapshot of the propagation channel. In other words, antennas are spatial filters which have inherent discrimination properties for the polarization, angle and phase characteristics of the propagation channel.

Ideally, the radiation (and reception) characteristics of antennas confined within a given aperture can be directly optimized for good performance against the propagation channel. The principle can be easily demonstrated via a deterministic Line-of-Sight (LoS) environment with no scatterer. In this case, it is obvious that the transmit and receive antennas that are optimally matched to the channel are co-polarized current source and field sensor. It will be of little use to ensure optimum match between the antenna and the RF circuit, when there is severe mismatch between the antenna and the channel, which in this LoS example can be transmit and receive antennas that are orthogonally polarized.

For practical multipath environments that are better described by their statistical properties, radiation and reception characteristics that are optimal for diversity gain can be derived given the APS of the environment [JQB08]. Practical considerations such as adding the constraint of nonoverlapping currents have also been investigated [DEJ09].

10.5 RELATED ISSUES AND FUTURE OUTLOOK

Two interesting developments that are closely related to the design and implementation of compact multiple antenna terminals are described in this section.

10.5.1 Correlation from S-parameters

The conventional approach of obtaining antenna correlation from 3D radiation patterns are both time-consuming and costly, since an anechoic chamber is required and the patterns are measured at each antenna port, one frequency at a time. For this reason, and due to the ease of obtaining the S-parameters of antennas using vector network analyzers (VNAs), it has become increasingly popular to apply (10.1) to obtain envelope correlation (see e.g., [COP07, AYL07, DLLT+08, GP08, ZWYH08]). Equivalently, the correlation expression can also be derived from antenna and load impedance parameters [DK04]. Nevertheless, it should be emphasized that two assumptions are implicit in the derived expression (10.1), namely:

1. The radiation efficiency is 100% for all antennas; and,
2. The environment is described by uniform 3D APS.

Therefore, the conventional technique of calculating correlation from radiation patterns should be used when these conditions are not met. The impact of losses on the accuracy of the expression is the subject of [Hal05, Stj05].

10.5.2 Over-the-Air Performance

Due to the expected large-scale rollout of multiple-antenna terminals in the near future, mobile phone manufacturers and network operators are scrambling to find simple yet accurate ways to characterize and differentiate between terminal products equipped with this feature. Part of this effort is focused on the development of test methodologies for radiated or Over-the-Air (OTA) performance evaluation of such devices. While the test methods for single-antenna terminals concern the metrics of Total Radiated Power (TRP) and Total Isotropic Sensitivity (TIS), they are no longer adequate for multiple-antenna devices, due to such devices' spatial filtering capability. The uniqueness of a multiple antenna system can be seen in the second-order statistics of its receive branches (or more commonly called *covariance matrix*). Apart from the diagonal values which are the average received powers of the antenna branches, which also appear in the special case of single antenna systems, the off-diagonal values (which are related to the correlation between the branches) play a vital role in determining whether the antenna system of the terminal device has the required spatial DoFs to support the expected performance improvement. The statistical independence between branch signals is a particularly challenging design criteria, considering the many constraints in designing the antenna system. It is also vitally important to consider the effect of user on the performance of such devices, since they are almost invariably used in close proximity with the hand and/or head of a user.

Currently, there is excellent coordination between the three main parties involved in the multiple-antenna test method development effort: 3GPP RAN4[12] (http://www.3gpp.org/), CTIA ERP Workgroup (http://www.ctia.org) and EU COST Action 2100 Subworking Group 2.2[13] (http://www.cost2100.org). Consistent to the expertise that exists within the different groups, it was agreed that COST2100 will focus on the anechoic chamber based test methods, e.g., [YSI+06], whereas CTIA is active in coordinating activities to develop reverberation chamber based test methods through its reverberation chamber subgroup (CTIA RCSG). Nevertheless, CTIA has also established a subgroup to look into anechoic chamber based methods in April 2009, which works in close liaison with COST2100 Subworking Group 2.2. The main requirement is that the test method should be good enough for its purpose of characterizing multiple antenna terminals and yet remain simple, cost-effective, backward compatible with single-antenna tests. Among the issues being addressed by the anechoic chamber based method include the performance metrics to measure, the required complexity of the multiple probe measurement system and the channel model to use. In order to coincide with the deployment of multiple antenna mobile systems, preliminary proposals are expected to be ready by the end of 2010.

[12]Working Group 4 under the Technical Specification Group (TSG) of radio access network (RAN) in 3GPP works on the RF aspects of UTRAN/E-UTRAN. RAN4 performs simulations of diverse RF system scenarios and derives the minimum requirements for transmission and reception parameters, and for channel demodulation.

[13]Subworking Group 2.2 has a long history of contributing to (single-antenna) anechoic chamber-based test method development through its predecessors in COST Action 273 and COST Action 259.

10.5.3 Future Outlook

Despite the research progress made in recent years, much work remains to be done for compact multiple antenna systems. Several promising areas of research are identified as follows:

- Most of the existing papers on multiple antennas for compact implementations simply make use of two or more "good" single antenna elements that are placed as far away from one another as possible for spatial diversity and/or with different orientations for polarization diversity. While this approach may suffice at high frequency and/or when the required number of antenna elements is small, a more rigorous design approach of the individual array elements is needed to achieve good overall array characteristics. This is a particularly challenging task when the size of the entire structure, including the ground plane, is electrically small [VOKK02, VOKV06], as is the case for the WCDMA850 band and the existing and upcoming cellular networks in the even lower 700 MHz and 450 MHz frequency bands [NSD+07].
- Innovative antenna solutions can be of interest, e.g., MIMO transmission with a single RF front end using a switched parasitic antenna with orthogonal bases [KKP08], multiport DRA arrays [TPL+09, TPLY10] and the use of metamaterials or metamaterial-inspired structures to reduce antenna size and coupling [MD09, ZE10].
- Practical constraints for terminal antennas are often not taken into consideration in many of the existing papers. In reality, apart from being physically small, terminal antennas are often required to be as unobstrusive as possible. This means that array antenna structures that are conformal to the terminal casing/components, e.g., [LJ09], would be more favorable.
- More accurate modeling of the transceiver chain is crucial for identifying key issues in the design of compact multiple antenna systems. For example, several recent contributions [MJ05, Gan06a, Gan06b, DHGL07, DHGL10] focus on the impact of different noise sources on the performance of receive antenna arrays, including external noise, antenna noise, amplifier noise, and downstream noise. One important result is that when amplifier noise dominates, the multiport conjugate match for maximum power transfer is no longer optimum for SNR. Instead, matching for the noise figure leads to better SNR than the maximum power match, highlighting that the noise behavior should also be carefully considered.

10.6 CONCLUSIONS

In this chapter, an overview of the current challenges and opportunities in compact multiple antenna terminals is presented. The focus is on the big picture, where arguably the most fundamental issue is the size-performance trade off of such devices. In order words, the performance of multiple antenna systems can increase with the number of antennas, due to more spatial samples (or information) being provided in the communication channel, but the possible increase of the number of antennas is severely constrained by the compact sizes of today's terminals and mutual coupling effects. An example compact multiband, multiple antenna prototype serves to highlight the problem of coupling at the WCDMA850 band, where the implementation of merely two antennas results in severe coupling, high correlation and significantly degraded diversity and capacity performance. Two different philosophies of alleviating the coupling problem are introduced—one considering only the antenna performance, whereas the other attempts to also include the propagation environment in its design criteria. Finally, even though multiple antenna systems in their various forms and applications have been around for decades, their application in compact terminals is a relatively new development, and therefore many exciting challenges remain to be addressed in the future.

10.7 ACKNOWLEDGMENT

The author would like to thank the following colleagues and collaborators for contributions to the research summarized in this chapter: Professor Jørgen Bach Andersen of Aalborg University, Denmark; Ms. Vanja Plicanic, Mr. Zhinong Ying, Mr. Thomas Bolin of Sony Ericsson Mobile Communications AB, Lund, Sweden; Mr. Ruiyuan Tian, Professor Andreas F. Molisch and Professor Gerhard Kristiansson of Lund University, Sweden; Dr. Anders Derneryd of Ericsson Research, Gothenburg; Dr. Yuanyuan Fei. The author also thank Professor Michael A. Jensen of Brigham Young University, USA, Assistant Professor Jon Wallace of Jacobs University Bremen, Germany, and three anonymous reviewers for their helpful feedback on the manuscript. Financial support from Vetenskapsrådet (Grant no. 2006-3012) and VINNOVA (Grant no. 2007-01377 and 2008-00970) is gratefully acknowledged.

CHAPTER

Conclusion: MIMO Roadmaps

11

Rodolphe Legouable and Luis Correia

11.1 SYSTEMS AND ROADMAPS

From an operator view, MIMO allows some expectations, especially for: increasing the throughput for inner-cell users thanks to the spatial multiplexing technique, increasing the throughput at the cell-edge by using beamforming or increasing: the cell coverage, the terminal/base station performance through generalization of using advanced receiver schemes and the robustness of control channels reception, among others. So, all these nice features will allow a better experience for the customers and a better capacity for the operators. This section aims at mainly focusing on the MIMO aspects for both radio mobile systems such as 3GPP/LTE and IEEE802.16m (WiMAX). First, we address the roadmaps of both systems and how they want to contribute to become a 4G system as specified by IMT-Advanced (IMT-A). We mainly focus on the basic MIMO configurations proposed in these standards and their perspectives of extension in ongoing standardization processes. Currently, the ITU-R wants to define a "4G system." Two main candidates that are the LTE-Advanced and the evolution of the IEEE802.16m will answer soon to ITU-R and to the IMT-Advanced recommendations for the definition of the 4G standard. Among the main evolutions, the MIMO component is the major technology that will allow the systems to achieve the target requirements. The number of transmit and receive antennas will of course be increased but also some new concepts such as COordinated MultiPoint (COMP), relaying, implementation of beamforming and spatial rejection will be carried out. So, in this part, these new concepts will be described. To conclude, some MIMO perspectives for the future evolutions of the IEEE802.11n, which are under definition via the IEEE802.11ac standardization process, and some wired systems such as for Digital Subscribe Layer (DSL), Power Line Current (PLC), and optics are introduced.

11.1.1 Toward 4G Radio Systems

This subsection aims at first defining the current 3GPP/LTE and IEEE802.16-2009 systems, especially in terms of roadmaps and MIMO specifications and then to introduce their future evolutions that will be used in the IMT-A proposal. Figure 11.1 gives

MIMO. DOI: 10.1016/B978-0-12-382194-2.00011-3

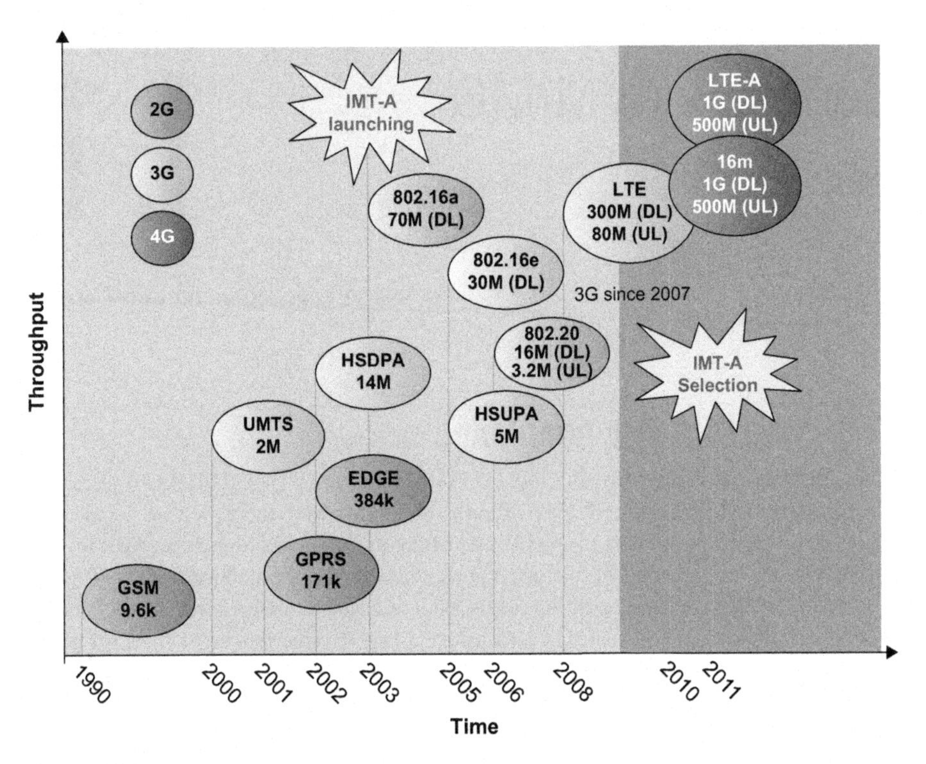

FIGURE 11.1

Positioning of radio cellular systems.

an overview of the situation of the cellular radio mobile systems since the beginning of the 1990s.

Currently the 3G network is complementary to the 2G one and it allows more services, more throughput and a better Quality-of-Service (QoS). At this time being, more than 400 millions of subscribers have a 3G access. However, despite the deployment of the High-Speed Packet Access (HSPA), the quick network congestion has been anticipated. So, in 2003, the International Telecommunication Union (ITU) has launched the International Mobile Telecommunications (IMT) advanced project that aims at defining the 4G network. The main requirement of such 4G system is to obtain throughputs of 1Gbit/s and 500 Mbit/s in down and up links, respectively. The target for IMT-A deployment is around 2013–2015. To anticipate the network congestion, the 3rd Generation Partnership Project (3GPP) has launched in 2004, an exploratory project called Long Term Evolution (LTE), which has defined new key features (new radio interfaces, new core network and so on). The first 3GPP/LTE standard definition has been published at the end of 2008 (in particular, through release 8 - R8, [3GP09b], which indicates an increasing stability as the rate of change requests is steadily decreasing) for first deployments expected in 2010–2011. In parallel to the

3GPP/LTE, the IEEE802.16/WiMax standard also proposes a radio cellular system that is a 4G candidate and which will answer to the IMT-A call for technology based on the IEEE802.16m system definition.

The 3GPP/LTE System

Figure 11.2 gives an overview of the planning of the 3GPP/LTE activities during the two years that allowed to edit the R8. This figure shows the schedule between the different Radio Access Network (RAN) groups that have defined the different layers of the system. Among these groups, RAN1 has the role to define layer 1 (L1) and more specifically the physical layer. At the L1 level, some MIMO schemes have been proposed either to increase the system's throughput by proposing spatial multiplexing schemes or to improve the robustness of the system. In the Down-Link (DL), multiantenna transmission with 2 or 4 transmit antennas is supported. The maximum number of code words is two, irrespective of the number of antennas with a fixed-mapping between code words to layers (for more details, see Chapter 9).

Spatial division multiplexing of multiple modulation symbols streams to a single User Equipment (UE) using the same time-frequency resource, also referred to as single-user MIMO is supported. When a MIMO channel is solely assigned to a single UE, it is known as *SU-MIMO*. Open and closed-loop spatial multiplexing techniques

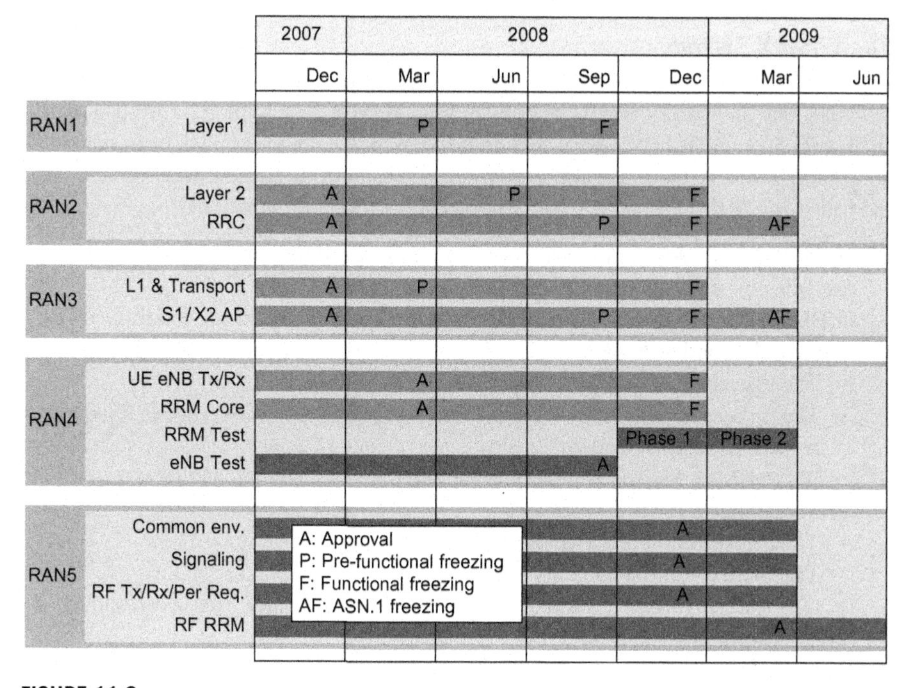

FIGURE 11.2

3GPP/LTE workplan.

are supported. Spatial division multiplexing of modulation symbol streams to different UEs using the same time-frequency resource, also referred to as multiuser MIMO, is also supported. There is semi-static switching between SU-MIMO and MU-MIMO per UE.

In addition, the following techniques are supported:

- Codebook based pre-coding with a single pre-coding feedback per full system bandwidth when the system bandwidth (or subset of resource blocks) is smaller or equal to 12RB and per 5 adjacent resource blocks or the full system bandwidth (or subset of resource blocks) when the system bandwidth is larger than 12RB.
- Rank adaptation with single-rank feedback referring to full system bandwidth. Node B can override rank report.
- Transmit diversity such as space-frequency Alamouti coding scheme.

In the Up-Link (UL), the baseline antenna configuration for UL MIMO is the MU-MIMO. To allow for MU-MIMO reception at the Node B, allocation of the same time and frequency resource to several UEs is supported, each of which transmits on a single antenna. Closed-loop transmit diversity-based on antenna selection shall also be supported for FDD (optional in UE).

The WiMAX System

Figure 11.3 gives an overview of the WiMAX evolution during the 4 last years. It shows the interaction between the mobile WiMAX technology forum and the IEEE 802.16 standard. We can notice that the WiMAX evolution has been constant, addressing in its Releases 1.0 and 1.5, which corresponds to the IEE802.16e standard, the following MIMO scenarios (see Chapter 8 for more explanations):

- In DL, two transmit and two receive antennas are implemented, whereas in UL, one transmit and two receive antennas are considered.
- In DL, MIMO Matrix A that corresponds to Space-Time Block-Coding (STBC) of type Alamouti, allows:
  - The transmission of redundant data streams;
  - The increase of coverage (or increase of capacity if the coverage is fixed) with respect to SISO coverage.
- In DL, MIMO Matrix B that corresponds to the Spatial Multiplexing (SM) technique, allows:
  - The transmission of independent data streams;
  - The increase of the capacity/throughput.
- In DL, the Advanced Antenna System (AAS) technique that corresponds to beamforming provides:
  - The generation of a single beam to a particular mobile station;
  - The increase of coverage (or increase of capacity if the coverage is fixed to non-AAS coverage).

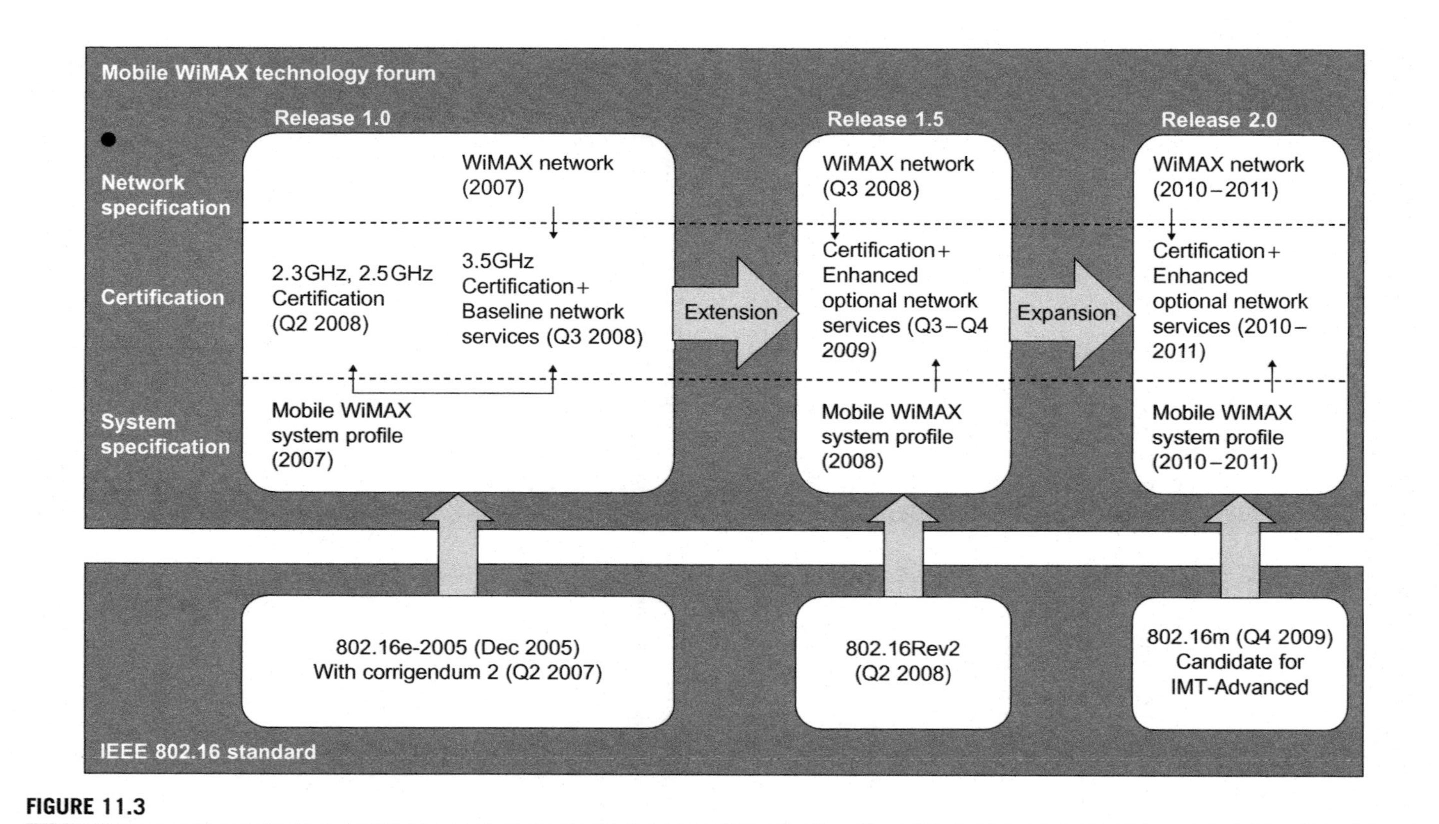

FIGURE 11.3

Mobile WiMAX releases and certification roadmap.

- In UL, the MU-MIMO technique that corresponds to a collaborative MIMO scheme, where the transmission of separate mobile stations with the same resource allocation is carried out.

The DL MIMO A scheme is able to bring a significant gain in terms of Signal-to-Interference plus Noise Ratio DL (SINR) and DL sector/user throughput (especially in an interference limited environment). Therefore, it allows to:

- Increase DL throughput (30%) wherever in the cell;
- Increase coverage with SINR gain in cell edge radio condition;
- Increase coverage with better indoor penetration;
- The greatest part of this gain is, however, brought by Rx diversity (15% of the gain only thanks to Alamouti block coding).

The DL MIMO B is able to double the initial sector throughput, but only when DL SINR is sufficiently high. However, the percentage of a cell that may have this radio condition is very small in realistic networks. In addition, it may not be exploitable in an unitary frequency reuse factor scheme for network deployment.

Consequently, we can remark that IEEE802.16Rev2 (named *IEEE802.16-2009* since its edition in May 2009 [IEE09a]), and 3GPP/LTE consider basically the same MIMO scenarios. Their first deployments are planned progressively from mid-2011 to 2015. However, both systems are currently candidates to the 4G radio mobile system definition, specified by the IMT-A.

LTE-Advanced and IEEE802.16m as Candidates to IMT-Advanced

Both organizations are planning to submit a candidate Radio Interface Technology (RIT) for September 2009 within ITU IMT-Advanced proposals framework. These proposals are not the standards specifications but the technology descriptions (including self-evaluation). To reach the performance targets set by the IMT-Advanced requirements and specified in Figure 11.4, the two groups are considering similar advanced features to bring to existing standards. These advanced features are mainly the following:

- Extension of the MIMO schemes (increased number of antennas, advanced multiantenna processing techniques);
- Support of wider bandwidth (through aggregation of two or more component carriers);
- Relaying–Coordinated multipoint transmission and reception (also called *Multi-BS MIMO*).

Some other features (enhanced multicast and broadcast services, location-based services support, femtocells support, self-organizing network support, and coexistence with other radio technologies) are addressed differently within the two bodies (while already included in existing releases and just requiring enhancements) and will not have a direct impact on systems performance. Then, if both systems are retained by the IMT-Advanced to be 4G systems, the WiMAX release 2.0 will define profiles

Parameter	Requirement		
		UL	DL
Peak spectrum efficiency[1]		6.75 bps/Hz	15 bps/Hz
Average spectrum efficiency[2]	Indoor	2.25 bps/Hz/cell	3 bps/Hz/cell
	Microcellular	1.8 bps/Hz/cell	2.6 bps/Hz/cell
	High speed	0.7 bps/Hz/cell	1.1 bps/Hz/cell
Cell edge spectrum efficiency[2], [3]	Indoor	0.07 bps/Hz/cell	0.1 bps/Hz/cell
	Microcellular	0.05 bps/Hz/cell	0.075 bps/Hz/cell
	Base coverage urban	0.03 bps/Hz/cell	0.06 bps/Hz/cell
	High speed	0.015 bps/Hz/cell	0.04 bps/Hz/cell

(1) Assuming 4×4 MIMO in DL & 2×4 MIMO in UL

(2) Assuming 4×2 MIMO in DL & 2×4 MIMO in UL

(3) Assumes 10 active users per cell

FIGURE 11.4

IMT-Advanced Requirements.

issued from the IEEE802.16m and LTE-A with its Release 10 should be deployed on the horizon of 2013–2015.

Figure 11.5 and 11.6 show the workplan of IEEE802.16m and the LTE-Advanced scheduling, respectively, with respect to the IMT-advanced (ITU-R) planning. At the end of September 2009, the letter ballot phase (term used in IEEE standardization) and study item (term used in 3GPP) that corresponds to the answer to the IMT-A were finished and listed the main technologies and processes that will allow to achieve the IMT-A requirements. Subsequently, both IEEE802.16m and 3GPP/LTE standardization bodies are currently entered in their sponsor ballot and work item phases, respectively. During these periods, the relative standards will be written. In parallel, the ITU-R will evaluate all the systems that answers to the proposal submission, in order to select the 4G system(s). With the most optimistic view, the whole 4G system(s) will be entirely specified during the first quarter of year 2011, targeting a network deployment in 2013–2015.

Among the main technologies that allows us to reach the IMT-Advanced requirements, we of course find the MIMO component. It allows us to offer the necessary coverage and/or capacity gains required by the network at any given time and

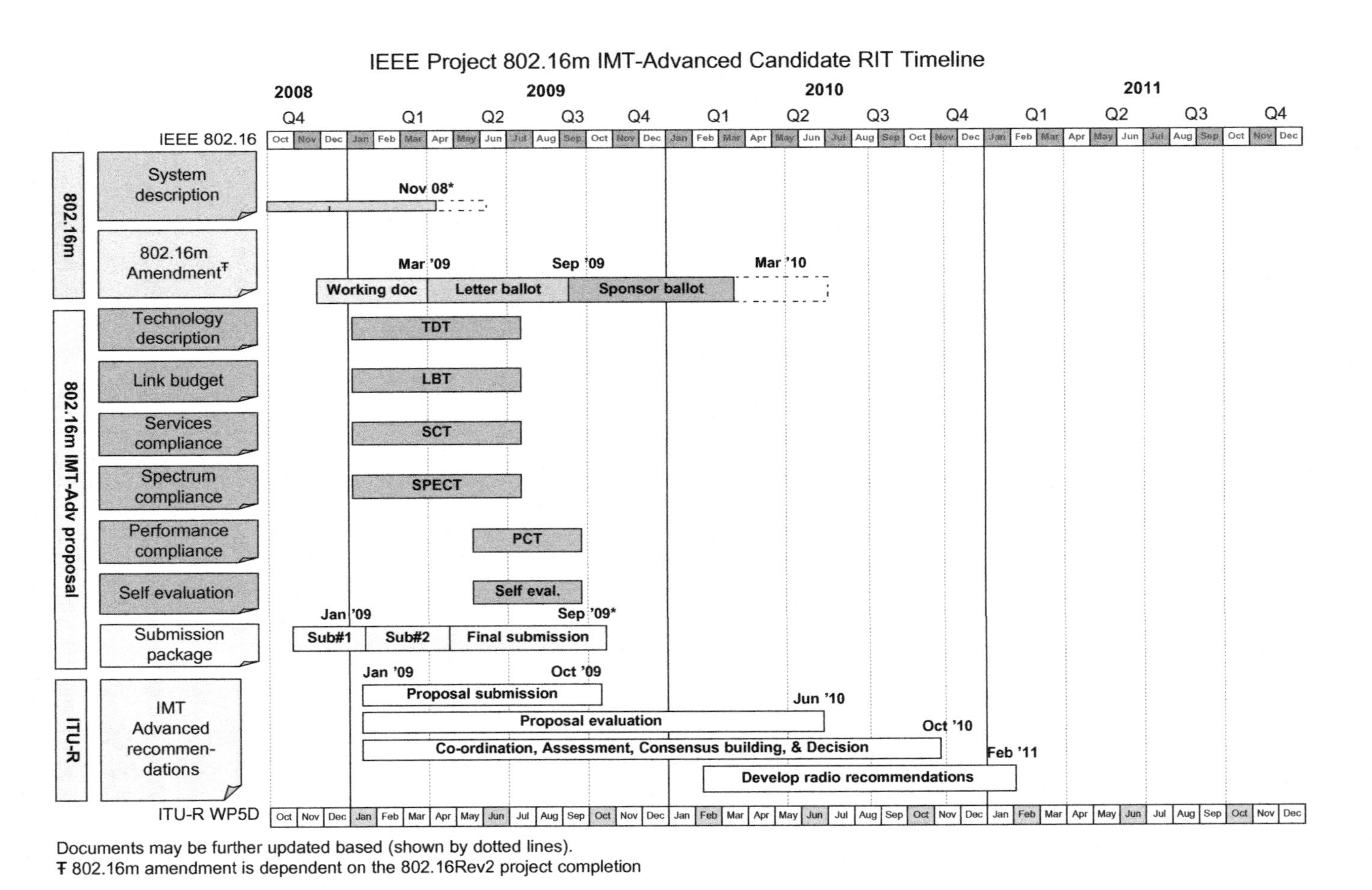

FIGURE 11.5

WiMAX Evolutions–IEEE 802.16m standard development workplan and IMT-Advanced.

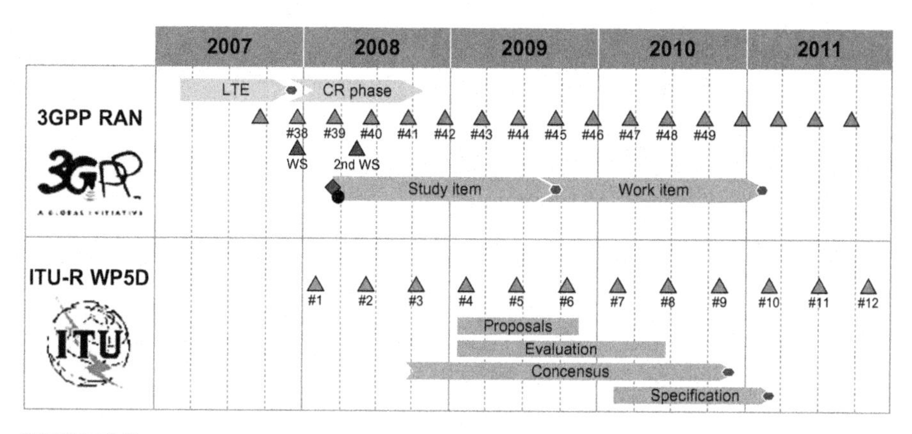

FIGURE 11.6

LTE-Advanced scheduling and IMT-Advanced.

location. Figure 11.7 depicts the generic architecture of the supported MIMO techniques in IEEE802.16m and LTE-A. The number of transmit and receive antennas will increase, with eight antennas at the base station, and four at the UE appearing to be the goal of both systems. Also, advanced multiantenna processing techniques are considered for higher order and multiuser scenarios. Then, the extension of spatial multiplexing up to 4 layers for both DL and UL is suggested. This indirectly implies that the implementation of advanced receivers will likely be unavoidable to achieve co-antenna interference cancellation. Enhanced multiuser MIMO schemes and rank mode adaptation capabilities are also proposed. In addition, the application of MIMO techniques is extended to the preamble transmission, the broadcast and multicast services, and also to the enabling of femtocells, and to the relaying and advanced interference mitigation techniques.

In the following, some advanced MIMO techniques promoted for IMT-A are presented such as COordinated MultiPoint (COMP), relaying and implementation of spatial rejection techniques.

11.1.2 Advanced MIMO Techniques Submitted in 4G

In this section, we describe in some detail the COMP, relays and advanced spatial rejection techniques that are pushed in the 4G system definition.

The Coordinated MultiPoint Technique

Coordinated MultiPoint transmission and reception is proposed for IMT-A and is intended to increase the spectrum efficiency and overall system throughput for both DL and UL at cell edge. The basic idea is that transmission of packets from one or more geographically separated sites to a single UE is dynamically coordinated in some way, relying on fast communication between the involved eNBs and UEs. Figure 11.8 represents a DL coordinated transmission between 3 BS and 3 UEs. The

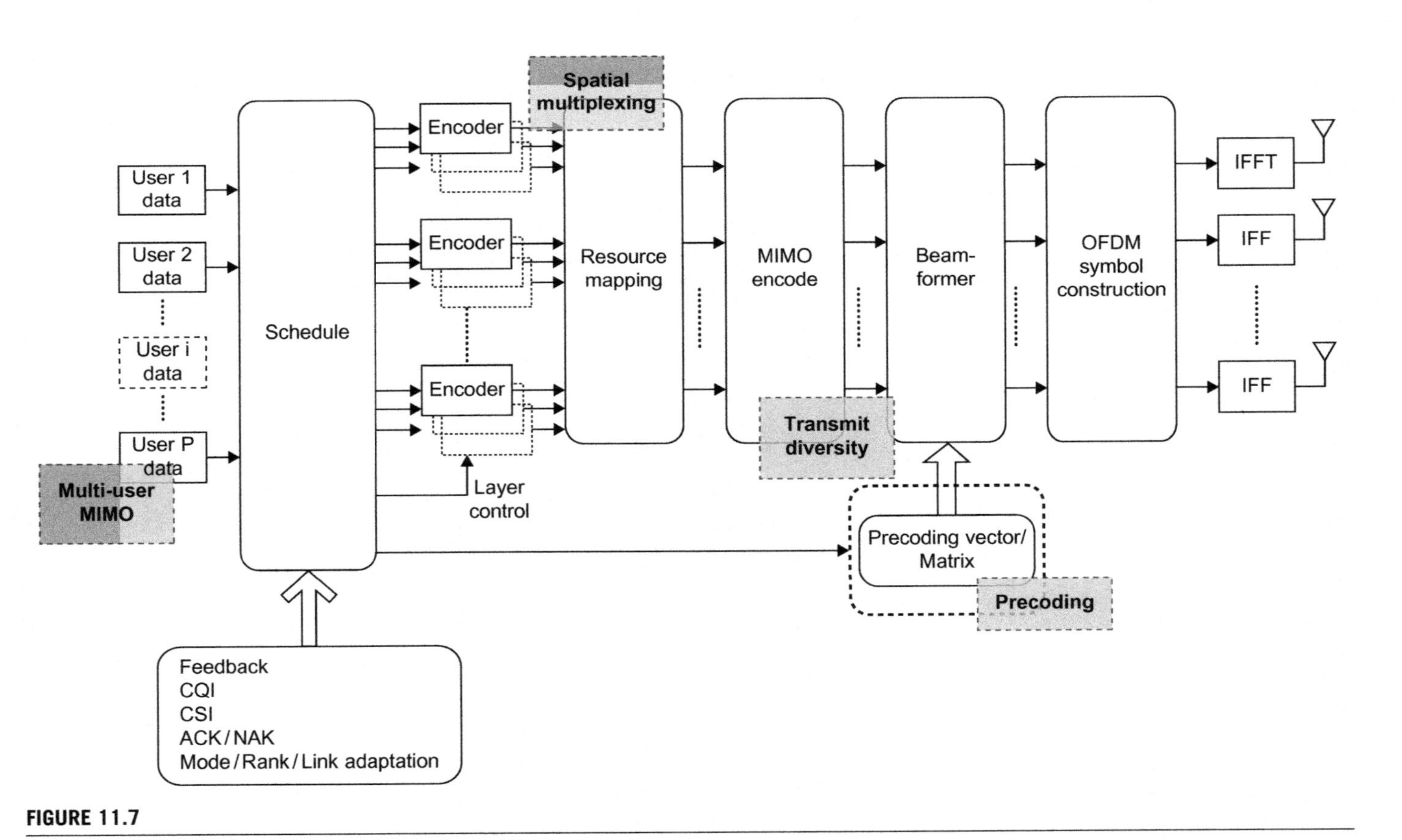

FIGURE 11.7
Support of MIMO techniques through generic architecture.

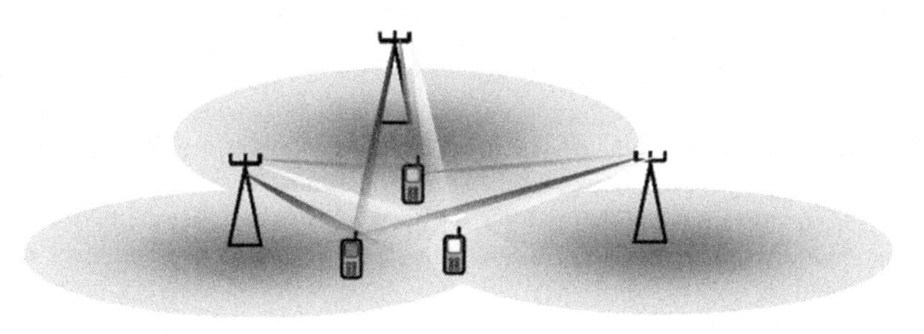

FIGURE 11.8

Example of DL COMP transmission.

main idea of COMP is to minimize the interference by attempting to avoid it. In the DL, the idea is to have distributed beamforming with or without Spatial Division Multiple Access (SDMA). In the UL, the main idea is to cancel the intercell interference.

DL CoMP

There are two forms of coordinated transmissions:

1. *The coordinated scheduling and/or beamforming*. In that configuration, data to a single UE is instantaneously transmitted from one of the transmission points and scheduling decisions are coordinated to control; for example, the interference generated in a set of coordinated cells;
2. *The joint processing/transmission*. In that case, data to a single UE is simultaneously transmitted from multiple transmission points; for example, in order to (coherently or noncoherently) improve the received signal quality and/or cancel actively interference for other UEs.

UL CoMP

Coordinated multipoint reception implies reception of the transmitted signal at multiple, geographically separated points. Uplink coordinated multipoint reception is expected to have very limited impact on the physical layer specifications largely because, for background compatibility reasons, LTE R8 already transmits user-specific reference signals. This indeed allows UE transmitted signals to be properly demodulated at several eNodeBs. How the eNodeBs then exchange information in order to allow a joint demodulation of the signal has yet to be defined. Scheduling decisions can also be coordinated among cells in order to control interference and may have some physical layer specification impact.

CoMP can be considered as a framework that exploits the tight coordination and sharing of information between eNB's. As such, it is an enabler for many new concepts that may not have been conceived or proposed yet. For instance, it could be used as a framework for more efficient inter-layer co-operation between eNB's in different layers, or for sharing of information enabling intercell interference cancellation, and so forth.

Clearly the exchange of control and user plane data between eNB's has some significant implications for the design and performance requirements of the backhaul network. At this time, it is not exactly clear what the timing requirements are for each of the various modes of CoMP, and this will come out of ongoing the performance evaluations in the LTE-A and IEEE802.16m feasibility studies. Some ideas about COMP could be:

- *Coordinated Scheduling.* In general, control data used for coordinated scheduling may comprise information about the active users in one cell that enables adjacent cells to co-ordinate their transmissions in a way which minimizes mutual interference. In principle it may also include information that could be broadcast to UE's in other cells to enable for instance intercell interference cancellation. However, this is not being seriously considered at the time of this writing. Typically, this information needs to be "timely," otherwise it will quickly be out dated and useless. Since users are typically scheduled every Time-to-Interval (TTI), this periodicity is probably the one required for "timely" information exchange between eNB's for supporting this CoMP mode. The volume of information exchanges between BS for this mode will be much lower than the user plane data rate. Therefore, it is the latency requirement that dominates the backhaul requirement for this mode.
- *Joint Processing/Transmission.* The coordinated simultaneous transmission of user packets from multiple sites to the same UE requires very tight time constraints on the backhaul, as well as time synchronization of the eNB's. Tight time constraints on packet delivery are required in order to minimize impact on user plane latency and on some physical layer procedures, e.g., HARQ (Hybrid Automatic Repeat reQuest). It seems likely that the requirement for backhaul delay could be similar to coordinated scheduling, i.e., every 1 ms. This mode requires sharing of user plane data between two or more cooperating eNB. Although this is likely to apply mainly for users at or close to cell edge locations the rates could nevertheless reach typically several Mbps for short bursts.

Relays

A relay is an equipment able to receive, process, and retransmit the received signals. The main expected benefits are: mitigation of coverage losses due to deployments in higher bands, and increase of cell edge throughput by a better distribution of the capacity within the cell. One main issue of relaying is that relays should be easy to deploy and be cost-effective. Figure 11.9 shows a representation of a transmission implementing a relay, together with an illustration showing a coverage increase owing to the relay.

Relays can be classified according to different system layers (L1/L2/L3) depending on the layer they operate on. L1 relays are evolved repeaters (with, for example, power control), L2 relays are decode, and forward relays and L3 relays are base

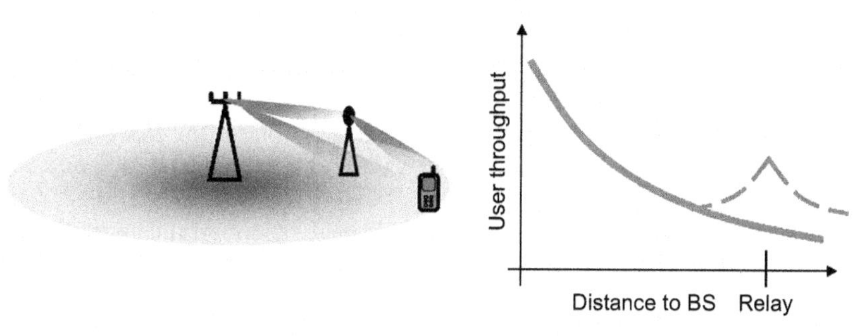

FIGURE 11.9

Illustration and impact of relay transmission.

stations with in-band backhaul. Following are more details, according to these kinds of relays.

L1 Relays

The term L1 relay refers to an Amplify and Forward relay, in which no decoding is carried out by the relay node. The simplest form of an L1 relay is an Radio Frequency (RF) repeater, which amplifies the input signal (desired signal, as well as thermal noise and interference) with a short delay (typically a fraction of a cyclic prefix, i.e., $\leq$1–2 μs). Note that simple L1 relays permit the direct signals from eNB and from the relay to be combined by the UE, provided that the total delay spread is less than the cyclic prefix. More advanced L1 relays (or "smart repeaters") have been considered, in which the relay only amplifies a portion of the overall bandwidth. However, this implies some digital processing (FFT/IFFT) to be done at the relay and involves an associated delay of one or more OFDM symbols. L1 repeaters are capable of increasing cell spectrum efficiency by boosting the throughput in otherwise poor coverage zones, which in turn, raises the cell average throughput and spectrum efficiency. However, since they generally transmit and receive at the same time on the same frequencies, they require high isolation between the transmit and receive antennas to avoid oscillation. It should be noted that echo cancellation techniques are already available to effectively increase the antenna isolation and significantly ease the deployment issues seen with legacy repeaters. If the delay in a smart repeater is greater than a subframe, then it becomes theoretically possible to avoid simultaneous reception and transmission, and antenna isolation is no longer a problem. However, if the delay is large, there are various issues that arise:

- *HARQ timeline.* The additional delay disrupts the HARQ process timeline, causing failure in on-time delivery of ack/nack messages;
- Disruption to channel estimation and measurement reporting;
- *Serving cell selection.* There is no way for the UE to differentiate between the eNB and the relay for measurement reporting to eNB. As a result, the eNB cannot

determine which UE's should be served by eNB directly and those to be served by the relay;
- *Power control.* Since UEs are unaware of the relay node as a separate entity, the relay node will not be able to independently control the power of UEs in its vicinity. This results in large power variations and potential jamming at the relay node;
- The additional delay (meaning well outside the cyclic prefix) causes additional interference in zones of overlapping coverage between L1 relay and serving eNB.

L2 Relays

An L2 relay typically refers to a Decode and Forward relay, which is not a fully-fledged eNB. This case can be further classified according to the level of functionality that resides in the relay node. In particular, we can think of the following subcases:

- HARQ operates end-to-end between the UE and the eNB. The relay may assist the eNB by simultaneously transmitting some of the data and/or HARQ signaling (PDCCH/ACK, etc) from the eNB to the UE in the DL, and from the UE to the eNB in the UL. This type of L2 relay is widely considered to have several issues to take into account (similar to L1 smart relay);
- Forwarding above HARQ. In this case, HARQ operates independently on both hops. The relay node could potentially carry out its own scheduling and rate adaptation. Alternatively, the serving eNB could carry out scheduling and/or rate adaptation on behalf of the relay and signal the corresponding information to the relay node;
- Forwarding above Medium Access Control (MAC) (de)multiplexing, corresponding to Radio Link Control (RLC) Protocol Data Unit (PDU);
- Forwarding above RLC on Packet Data Convergence Protocol (PDCP) PDUs.

The other types are more similar to L3 relays from Physical and MAC layer, perspectively, but may not be as efficient. While the relay node may not be a full-fledged eNB in these cases, it transmits its own cell IDentification (ID), pilot and reference symbols, makes its own scheduling decisions and transmits/receives the corresponding control channels, and transmits/receives data according to these scheduling decisions. However, the relay scheduler is more constrained in these cases as compared to the case of an L3 relay.

L3 Relays

An L3 relay transmits synchronization signals; it is indistinguishable from a regular cell. IP packets are transported on the relay backhaul link (between relay and serving eNB) as opposed to PDCP, RLC, or MAC packets in the case of L2 relays. From a UE perspective the L3 relay is a fully fledged eNB, and from an eNB perspective it appears much like a UE. This minimizes the impact on UE and eNB specifications and functionalities. The L3 relay also includes full Radio Resource Control (RRC) functionality, which means there is no impact on existing specifications and also overcomes one of the main drawbacks of L2 relays. The fact that full IP packets are visible at the relay also means that queue management can be performed by the relay, eliminating the need for a flow control procedure between the eNB and the relay.

Advanced MIMO Transmission and Reception Techniques

In this subsection, we propose an overview of advanced Tx beamforming and Rx spatial rejection techniques.

Tx Beamforming

The use of smart antennas for mitigating intercell interference seems to be an issue in 4G systems. The MIMO scheme family is based on beamforming and could be implemented either for the DL or the UL. As depicted in Figure 11.10, the energy is focused toward the user, which allows an enhancement of the received useful signal power. In addition, the system is almost exempted from interference, under the condition of no beam collision. When collisions occur, interference is often dominated by a single source, which is easy to mitigate in an interference cancellation detector. Beamforming can be carried out in different ways; at the highest level we distinguish adaptive and fixed-beamforming. In adaptive beamforming, the antenna systems are designed to optimize antenna patterns. In fixed-beamforming, or Grid of Beam (GoB) approaches, a certain number of pre-defined beams are used and the beamforming problem reduces to beam selection, which requires less feedback information than adaptive approaches. For both adaptive and fixed-beamforming, correlated antennas are preferred; for example, with half a wavelength element separation. Transmit

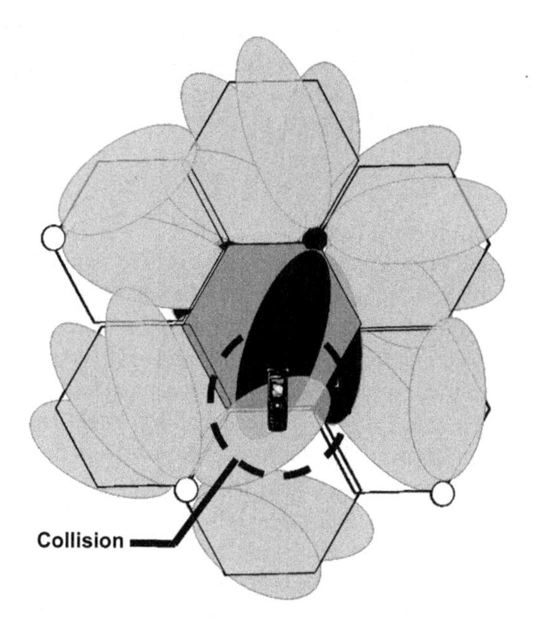

FIGURE 11.10

Tx beamforming representation.

beamforming is in principle applicable both at BS and UE, but in practice it will most probably be limited to the BS. It is also possible to implement SDMA as further multiple access component on top of beamforming, in order to serve several users at the same time on the same resources, owing to spatial UE separation.

Adaptive Beamforming

In adaptive beamforming, the antenna pattern is adapted in order to provide optimal reception for the scheduled user. This optimization can be done according to different criteria. The adaptation is based on channel knowledge, which requires a signaling overhead that has to be feedback periodically as the result of channel temporal variations. For simplicity, fixed-beamforming may be suitable approach in terms of performance to complexity ratio.

Fixed-beamforming or GoB

Adaptive beamforming as described previously adapts the transmit patterns for each user to long term Channel State Information (CSI) at the transmitter, such as the transmit covariance matrix. A low-complex approach to such adaptive beamforming is to use a finite set of fixed-antenna patterns, which will generate a set of beams matched to the long term transmit covariance matrices of different parts of the coverage area. All users in the coverage area share the set of beams, and the problem of beamforming is reduced to beam selection. The amount of channel knowledge required at the transmitter, once the beams have been designed, is then small. Since the adaptation is typically done over long term CSI, the required amount of feedback signaling is low or even none in case uplink received signals are used to determine which beam is the best. This spatial processing scheme seems very interesting for the wide area scenario and is intended to be used for single stream transmission (i.e., no SDMA) in a triple sector site.

Rx Spatial Rejection Techniques

By providing the radio receivers with multiple receive antennas, it is possible to implement different combining schemes in the baseband signal processing. Since the radio channels from the transmit antenna to the various receive antennas tend to fade differently, multiantenna receivers provide receive diversity both for the signal of interest and for the interference. With appropriate selection of the antenna; for example, in order to account for radio channel fading, interference power and spatial structure of the interference, such multiantenna receivers may provide increased robustness to both fading and interference. This, in turn, may improve the radio network coverage, capacity and user data rates. Maximum Ratio Combining (MRC) and Interference Rejection Combining (IRC) are two well-known combining schemes. Their impact is illustrated in Figure 11.11. With MRC the combining antenna weights are selected according to the channel quality of the desired signal, the noise power and the interference power at the different receive antennas. IRC is an extension of MRC, which also takes the spatial characteristics of the receive signals into account,

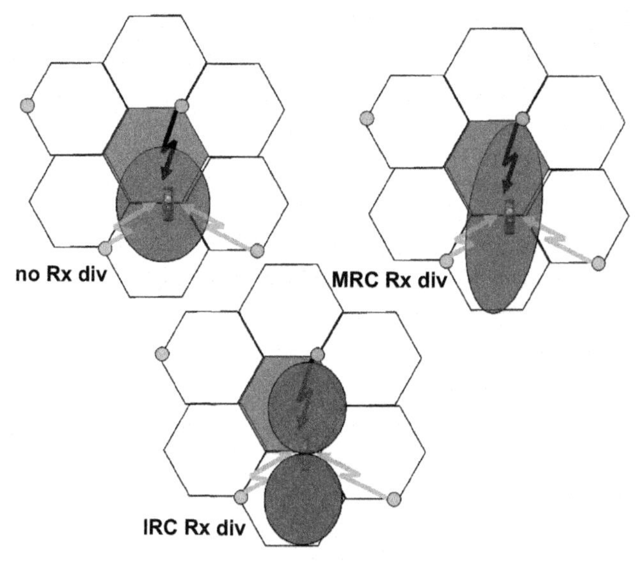

FIGURE 11.11

Rx spatial rejection representation.

therefore, enables interference suppression at the receiver. IRC determines the combining weights based on the channel and the (spatial) noise and interference. MRC and IRC can be applied both at the BS receiver and at the UE receiver and are consequently possible in order to use to improve both DL and UL performance.

Maximum Ratio Combining (MRC)

MRC combines coherently the signals output by the various sensors, by applying weights depending on the signal-to-noise ratio at each antenna output. This means that MRC can be seen as receive beamforming, since the effective antenna pattern resulting from application of the antenna weights is a beam toward the desired signal. As the true MRC requires the estimation of the noise power, the weights are generally approximated by being only proportional to the strength of the desired signal on each antenna, provided by the corresponding channel estimate. MRC provides diversity gain only when the fades experimented on each receive antenna are sufficiently decorrelated. In addition, MRC provides coherent combining gain as the desired signals are coherently added, whereas the interference signals are not. Note that when interference is spatially white, MRC provides optimal performance according to the maximization of the SINR value.

Interference Rejection Combining (IRC)

IRC is an extension of traditional MRC that also accounts for the spatial structure of the interference, which allows the interference to be partly rejected in the spatial domain [Vau88]. If we again make the analogy to beamforming, this means that IRC can be seen as receive beamforming with null steering, since the effective antenna pattern now also has nulls in the direction of the interferer(s). A receiver equipped with M_T receive antennas can perfectly reject $M_T - 1$ interfering sources, provided the interfering signals are received with different spatial signatures, that is, different directions of arrival, compared to the signal of interest. When the number of sources are greater than $M_T - 1$, which is usually the case in real-world systems, the IRC rejects as much interference as possible with a linear processing, in a way that maximizes the SINR at the receiver output. The main drawback of the IRC technique is the need to have accurate channel and interference estimates, which is essential for a proper performance of the IRC technique.

11.1.3 Other Systems

In direct link with the IEEE802.11n system (see Chapter 7), a new standard called IEEE802.11ac is under definition. The optimistic roadmap of this standard is to propose system deployment at the end of 2011. The aim of this standard is to achieve 1 Gbit/s at the MAC throughput for indoor transmission. For this achievement, the possibility of using 8 or 16 antennas with SDMA technique at each access point is under consideration, but the cost still needs to be evaluated. The interests envisioned by the system are:

- The possibility to continue improving the throughputs without increasing the number of antennas at the UE;
- A better exploitation of the resources with lower constraints on frequency bandwidth increase;
- The possible change request of the MAC protocol for improving the MAC efficiency by decreasing its overhead induced by signaling, which is strongly falling down since the IEEE802.11n;
- The improvement of every link quality by beamforming;
- The possibility of a better sharing of the interference between cells.

Most designers of communications systems, not all of them wireless, have well understood that the MIMO component is the main upcoming technology, providing a better robustness and a higher throughput in transmission. Thus, it now appears that for some wired systems such as DSL, PLC, or optics it is investigated whether it is possible to achieve gain through MIMO techniques in the corresponding transmission medium. Significant progress is expected in the coming years in this area as discussed below.

Wired Systems

Using a multiuser channel model for DSL transmission, as illustrated in Figure 11.12, can be very promising to enhance the throughput on the copper wire pairs. Most of the currently deployed DSL technologies do not take into account the crosstalk interference emanating from their own environment, whereas the channel capacity can be drastically improved if this limitation is overcome. Recent research in DSL transmission focuses on coordination between users at the access node. Vectored transmissions involve multiuser signal processing such as crosstalk pre-compensation or cancellation [GJ02]. The multiuser channel is here considered as a MIMO channel. These techniques are proven to approach the performance of perfect crosstalk suppression [Cen04]. The main difficulty for implementing vectoring is the knowledge of the crosstalk coupling coefficients between users. Another challenge is the need to manage the complexity when a high number of users have to be coordinated. Different techniques are currently under study to estimate the crosstalk coupling coefficients [ALMH08] and to provide low complexity vectored DSL transmissions. MIMO techniques can also be implemented in powerline communications. First, they were proposed for 4-wire cables relying on differential transmissions and a 2×2 channel model [GHF05]. A further study allows for a 3×3 modeling for the same 4-wire cable by considering common-mode propagation [HG07]. There is the need of a sufficient number of uncorrelated propagation modes to implement MIMO transmission on PLC systems. For indoor domestic PLC, the use of the three wires of the household cabling might provide an opportunity for the implementation of MIMO techniques. Recently, the use of such techniques has also been proposed in optical transmissions and has been applied in multimode fibre links [TAB07], through the excitation of distinct transmission modes in the fibre and the use of multimode receivers. The main difficulty is the feeding and detection of light in order to obtain sufficiently independent propagation modes, therefore, maximizing the MIMO capacity. Another strategy is to benefit from the polarization diversity in order to implement a MIMO system [MSY08]. It is possible in this case to build a 2×2 system where each polarization branch can carry a stream. Spatial multiplexing is fulfilled in this way. Either OFDM or single-carrier modulation can be used. 2×2 MIMO systems using QPSK and polarization diversity (DP-QPSK) have been built, where the multiplexing element is a polarization combiner and the de-multiplexing element is a polarization splitter. The receiver consists of a time-domain equalization filter applied on each stream, which cancels the chromatic dispersion effect. A second MIMO equalizer is dedicated to the compensation of the polarization-mode

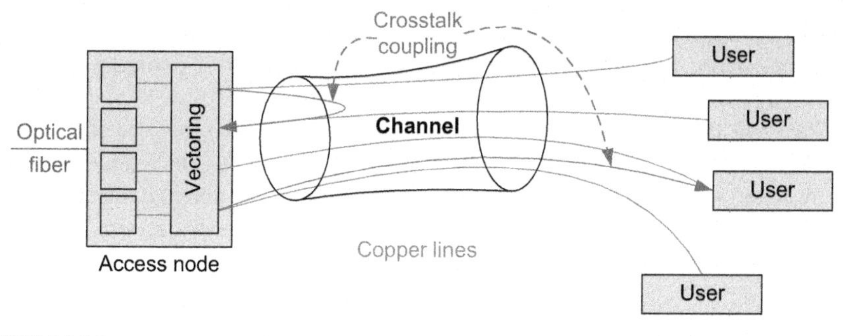

FIGURE 11.12

Representation of the channel in a vectored DSL transmission scheme.

dispersion, which can limit drastically the transmission of high throughput on the long-haul network.

In conclusion of this part, we have seen that the MIMO component is and will be more and more implemented in most transmission systems under development, with a valuable benefit in terms of robustness increase and/or data throughput.

11.2 A BIRD'S EYE VIEW ON CURRENT AND FUTURE PROSPECTS FOR MIMO

It is part of the "natural" evolution of systems that a first version is standardized and implemented, further versions following, improving specific aspects of the standard, including additional features, toward an increase of performance in various directions. These improvements can be dealing with very specific aspects of the system, hence, not being directly usable for other systems, or can be the result of integrating some general technology into the system, thus, taking advantage of a broader knowledge and of a larger number of suppliers. The use of MIMO in cellular networks and WLANs is clearly in the latter case.

Sometimes, the implementation of features in a system requires only the effort of infrastructure manufacturers, in the sense that these features do not affect mobile terminals; the use of diversity in a base station is a good example of this situation. However, many other times, a joint work and collaboration between infrastructure and mobile terminals manufacturers is required and essential, in order to achieve the desired target. Obviously, the latter situation increases the complexity of system improvement. In the case of MIMO, the latter rule also applies, since it requires the use of multiple antennas in both base station/access point and mobile terminal.

The increase of system performance can also be seen from two different perspectives. On one hand, some features are totally independent of users, being fully controlled by operators; the use of half-rate in GSM is a good example, among many others. On the other hand, many performance parameters do depend on the type of terminal users have, and on how they use it; of course, this creates problems in taking full advantage of the technology and in achieving the goals of optimum/maximum performance. Again, MIMO is the latter perspective.

It is clear that, although MIMO is a very powerful technology for the increase of performance (i.e., capacity, data rate, quality of service, etc.) of mobile and wireless systems, really profiting from it can be a complex and nonguaranteed process. Still, the potential of MIMO is so high that, regardless of these barriers, it is being implemented in current upgrades of these systems; for example, LTE, WiMAX, and WiFi.

Nowadays, major manufacturers for cellular networks infrastructure already deliver MIMO products for base stations; four antennas are already made available, and trials with eight antennas have also been announced. However, the situation is quite different as far as mobile terminals are concerned: at this end of the link, terminals are severely size constrained and the space available to locate multiple antennas is often desperately limited. Fortunately, there is a lot of work being done on co-location of multiple (e.g., 2) antennas in mobile phones with a relatively low correlation, especially for orthogonal linear polarizations (as addressed in one of the Chapter 10 of this book). On the other hand, one should not look into this problem by taking only (current) mobile phones as user's terminals, other devices being real alternatives as well; personal computers (to be used not only for WiFi, but also for cellular networks) may include two or four antennas in a relatively decoupled way (there are a number of publications on this topic), and even other hypothesis that do not present space problems can be considered; for example, the use of cars as the "interface" between the mobile phone and wave propagation (this is already a reality today, for high-end cars).

So, although there may currently be some real problems in the implementation of MIMO in mobile and wireless systems, by taking advantage of multiple antennas at both ends of the link, the conditions for the efficient deployment of MIMO systems will definitely be favorable in the near future.

Chapters 1–5 and 6–10 are structured in such a way that the first one presents the general models and fundamental aspects of MIMO, which can be applicable to any system, while the second deals with the application to several mobile and wireless systems. Given the interest in the implementation aspects, in what follows, one summarizes the main features and benefits of MIMO in these systems.

The implementation of MIMO in IEEE 802.11 WLANs, i.e., WiFi, is addressed in one Chapter 7. It has been introduced in the "802.11n version" of the standard [IEE09b], enabling (at the physical layer) data rates up to 600 Mbps with a 40 MHz bandwidth, in a 4×4 MIMO configuration. Changes are described in the transmitter, at the Physical Layer Convergence Procedure and frame format, and at the signal processing architecture, in the receiver, where system performance depends on the implemented algorithms and the solution is not unique, looking into signal detection, synchronization, carrier frequency offset estimation, channel estimation, and decoder. A comparison of performance results is done, with and without space time block coding, using combinations of one and two antennas at both transmitter and receiver (i.e., from SISO to MIMO); MIMO does improve system performance, enabling gains in signal to noise ratio, frame error ratio, data rate, or other parameters, depending on the perspective used for the analysis. Gains can be as high as 10 dB in signal to noise ratio, or the increase in data rate of almost an order of magnitude, when comparing MIMO with SISO.

Another Chapter 8 deals with the implementation of MIMO in IEEE 802.16 cellular networks, that is, WiMAX. In the recent standard [IEE09a], up to four antennas are foreseen, but most of the proposed schemes apply only to two, like the two schemes that are mandatory for use in downlink, but are optional for uplink, i.e., Space Time Block Coding and Spatial Multiplexing; still, a combination of these two schemes is being considered. Again, MIMO improves system performance, as expected. Additionally, a brief analysis of the transceiver architecture is performed for mobile terminals, showing that it is possible to incorporate a second antenna, together with its circuitry, given both size and cost issues.

The implementation of MIMO in LTE is considered within more than one Chapter 9. LTE has been designed having MIMO in mind from the beginning [3gp09]. The effort in standardisation was focused on specifying efficient schemes for downlink single user MIMO (that is, no indications are given for a transmission to multiple users such as broadcast), while for uplink the system was designed on the assumption that the mobile terminal front-end comprises a single signal chain; hence, for the latter, the support of single user MIMO is limited to adaptive transmit antenna switching; multiuser schemes are not excluded but are left to manufacturers implementation. The techniques and schemes considered in LTE for MIMO are not that much different from the ones taken in the previous systems, namely because they all share basic aspects of the multiple access technique. Detailed information is given on several implementation aspects, and LTE-Advanced is considered as well. Additionally, the use of Space Division Multiplexing together with MIMO is addressed, namely the problems related to detection algorithms and their mapping in the context of the standard. It is concluded that the gap between theoretical algorithms and practical implementations is huge, but also that there are ways to bridge this gap. Also, the importance of scalable or flexible solutions is stressed, since worst-case design is not an acceptable solution for complex algorithms implementation in mobile terminals; a scalable architecture can support a high throughput by relying on simple algorithms when the channel is favorable, and may be able to be reconfigured in order to exploit much more complex detection strategies when the channel is worse.

However, MIMO development and usage is not going to be finished by the standardization of the previously mentioned systems, and one may see quite a number of developments in the future, some of them listed in the following text.

Due to the fact that the UHF band has excellent propagation characteristics for mobile and wireless communications, the race on higher data rates has not implied (yet) the move to higher frequency bands, and basically all mobile and wireless systems working today use this band. Moreover, with the digital dividend (the use of part of the spectrum released from broadcast systems to mobile and wireless ones), the pressure to go higher in frequency has decreased. The achievement of higher data rates are usually obtained by more efficient multiple access techniques and modulation schemes, better management of the radio resources, and so on, rather than by investigating in the area of propagation and channels. MIMO was the exception confirming the rule, i.e., by taking advantage of the randomness of the propagation channel, it was possible to profit from the "parallel channels," and therefore, to enable higher data rates.

Nevertheless, one cannot avoid going up in frequency, if higher data rates are to be obtained, especially because efficiency and performance boundaries are being attained today. The knowledge of propagation and channels up to the UHF band is almost complete, but beyond that (including mm waves and up to the THz band) there is still a lot of work to be done, and a thorough characterization of MIMO in these high frequency bands is still to be performed.

Mobile communications gave rise many years ago to the fears from (electromagnetic) radiation. Basically, the extremely fast introduction of a new technology in the mass market, without a proper explanation of the system behavior, generated a lot of health concerns, which are still present today. One can currently envision that environmental concerns, which are already of key importance nowadays ("green communications" is in the agenda of many initiatives, projects and fora), will become even more important, with a huge impact on systems and networks, concerning their development, deployment and operation. Therefore, energy efficiency has to be (is already being) taken into account in mobile and wireless communications. MIMO can play an important role in this matter, through a better exploitation of the spatial/polarization degrees of freedom toward more efficient communications, at a lower transmission power. Moreover, by combining MIMO with beamforming, one can further decrease the transmission power (given the increase in the corresponding antenna gain); hence, providing additional contribution to increase energy efficiency.

Another trend regards mobile terminals. The evolution has been such that one can foresee that users may carry just an RF SIM card for their identification by the network, using any terminal/device at hand for communication, ranging from a PC to a "TV screen," encompassing terminals embedded in cars and in the office, and reaching spectacles as the replacement of today's mobile phones. Therefore, short range communications, in the vicinity of a person's body, will play a major role. The exploration of MIMO in such distance ranges has not deserved much attention, most probably due to the low randomness of the propagation channel, as well as to the difficulties in having more than one antenna available at either end of the link. Still, given the development of technology (in antenna design, signal processing, nano-devices, etc.), together with an increase in frequency band, one can easily envision that the conditions for deploying MIMO in these kind of applications may arise in the near future.

Propagation in not so usual environments needs to be studied as well, for example: several indoor scenarios, with a differentiation between business and residential environments; for personal systems, in-, on- and off-body propagation; for car communications, propagation in between cars, from a heavy traffic situation in a motorway (addressing high speed as well) to a city street; public transport scenarios, including buses, trains, planes, and boats, among others. The characterization of all these scenarios will allow one to estimate how far they are appropriate for MIMO technologies, and how to optimize the location of antennas.

Location based services are already popular nowadays, where a user can know his/her location via the mobile phone, and take advantage of them to navigate or to find a nearest shop, among other applications. Basically, this means that the user is accessing the network for location purposes. However, one can invert this concept, that is, the network knowing where users are, and take advantage of it. For example, by exploiting users' terminals (and other devices) as channel sensors, the network can establish a geographical map of channel conditions, hence, of channel quality, and with that information forecast services availabilities in an efficient way (specifically for each user). The creation of such geographical channel mapping for better

MIMO usage (i.e., enabling the system to know where and when to use MIMO, and then, increasing data rate accordingly), in a given scenario (from indoors to streets, including bodies and cars, and many others), would definitely increase the overall system performance.

Machine-to-machine communications (recently renamed as the *Internet of Things*) no longer requires a dedicated introduction. The increasing number of applications of this type of communications, where the user has no intervention, is obvious, and definitely will surpass human communications (voice or data) in terms of traffic volume, as data has already surpassed voice in terms of exchanged number of information bits. This also extends to car-to-car and car-to-infrastructure communications, and other types of systems. This means that one will have devices all over the surroundings, at any location, many times with low power consumption requirements, communicating not only among each other but also to a control or information network. Again, MIMO can play an important role in this type of communication, and conditions for deployment should be explored.

Sensors can be considered as subset of the *Internet of Things*, not only for personal use (e.g., on the body), but also through myriad devices that start being implanted in our surroundings (e.g., in cars or houses). Being a subset of the *Internet of Things*, the same type of problems apply, but one can envision in this case that power consumption and efficiency of communications will play enhanced roles. This opens a wide range of study cases, by extending the MIMO concept to the use of different multiple antennas, located at somehow random locations (e.g., buttons on clothes); hence, creating the need to analyze system behavior under those conditions.

In summary, the possibilities for deploying and amplifying MIMO technologies in the future are immense, bringing to the conclusion that the exploitation of these technologies is just in its infancy.

List of Symbols

$\triangleq$	variable definition
$\mathcal{E}$	expectation operator
$\otimes$	Kronecker product
$\odot$	Hadamard (element-wise) product
$\star$	convolution product
$\doteq$	exponential equality, $f(x) \doteq x^n \Leftrightarrow \lim_{x \to \infty} \frac{\log_2(f(x))}{\log_2(x)} = b$
$\mathbb{R}$	real field
$\mathbb{Z}$	integer field
$\mathbb{C}$	complex field
a, α	scalar
$\mathbf{a}$	column or row vector
$\mathbf{A}$	matrix
$\mathbf{A}(m,n)$	element in m^{th} row and n^{th} column of matrix $\mathbf{A}$
$\mathbf{A}(m,:)$	m^{th} row of matrix $\mathbf{A}$
$\mathbf{A}(:,n)$	n^{th} column of matrix $\mathbf{A}$
$\mathbf{A}^T$	transpose
$\mathbf{A}^*$	conjugate
$\mathbf{A}^H$	conjugate transpose (Hermitian)
$\mathbf{A}^\dagger$	pseudo-inverse
$\mathbf{A} \succcurlyeq 0$	means that $\mathbf{A}$ is positive semidefinite
$\det(\mathbf{A})$	determinant of $\mathbf{A}$
$\mathrm{Tr}\{\mathbf{A}\}$	trace of $\mathbf{A}$
$\Re[\mathbf{A}]$	real part of $\mathbf{A}$
$\Im[\mathbf{A}]$	imaginary part of $\mathbf{A}$
$\|a\|$	absolute value of scalar a
$\\|\mathbf{A}\\|_F$	Frobenius norm of $\mathbf{A}$
$\|\mathbf{A}\|$	element-wise absolute value of matrix $\mathbf{A}$
$\\|\mathbf{a}\\|$	norm of $\mathbf{a}$
$\mathrm{vec}(\mathbf{A})$	stacks $\mathbf{A}$ into $mn \times 1$ vector columnwise
$\mathrm{diag}\{a_1, a_2, \ldots, a_n\}$	$n \times n$ diagonal matrix with $(m,m)^{th}$ element $= a_m$
$\mathbf{0}_{m \times n}$	$m \times n$ matrix of zeros
$\mathbf{1}_{m \times n}$	$m \times n$ matrix of ones
$\mathbf{I}_m$	$m \times m$ identity matrix
$\lceil x \rceil$	$\lceil x \rceil = \begin{cases} 1 & \text{if } x > 0 \\ 0 & \text{if } x = 0 \end{cases}$
$\mathrm{mod}\ x$	modulo x operation
$\delta[x]$	Dirac delta (unit impulse) function
$\sharp\tau$	cardinality of the set τ
$\mathcal{Q}(x)$	$\mathcal{Q}$-function, $\mathcal{Q}(x) = 1/\sqrt{2\pi} \int_x^\infty e^{-t^2/2} dt$
$E_p(x)$	exponential integral of order p, $E_p(x) = \int_1^\infty e^{-xu} u^{-p} du$,

	$\Re[x] > 0$
$\Gamma(x,y)$	incomplete gamma function, $\Gamma(x,y) = \int_y^{\infty} u^{x-1}\, e^{-u}\, du$, $\Re[x] > 0$
$\Gamma(x)$	gamma function, $\Gamma(x) = \Gamma(x,0) = \int_0^{\infty} u^{x-1}\, e^{-u}\, du$, $\Re[x] > 0$
$p_x(x)$	probability density function of random variable x
$p_{x_1,\ldots,x_n}(x_1,\ldots,x_n)$	joint probability density function of random variables $x_1,\ldots,x_n$

List of Acronyms

ADPS Angular Delay Power Spectrum
AoA Angle of Arrival
AoD Angle of Departure
APP A Posteriori Probability
APS Angular Power Spectrum (also Azimuth Power Spectrum)
AS Angular Spread
AWGN Additive White Gaussian Noise
BC Broadcast Channel
BCJR BCJR algorithm
BD Block Diagonalization
BER Bit Error Rate
BICM Bit-Interleaved Coded Modulation
BLAST Bell Labs Layered Space-Time
BPSK Binary Phase Shift Keying
CA Code-Aided
CAS Cluster (or Component) Angular Spread
CDD Cyclic Delay Diversity
CDF cumulative distribution function
CDMA Code Division Multiple Access
CL Closed Loop
CMD Correlation Matrix Distance
COMP COordinated MultiPoint
COST European Cooperation in Science and Technology
CQI Channel Quality Indicator
CRB Cramer Rao Bound
CSI Channel State Information
CSIR Channel State Information at the Receiver
CSIT Channel State Information at the Transmitter
CTC Convolutional Turbo Codes
DA Data-Aided
D-BLAST Diagonal Bell Labs Space-Time
DFE Decision Feedback Equalization

DFT Discrete Fourier Transform

DL Down Link

DMC Dense Multipath Component

DMT Discrete MultiTone

DM-RS DeModulation Reference Signals

DoA Direction of Arrival

DoD Direction of Departure

DoF Degrees of Freedom

DPC Dirty Paper Coding

DRA Dielectric Resonator Antenna

DSL Digital Subscriber Line

EM Expectation-Maximization

ESPRIT Estimation of Signal Parameters via Rotational Invariance Techniques

EU European Union

FDD Frequency-Division Duplex

FEC Forward Error Correction

FFT Fast Fourier Transform

FIR Finite Impulse Response

FTW Forschungszentrum Telekommunikation Wien

GSCM Geometry-Based Stochastic Channel Model

GSM Global System for Mobile Communications

HARQ Hybrid Automatic Repeat reQuest

HEM Half biased Expectation Maximisation

HQF-GRV Hermitian Quadratic form of complex Gaussian random variables

HSPA High Speed Packet Access

IC Interference Channel

i.i.d. independent and identically distributed

IMT International Mobile Telecommunications

IOs Interacting Objects

IP Internet Protocol

IRC Interference Rejection Combining

ISI Inter Symbol Interference

IST Information Society Technologies

ITU-R International Telecommunications Union - Radio

JT Joint Transmission

LOS Line Of Sight
LS Large Scale
LST Layered Space-Time
LTE Long Term Evolution
MAC Medium Access Control
MACh Multiple Access Channel
MCS Modulation and Coding Scheme
MEG Mean Effective Gain
MI Mutual Information
MIMO Multiple-Input Multiple-Output
MISO Multiple-Input Single-Output
ML Maximum Likelihood
MMSE Minimum Mean Squared Error
MPC Multipath Components
MRC Maximal Ratio Combining
MSE Mean Squared Error
MUD Multi User Detection
MUSIC Multiple Signal Classification
MU-MIMO Multi User MIMO
NDA Non-Data-Aided
NLOS Non Line Of Sight
OFDM Orthogonal Frequency Division Multiplexing
OFDMA Orthogonal Frequency Division Multiplexing Access
OL Open Loop
OSIC Ordered Successive Interference Cancellation
O-STBC Orthogonal Space-Time Block Codes
OTA Over The Air
PAM Pulse Amplitude Modulation
PAS Power Angular Spectrum
PBCH Physical Broadcast Channel
PCB Printed Circuits Board
PDCCH Physical Downlink Control Channel
PDCP Packet Data Convergence Protocol
PDF Probability Density Function
PDP Power Delay Profile

PDSCH Physical Downlink Shared Channel
PDU Protocol Data Unit
PER Packet Error Rate
PIFA Planar Inverted F Antenna
PLC Power Line Communication
PMCH Physical Multicast Channel
PMI Preferred Matrix Indicator
PSK Phase Shift Keying
PUCCH Physical Uplink Control CHannel
PUSC Partial Usage of Subcarriers
PUSCH Physical Uplink Shared CHannel
QAM Quadrature Amplitude Modulation
QoS Quality of Service
QO-STBC Quasi-Orthogonal Space-Time Block Codes
QPSK Quadrature Phase Shift Keying
RB Resource Block
RCM Random Cluster Model
RE Resource Element
RF Radio Frequency
RI Rank Indicator
RLC Radio Link Control
rms root mean square
RRC Radio Resource Control
RRH Remote Radio Heads
Rx Receiver
SAGE Space-Alternating Generalized Expectation maximization
SB Scheduling Block
SCM Spatial Channel Model
SC-FDMA Single Carrier Frequency Division Multiplexing Access
SDD System Description Document
SDMA Spece Division Multiplexing Access
SFBC Space Frequency Block Coding
SIC Successive Interference Cancellation
SIMO Single-Input Multiple-Output
SINR Signal to Interference plus Noise Ratio
SIR Signal to Interference Ratio

SISO Single-Input Single-Output
SI-SO Soft-Input Soft-Output
SLNR Signal to Leakage plus Noise Ratio
SM Saptail Multiplexing
SNR Signal-to-Noise Ratio
SRS Sounding Reference Signals
STC Space-Time Code
ST-BICM Space Time Bit-Interleaved Coded Modulation
SU-MIMO Single User MIMO
SVD Singular Value Decomposition
TB Transport Block
TDD Time-Division Duplex
TDMA Time Division Multiplexing Access
TIS Total Isotropic Sensitivity
TRP Total Radiated Power
TSD Tile Switched Diversity
TTI Transmission Time Interval
TU Typical Urban
Tx Transmitter
UE User Equipment
UEM Unbiased Expectation Maximisation
UHF Ultra High Frequency
UL Uplink
ULA Uniform Linear Array
UTA Uniform Triangular Array
V-BLAST Vertical Bell Labs Space-Time
WCDMA Wideband Code Division Multiplexing Access
WiMAX Worldwide Interoperability for Microwave Access
WINNER Wireless World Initiative New Radio
WLAN Wireless Local Area Network
WMF Whitened Matched Filter
XPD Cross-Polarization Discrimination
ZF Zero Forcing
3G Third Generation
3GPP 3rd Generation Mobile Group
4G Fourth Generation

References

[3gp03] Spatial channel model for multiple input multiple output (MIMO) simulations. Technical Report 3GPP TR 25.996 V6.1.0, September 2003. Available: http://www.3gpp.org/ftp/Specs/html-info/25996.htm.

[3gp09] 3GPP. Evolved Universal Terrestrial Radio Access (E-UTRA) and Evolved Universal Terrestrial Radio Access Network (E-UTRAN) Overall Description. Stage 2 (Release 9), Technical Specification Group Radio Access Network, TS 36.300 v9.2.0, December 2009. Available: http://www.3gpp.org/ftp/Specs/archive.

[3GP07] CDD precoding for 4Tx antennas. Technical Report R1-072019, 3GPP, 2007.

[3GP09] 3GPP. Technical specification group radio access network; requirements for further advancements for E-UTRA (LTE-Advanced). Available: http://www.3gpp.org/ftp/Specs/archive/36_series/36.913/36913-900.zip, 2009.

[3GP09a] Overview of 3GPP Release 8. Technical report, 3GPP, 2009.

[3GP10a] Report of 3GPP TSG RAN WG1 #59bis. Technical report, 3GPP, 2010.

[3GP10b] Summary of evaluation results for 3GPP requirements. Technical Report R1-100823, 3GPP, 2010.

[3GPP07] 3GPP. Evolved Universal Terrestrial Radio Access (E-UTRA); Physical channels and modulation (Release 8), TS 36.211 V8.0.0, 2007. Available: http://www.3gpp.org/ftp/Specs/archive/36_series/36.211/36211-800.zip.

[A.M02] A. M. Sayeed. Deconstructing multiantenna fading channels. *IEEE Trans. Signal Process.*, 50(10):2563–2579, October 2002.

[AB08] A. Burr. Generalised linear precoding for MIMO using tensor notation. COST 2100 TD(08)452, 4th MCM, Wroclaw, Poland, 6–8 February 2008.

[Ada09] Adaptive codebook designs for MU-MIMO. Technical Report R1-091282, 3GPP, 2009.

[ADC95] N. Al-Dahir and J. M. Cioffi. MMSE decision-feedback equalizers: finite-length results. *IEEE Trans. Inf. Theory*, pages 961–975, July 1995.

[AdN05] J. Bach Andersen and J. Ødum Nielsen. Modelling the full MIMO matrix using the richness function. *ITG/IEEE Workshop on Smart Antennas*, Duisburg, Germany, 2005.

[AFRGAP09] M. Alvarez-Folgueiras, J. A. Rodriguez-Gonzalez, and F. Ares-Pena. Low-sidelobe patterns from small, low-loss uniformly fed linear arrays illuminating parasitic dipoles. *IEEE Trans. Antennas Propag.*, 57(5):1584–1586, May 2009.

[AL06] J. B. Andersen and B. K. Lau. On closely coupled dipoles in a random field. *IEEE Antennas Wireless Propag. Lett.*, 5:73–75, 2006.

[Ala98] S. M. Alamouti. A simple transmit diversity technique for wireless communications. *IEEE J. Select. Areas Commun.*, 16(8):1451–1458, October 1998.

[ALMH08] A. Amayreh, J. Le Masson, and M. Hélard. Blind crosstalk channel identification in dmt-based dsl systems. *Globecom*, New Orleans, LA, USA, December 2008.

[AME02] A. Kuchar, M. Tangemann, and E. Bonek. A real-time DOA-based smart antenna processor. *IEEE Trans. Veh. Technol.*, 51:1279–1293, November 2002.

[AMI07] G. Acosta-Marum and M. A. Ingram. Six time- and frequency-selective empirical channel models for vehicular wireless lans. *1st IEEE International Symposium Wireless Vehicular Communications*, Baltimore, MD, May 2007.

[AMM+02] A. F. Molisch, M. Steinbauer, M. Toeltsch, E. Bonek, and R. S. Thomä. Capacity of MIMO systems based on measured wireless channels. *IEEE J. Select. Areas Commun.*, 20(3):561–569, April 2002.

[AMY+09] A. Abe, N. Michishita, Y. Yamada, J. Muramatsu, T. Watanabe, and K. Sato. Mutual coupling reduction between two dipole antennas with parasitic elements composed of composite right-/left-handed transmission lines. *International Workshop Antenna Technology (IWAT)*, Santa Monica, CA, USA, March 2009.

[AR76] J. B. Andersen and H. H. Rasmussen. Decoupling and descattering networks for antennas. *IEEE Trans. Antennas Propag.*, AP-24(6):841–846, November 1976.

[AS84] N. Amitay and J. Salz. Linear equalization theory in digital data transmission over dually polarized fading radio channels. *Bell Syst. Tech. J.*, 63:2215–2259, December 1984.

[Ave07] J. Avendal. Multiband diversity antennas. Master's thesis, Lund University, Lund, Sweden, January 2007. Available: http://www.es.lth.se/teorel/Publications/TEAT-5000-series/TEAT-5084.pdf.

[AYL07] J. Avendal, Z. Ying, and B. K. Lau. Multiband diversity antenna performance study for mobile phones. In *Proceedings of the International Workshop Antenna Technology (IWAT)*, pages 193–196, Cambridge, UK, March 2007.

[B01] A. G. Burr. Modulation and Coding for Wireless Communications. Prentice-Hall, Harlow, Essex, UK, 2001.

[B08] A. G. Burr. Tensor-based linear precoding models for frequency domain MIMO systems, in 1st COST2100 Workshop MIMO and Cooperative Communications Trondheim, Norway, June 3rd–4th 2008.

[B09] A. G. Burr. Multiplexing gain of multiuser MIMO on finite scattering channel. COST 2100 TD(09)871, 8th MCM, Valencia, May 18–19, 2009.

[BBW+05] A. Burg, M. Borgmann, M. Wenk, M. Zellweger, W. Fichtner, and H. Bolcskei. VLSI implementation of MIMO detection using the sphere decoding algorithm. *IEEE J. Solid-State Circuits*, 40(7):1566–1577, July 2005.

[BCJR74] L. Bahl, J. Cocke, F. Jelinek, and J. Raviv. Optimal decoding of linear codes for minimizing symbol error rate. *IEEE Trans. Inf. Theory*, 20(2):284–287, 1974.

[BDMP98] S. Benedetto, D. Divsalar, G. Montorsi, and F. Pollara. Serial concatenation of interleaved codes: performance analysis, design, and iterative decoding. *IEEE Trans. Inf. Theory*, 44(3):909–926, May 1998.

[BDSR+08] B. Bougard, B. De Sutter, S. Rabou, D. Novo, O. Allam, S. Dupont, and L. Van der Perre. A Coarse-Grained Array based Baseband Processor for 100Mbps+ Software Defined Radio. In *Proceedings of the Design, Automation and Test in Europe (DATE)*, pages 716–721, March 2008.

[BDSV+08] B. Bougard, B. De Sutter, D. Verkest, L. Van der Perre, and R. Lauwereins. A coarse-grained array accelerator for software-defined radio baseband processing. *IEEE Micro*, 28(4):41–50, July 2008.

[BG96] C. Berrou and A. Glavieux. Near optimum error correcting coding and decoding: turbo-codes. *IEEE Trans. Commun.*, 44(10):1261–1271, 1996.

[BGPvdV06] H. Bolcskei, D. Gesbert, C. B. Papadias, and A. J. van der Veen. Space-Time Wireless Systems: From Array Processing to MIMO Communications. Cambridge University. Press, July 2006.

[BKH97] G. Bauch, H. Khorram, and J. Hagenauer. Iterative equalization and decoding in mobile communications systems. In *Proceedings of the 2nd EPMCC'97 together with 3rd ITG-Fachtagung "Mobile Kommunikation,"* Bonn, Germany, October 1997.

[BL10] Bell Labs. Bell Labs scientists shatter limit on fixed wireless transmission from Bell Labs press release. Available: http://www.alcatel-lucent.com/wps/portal/!ut/p/kcxml/04_Sj9SPykssy0xPLMnMz0vM0Y_QjzKLd4y3cDcFSYGZzgH6kShiBvGOCBFfj_zcVP0gf W_9AP2C3NCIckdHRQAzW18m/delta/base64xml/L3dJdyEvd0ZNQUFzQUMvNElVRS82X0FfN01U?LMSG_CABINET=Bell_Labs&LMSG_CONTENT_FILE=News_Features/News_Feature_Detail_000123&lu_lang_code=en_WW, accessed 24th June 2010.

[BM96] S. Benedetto and G. Montorsi. Serial concatenation of block and convolutional codes. *Electron. Lett.*, 32:887–888, May 1996.

[BO01] M. Bengtsson and B. Ottersten. *Handbook of Antennas in Wireless Communications*, chapter 18: Optimal and Suboptimal Transmit Beamforming. CRC press, August 2001.

[BP31] H. H. Beverage and H. O. Peterson. Diversity receiving system of RCA Communications, Inc., for radiotelegraphy. *Proc. IRE*, 19(4):529–561, April 1931.

[BRC03] S. Blanch, J. Romeu, and I. Corbella. Exact representation of antenna system diversity performance from input parameter description. *Elect. Lett.*, 39(9):705–707, May 2003.

[BRV05] J. Belfiore, G. Rekaya, and E. Viterbo. The Golden code: A 2×2 full rate space-time code with non vanishing determinants. *IEEE Trans. Inf. Theory*, 51(4):1432–1436, 2005.

[BS02] H. Boche and M. Schubert. A general duality theory for uplink and downlink beamforming. In *Proceedings of the IEEE Vehicular Technology Conference (VTC) fall*, Vancouver, Canada, pages 87–91 Vol. 1, September 2002.

[BS04] H. Boche and S. Stańczak. Convexity of some feasible QoS regions and asymptotic behavior of the minimum total power in CDMA systems. *IEEE Trans. Comput.*, 52(12):2190–2197, December 2004.

[BS05] H. Boche and S. Stańczak. Log-convexity of the minimum total power in CDMA systems with certain quality-of-service guaranteed. *IEEE Trans. Inf. Theory*, 51(1):374–381, January 2005.

[BS08a] H. Boche and M. Schubert. A calculus for log-convex interference functions. *IEEE Trans. Inf. Theory*, 54(12):5469–5490, December 2008.

[BS08b] H. Boche and M. Schubert. Concave and convex interference functions – general characterizations and applications. *IEEE Trans. Signal Process.*, 56(10):4951–4965, October 2008.

[BS08c] H. Boche and M. Schubert. The structure of general interference functions and applications. *IEEE Trans. Inf. Theory*, 54(11):4980–4990, November 2008.

[BS08d] H. Boche and M. Schubert. A superlinearly and globally convergent algorithm for power control and resource allocation with general interference functions. *IEEE/ACM Trans. Netw.*, 16(2):383–395, April 2008.

[BS09] H. Boche and M. Schubert. Perron-root minimization for interference-coupled systems with adaptive receive strategies. *IEEE Trans. Comput.*, 57(10):3173–3164, October 2009.

[BS10] H. Boche and M. Schubert. A unifying approach to interference modeling for wireless networks. *IEEE Trans. Signal Process.*, 58(6), June 2010.

[BSG04] M. Biguesh, S. Shahbazpanahi, and A. B. Gershman. Robust downlink power control in wireless cellular systems. *EURASIP J. Wirel. Commun. Netw.*, (2):261–272, 2004.

[BT06] E. Biglieri and G. Taricco. Fundamentals of MIMO channel capacity. In H. Bölcskei et al. (eds.), *Space-Time Wireless Systems*, Cambridge University Press, 2006.

[Bur03] A. G. Burr. Capacity Bounds and Estimates for the Finite Scatterers MIMO Wireless Channel. *IEEE J. Select. Areas Commun.*, 21(5):812–818, June 2003.

[BUV01] A. Berthet, B. Ünal, and R. Visoz. Iterative decoding of convolutionally encoded signals over multipath rayleigh fading channels. *IEEE J. Sel. Areas Commun.*, 19:1729–1743, September 2001.

[BVJY03] B. Vucetic and J. Yuan. *Space-Time Coding*. John Wiley, Chichester, 2003.

[BW06] E. Bonek and W. Weichselberger. What we can learn from multi-antenna measurements. In H. Bölcskei et al. (eds.), *Space-Time Wireless Systems*, Cambridge University Press, 2006.

[BW74] L. H. Brandenburg and A. D. Wyner. Capacity of the Gaussian channel with memory: The multivariate case. *Bell Syst. Tech. J.*, 53:745–778, May/June 1974.

[BZH04] A. G. Burr, L. Zhang, and S. Hirst. Space-time signal processing for real-world channels. In *Proceedings of the Euro-Microwave Conference*, Amsterdam, October 2004.

[CA04] C. Oestges and A. J. Paulraj. Beneficial impact of channel correlations on MIMO capacity. *Elect. Lett.*, 40(10):606–608, May 2004.

[CA08] V. Chandrasekhar and J. G. Andrews. Femtocell networks: A survey. *IEEE Commun. Mag.*, pages 59–67, September 2008.

[CBH+06] N. Czink, E. Bonek, L. Hentilä, J.-P. Nuutinen, and J. Ylitalo. Cluster-based MIMO channel model parameters extracted from indoor time-variant measurements. *Globecom IEEE Global Telecommunications Conference* San Francisco, CA, USA, 2006.

[CBVV+08] N. Czink, B. Bandemer, G. Vazquez-Vilar, L. Jalloul, C. Oestges, and A. Paulraj. Spatial separation of multi-user MIMO channels. Technical Report COST 2100 TD(08)622, 2008.

[CCMR07] C. Y. Chiu, C. H. Cheng, R. D. Murch, and C. R. Rowell. Reduction of mutual coupling between closely-packed antenna elements. *IEEE Trans. Antennas Propag.*, 55(6):1732–1738, June 2007.

[CDEF95a] J. M. Cioffi, G. P. Dudevoir, M. V. Eyuboglu, and G. D. Forney. MMSE decision feedback equalizers and coding – Part I: Equalization results. *IEEE Trans. Commun.*, pages 2582–2594, October 1995.

[CDEF95b] J. M. Cioffi, G. P. Dudevoir, M. V. Eyuboglu, and G. D. Forney. MMSE decision feedback equalizers and coding- Part II: Coding results. *IEEE Trans. Commun.*, pages 2595–2604, October 1995.

[CDS96] L. J. Cimini, Jr., B. Daneshrad, and N. R. Sollenberger. Clustered OFDM with transmitter diversity and coding. In *Proceedings of the Globecom'96*, pages 703–707, London, UK, November 1996.

[Cen04] R. Cendrillon. *Multi-user Signal and Spectra Coordination for Digital Subscriber Lines*. PhD thesis, Université catholique de Louvain, 2004.

[CH08] N. Costa and S. Haykin. A novel wideband MIMO channel model and experimental validation. *IEEE Trans. Antennas Propag.*, 56:550–562, February 2008.

[CH10] N. Costa and S. Haykin. Multiple-Input Multiple Output Channel Models, *Theory and Practice*. Wiley, 2010.

[Chi02] M. Chiani. Evaluating the capacity distribution of MIMO rayleigh fading channels. *IEEE International Symposium on Advances in Wireless Communications (ISWC)*, pages 3–5, Victoria, CA, September 2002.

[CHL02] J. M. Chaufray, W. Hachem, and Ph. Loubaton. Asymptotical Analysis of Optimum and Sub-Optimum CDMA Downlink MMSE Receivers. In *IEEE International Symposium on Information Theory*, Lausanne, Switzerland, July 2002.

[Chu48] L. J. Chu. Physical limitations of omni-directional antennas. *J. Appl. Phys.*, 19(12):1163–1175, December 1948.

[Chu72] D. C. Chu. Polyphase codes with good periodic correlation properties. *IEEE Trans. Inform. Theory*, 18(4): 531–532, July 1972.

[CIM04] D. Catrein, L. Imhof, and R. Mathar. Power control, capacity, and duality of up- and downlink in cellular CDMA systems. *IEEE Trans. Comput.*, 52(10):1777–1785, 2004.

[Cla68] R. H. Clarke. A statistical theory of mobile radio reception. *Bell Syst. Tech. J.*, 47(6):957–1000, July/August 1968.

[CLD$^+$08] A. Chebihi, C. Luxey, A. Diallo, P. Le Thuc, and R. Staraj. A novel isolation technique for closely spaced PIFAs for UMTS mobile phones. *IEEE Antennas Wireless Propag. Lett.*, 7:665–668, 2008.

[CLW$^+$03] D. Chizhik, J. Ling, P. W. Wolniansky, R. A. Valenzuela, N. Costa, and K. Huber. Multiple-input-multiple-output measurements and modeling in manhattan. *IEEE J. Select. Areas Commun.*, 21(3):321–331, April 2003.

[CM01] G. Caire and U. Mitra. Structured multiuser channel estimation for block-synchronous DS/CDMA. *IEEE Trans. Commun.*, 49:1605–1617, September 2001.

[CMC08] Cubic metric comparison of OFDMA and Clustered-DFTS-OFDM/NxDFTS-OFDM. Technical Report R1-084469, 3GPP, 2008.

[Com09] Best companion reporting for improved single-cell MU-MIMO pairing. Technical Report R1-091307, 3GPP, 2009.

[COP07] S. H. Chae, S. Oh, and S. Park. Analysis of mutual coupling, correlations, and TARC in WiBro MIMO array antenna. *IEEE Antennas Wirel. Propag. Lett.*, 6:122–125, 2007.

[Cor01] L. M. (ed.) Correia. *Wireless Flexible Personalised Communications, COST 259: European Co-operation in Mobile Radio Research*. Wiley, 2001.

[Cor06] L. M. (ed.) Correia. *Mobile Broadband Multimedia Networks*. Academic Press, 2006.

[CRB$^+$08] N. Czink, A. Richter, E. Bonek, J.-P. Nuutinen, and J. Ylitalo. Including diffuse multipath parameters in MIMO channel models. *IEEE Vehicular Technology Conference Fall*, Calgary, Canada, 2008.

[CSI09] Consideration on CSI-RS design for CoMP and text proposal to 36.814. Technical Report R1-093031, 3GPP, 2009.

[CS00] G. Caire and S. Shamai (Shitz). On achievable rates in a multi-antenna broadcast downlink. In *Proceedings of the Allerton Conference on Communications, Control and Computing, Monticello*, USA, October 2000.

[CS03] G. Caire and S. Shamai (Shitz). On the achievable throughput of a multi-antenna Gaussian broadcast channel. *IEEE Trans. Inform. Theory*, 49(7):1691–1706, July 2003.

[CT91] T. M. Cover and J. A. Thomas. *Elements of Information Theory*. Wiley, 1991.

[CTB98] G. Caire, G. Taricco, and E. Biglieri. Bit-interleaved coded modulation. *IEEE Trans. Inform. Theory*, 44(3):927–946, May 1998.

[CTKV02] C. Chuah, D. Tse, J. Kahn, and R. Valenzuela. Capacity scaling in MIMO wireless systems under correlated fading. *IEEE Trans. Inf. Theory*, 48(3):637–650, March 2002.

[CTP+07] M. Chiang, C. W. Tan, D. Palomar, D. O'Neill, and D. Julian. Power control by geometric programming. *IEEE Trans. Wirel. Commun.*, 6(7):2640–2651, July 2007.

[CW01] A. M. Chan and G. W. Wornell. A Class of Asymptotically Optimum Iterated-decision Multiuser Detectors. In *IEEE International Conference on Acoustics, Speech, and Signal Processing*, Salt Lake City, Utah, May 2001.

[CW04] H. J. Chaloupka and X. Wang. Novel approach for diversity and MIMO antennas at small mobile platforms. In *Proceedings of the IEEE International Symposium on Personalized, Indoor and Mobile Radio Communications (PIMRC)*, volume 1, pages 637–642, Barcelona, Spain, September 2004.

[CWZ03] M. Chiani, M. Z. Win, and A. Zanella. On the capacity of spatially correlated MIMO channels. *IEEE Trans. Inf. Theory*, 49(10):2363–2371, October 2003.

[CY08a] J. C. Coetzee and Y. Yu. Closed-form design equations for decoupling networks of small arrays. *Elect. Lett.*, 44(25):1441–1442, December 2008.

[CY08b] J. C. Coetzee and Y. Yu. New modal feed network for a compact monopole array with isolated ports. *IEEE Trans. Antennas Propag.*, 56(12):3872–3875, December 2008.

[CY08c] J. C. Coetzee and Y. Yu. Port decoupling for small arrays by means of an eigenmode feed network. *IEEE Trans. Antennas Propag.*, 56(6):1587–1593, June 2008.

[CYz+07] N. Czink, X. Yin, H. Özcelik, M. Herdin, E. Bonek, and B. Fleury. Cluster characteristics in a MIMO indoor propagation environment. *IEEE Trans. Wirel. Commun.*, 6(4):1465–1475, 2007.

[CZ07] S. Chen and T. Zhang. Low Power Soft-Output Signal Detector Design for Wireless MIMO Communications Systems. In *Proc. of the International Symposium on Low Power Electronics and Design*, pages 232–237, August 2007.

[CZN+09] N. Czink, T. Zemen, J.-P. Nuutinen, J. Ylitalo, and E. Bonek. A time-variant MIMO channel model directly parametrised from measurements. *EURASIP J. Wireless Communications and Networks*, 2009. Special Issue on Advances in Propagation Modeling for Wireless Systems.

[CZX07] S. Chen, T. Zhang, and Y. Xin. Relaxed K-Best MIMO signal detector design and VLSI implementation. *IEEE Trans. Very Large Scale Integ. (VLSI) Systems*, 15(3):328–337, March 2007.

[DBR04] S. Dossche, S. Blanch, and J. Romeu. Optimum antenna matching to minimise signal correlation on a two-port antenna diversity system. *Elect. Lett.*, 40(19):1164–1165, September 2004.

[DBR05] S. Dossche, S. Blanch, and J. Romeu. Decorrelation of a closely spaced four element antenna array. In *Proceedings of the IEEE Antenna Propagation Society International Symposium*, volume 1B, pages 803–806, Washington, DC, USA, July 2005.

[DCAB08] D. Chen and A. Burr. Adaptive stream mapping in MIMO-OFDM with linear precoding. COST 2100 TD(08)643, 6th MCM, Lille, France, 6–8 October 2008.

[dCDNL07] K. Q. da Costa, V. Dmitriev, D. C. Nascimento, and J. Cd. S. Lacava. Broadband L-probe fed patch antenna combined with passive loop elements. *IEEE Antennas Wirel. Propag. Lett.*, 6:100–102, 2007.

[DEJ09] D. D. Evans and M. A. Jensen. Near-optimal radiation characteristics for diversity antenna design. *International Workshop Antenna Technology (IWAT)*, Santa Monica, CA, USA, March 2009.

[DGMJ00] Da-Shan Shiu, G. J. Foschini, M. J. Gans, and J. M. Kahn. Fading correlation and its effect on the capacity of multielement antenna systems. *IEEE Trans. Commun.*, 48(3):502–513, March 2000.

[DHGL07] C. P. Domizioli, B. L. Hughes, K. G. Gard, and G. Lazzi. Receive diversity revisited: correlation, coupling and noise. *IEEE Global Telecommunications Conference (Globecom)*, pages 3601–3606, Washington, DC, USA, November 2007.

[DHGL10] C. P. Domizioli, B. L. Hughes, K. G. Gard, and G. Lazzi. Noise correlation in compact diversity receivers. *IEEE Trans. Commun.*, 58(5):1426–1436, May 2010.

[DHLdC03] M. Debbah, W. Hachem, P. Loubaton, and M. de Courville. MMSE analysis of certain large isometric random precoded systems. *IEEE Trans. Inf. Theory*, 49(5):1293–1311, May 2003.

[Die08] F. Dietrich. *Robust Signal Processing for Wireless Communications*. Springer, 2008.

[DJB95] C. Douillard, M. Jezequel, and C. Berrou. Iterative correction of intersymbol interference: Turbo-equalization. *Eur. Trans. Telecommun.*, 6(5):507–511, September-October 1995.

[DK04] A. Derneryd and G. Kristensson. Antenna signal correlation and its relation to the impedance matrix. *Elect. Lett.*, 40(7):401–402, April 2004.

[DLLT$^+$06] A. Diallo, C. Luxey, P. Le Thuc, R. Staraj, and G. Kossiavas. Study and reduction of the mutual coupling between two mobile phone PIFAs operating in the DCS1800 and UMTS bands. *IEEE Trans. Antennas Propag.*, 54(11):3063–3074, November 2006.

[DLLT$^+$08] A. Diallo, C. Luxey, P. Le Thuc, R. Staraj, and G. Kossiavas. Enhanced two-antenna structures for universal mobile telecommunications system diversity terminals. *IET Proc. Microw. Antennas Propag.*, 2(1):93–101, February 2008.

[DLR77] A. P. Dempster, N. M. Laird, and D. B. Rubin. Maximum-likelihood from incomplete data via the em algorithm. *J. Roy. Stat. Soc.*, 39(1):1–38, January 1977.

[DLT09] DL MIMO for LTE-Advanced. Technical Report R1-092028, 3GPP, 2009.

[DLTK08] C. Diallo, A. Luxey, R. Le Thuc, P. Staraj, and G. Kossiavas. Diversity performance of multiantenna systems for UMTS cellular phones in different propagation environments. *Int. J. Antennas Propag.*, 2008.

[DM05] M. Debbah and R. Müller. MIMO channel modeling and the principle of maximum entropy. *IEEE Trans. Inf. Theory*, 51(5), May 2005.

[Don07] Lu Dong. *MIMO Selection and Modeling Evaluations for Indoor Wireless Environments*. PhD thesis, School of Electrical and Computer Engineering, Georgia Institute of Technology, 2007.

[DPS08] E. Dahlman, S. Parkvall, and J. Sköld. *3G Evolution: HSPA and LTE for Mobile Broadband*. Academic Press, London, 2008.

[DRJ+06] S. Dossche, J. Rodriguez, L. Jofre, S. Blanch, and J. Romeu. Decoupling of a two-element switched dual band patch antenna for optimum MIMO capacity. In *Proceedings of the IEEE Antenna Propagation Society International Symposium*, pages 325–328, Albuquerque, NM, USA, July 2006.

[Dun44] O. E. Dunlap. *Radio's 100 Men of Science*. Harper and Brothers, New York and London, 1944.

[DV02] A. Dejonghe and L. Vandendorpe. Turbo-equalization for multilevel modulation: an efficient low-complexity scheme. In *Proceedings of the ICC'2002 – IEEE International Conference on Communications*, volume 3, pages 1863–1867, New York, USA, April 2002.

[DV05] P. Dayal and M. Varanasi. An optimal two transmit antenna space-time code and its stacked extensions. *IEEE Trans. Inf. Theory*, 51(12):4348–4355, 2005.

[EBF09] S. Eberli, A. Burg, and W. Fichtner. Implementation of a 2×2 MIMO-OFDM receiver on an application specific processor. *Microelectron. J.*, 40(11):1642–1649, 2009.

[Eea04] V. Erceg and et al. TGn channel models. Technical report, IEEE P802.11 Wireless LANs, May 2004. Available: grouper.ieee.org/groups/802/.

[Eff10] Effect of subspace information accuracy on MU-MIMO. Technical Report R1-100191, 3GPP, 2010.

[ESBC04] V. Erceg, P. Soma, D. S. Baum, and S. Catreux. Multiple-input multiple-output fixed wireless radio channel measurements and modeling using dual-polarized antennas at 2.5 GHz. *IEEE Trans. Wirel. Commun.*, 3(6):2288–2298, November 2004.

[ETM06] G. Eriksson, F. Tufvesson, and A. F. Molisch. Potential for MIMO systems at 300 MHz – measurements and analysis. *Antenn '06*, Linköping, Sweden, May 2006.

[ETW08] R. H. Etkin, D. N. C. Tse, and H. Wang. Gaussian interference channel capacity to within one bit. *IEEE Trans. Inf. Theory*, 54(12):5534–5562, December 2008.

[EWA10a] E. W. Weisstein. Incomplete gamma function from MathWorld—A Wolfram web resource. Available: http://mathworld.wolfram.com/IncompleteGammaFunction.html, accessed 3rd August 2010.

[EWA10b] E. W. Weisstein. Gamma function from MathWorld—A Wolfram web resource. Available: http://mathworld.wolfram.com/GammaFunction.html, accessed 3rd August 2010.

[EWJ10] E. W. Weisstein. Chi-Squared distribution from MathWorld—A Wolfram web resource. Available: http://mathworld.wolfram.com/Chi-SquaredDistribution.html, accessed 24th June 2010.

[FFLT08] Y. Fei, Y. Fan, B. K. Lau, and J. S. Thompson. Optimal single-port matching impedance for capacity maximization in compact MIMO arrays. *IEEE Trans. Antennas Propag.*, 56(11):3566–3575, November 2008.

[FG98] G. J. Foschini and M. J. Gans. On limits of wireless communications in a fading environment when using multiple antennas. *Wirel. Personal Commun.*, 6: 311–335, March 1998.

[FGAR08] P. J. Ferrer, J. M Gonzalez-Arbesu, and J. Romeu. Decorrelation of two closely spaced antennas with a metamaterial AMC surface. *Microw. Opt. Technol. Lett.*, 50(5):1414–1417, May 2008.

[FGVW99] G. J. Foschini, G. D. Golden, R. A. Valenzuela, and P. W. Wolniansky. Simplified processing for high spectral efficiency wireless communications employing multi-element arrays. *IEEE J. Select. Areas Commun.*, 17:1841–1852, November 1999.

[Fin10] Finalizing 4Tx codebook for UL SU-MIMO. Technical Report R1-101641, 3GPP, 2010.

[FJS02] B. Fleury, P. Jourdan, and A. Stucki. High-resolution channel parameter estimation for MIMO applications using the SAGE algorithm. *International Zurich Seminar on Broadband Communications*, pages 30-1–30-9, Zürich, Swiss, February 2002.

[Fle00] B. H. Fleury. First- and second-order characterization of direction dispersion and space selectivity in the radio channel. *IEEE Trans. Inform. Theory*, 46(6):2027–2044, September 2000.

[Fle10] A flexible feedback concept. Technical Report R1-100051, 3GPP, 2010.

[FLS+06] Y. Fei, B. K. Lau, A. Sunesson, J. B. Andersen, A. J. Johansson, and J. S. Thompson. Experiments of closely coupled monopoles with load matching in a random field. In *Europ. Conf. Antennas Propag., (EUCAP)*, Nice, France, November 2006.

[FM97] B. J. Frey and D. J. C. MacKay. A revolution: Belief propagation in graphs with cycles. *Advances in Neural Information Processing Systems 10*, December 1997.

[FN98] C. Farsakh and J. A. Nossek. Spatial covariance based downlink beamforming in an SDMA mobile radio system. *IEEE Trans. Comput.*, 46(11):1497–1506, November 1998.

[FOHD] F. Quitin, C. Oestges, F. Horlin, and P. De Doncker. Clustered channel characterization for indoor polarized MIMO systems. In *IEEE 20th International Symposium on Personal, Indoor and Mobile Radio Communications, PIMRC '09*, Tokyo, Japan, September 13–16, 2009.

[Fos96] G. J. Foschini. Layered space-time architecture for wireless communication in a fading environment when using multiple antennas. *Bell Labs Syst. Tech. J.*, pages 41–59, Autumn 1996.

[FWRH10] J. Furuskog, K. Werner, M. Riback, and B. O. Hagerman. Field trials of LTE with 4×4 MIMO. *ERICSSON Review*, 1, February 2010.

[Gan06a] M. J. Gan. Channel capacity between antenna arrays – part I: Sky noise dominates. *IEEE Trans. Commun.*, 54(9):1586–1592, September 2006.

[Gan06b] M. J. Gan. Channel capacity between antenna arrays – part II: Amplifier noise dominates. *IEEE Trans. Commun.*, 54(11):1983–1992, November 2006.

[GBJS08] D. Guha, S. Biswas, T. Joseph, and M. T. Sebastian. Defected ground structure to reduce mutual coupling between cylindrical dielectric resonator antennas. *Elect. Lett.*, 44(14):836–837, July 2008.

[GCYP07] Y. Gao, X. Chen, Z. Ying, and C. Parini. Design and performance investigation of a dual-element PIFA array at 2.5 GHz for MIMO terminal. *IEEE Trans. Antennas Propag.*, 55(12):3433–3441, December 2007.

[GDF03] G. D. Forney Jr. On the Role of the MMSE Estimation in Approaching the Information-Theoretic Limits of Linear Gaussian Channels: Shannon meets Wiener. In *41th Annual Allerton Conference on Communications Control and Computer*, Monticello, IL, October 2–4, 2003.

[GFVW99] G. Golden, C. Foschini, R. Valenzuela, and P. Wolniansky. Detection algorithm and initial laboratory results using V-BLAST space-time communication architecture. *Electron. Lett.*, 35(1):14–16, January 1999.

[GHF05] C. Giovaneli, B. Honary, and P. Farrell. Space-frequency coded OFDM system for multi-wire power line communications. *ISPLC*, Vancouver, Canada, 2005.

[GJ02] G. Ginis and J. Cioffi Vectored transmissions for digital subscriber line systems. *J. Sel. Areas Commun.*, 20(5):1085–1104, June 2002.

[GJJV03] A. Goldsmith, S. A. Jafar, N. Jindal, and S. Vishwanath. Capacity limits of MIMO channels. *IEEE J. Sel. Areas Commun.*, 21(5):684–702, 2003.

[GN06] Z. Guo and P. Nilsson. Algorithm and implementation of the K-best sphere decoding for MIMO detection. *IEEE J. Sel. Areas Commun.,* 24(3):491–503, March 2006.

[GN07] M. Gustafsson and S. Nordebo. On the spectral efficiency of a sphere. *Prog. Electromagnet. Res. (PIER)*, 67:275–296, 2007.

[Gol05] A. Goldsmith. *Wireless Communications*. Cambridge University Press, 2005.

[GP08] R. Glogowski and C. Peixeiro. Multiple printed antennas for integration into small multistandard handsets. *IEEE Antennas Wirel. Propag. Lett.*, 7:632–635, 2008.

[GP96] D. Gerlach and A. Paulraj. Base station transmitting antenna arrays for multipath environments. *Signal Processing (Elsevier Science)*, 54:59–73, 1996.

[Gra02] A. Grant. Rayleigh fading multiantenna channels. *EURASIP J. Appl. Signal Process., special issue on Space-Time coding*, 3:316–329, 2002.

[GS01] G. Ganesan and P. Stoica. Space-time block codes: A maximum SNR approach. *IEEE Trans. Inf. Theory*, 47:1650–1656, 2001.

[GSK09] M. Gustafsson, C. Sohl, and G. Kristensson. Illustrations of new physical bounds on linearly polarized antennas. *IEEE Trans. Antennas Propag.*, 57(5):1319–1327, May 2009.

[GSS+03] D. Gesbert, M. Shafi, D. Shiu, P. J. Smith, and A. Naguib. From theory to practice: an overview of MIMO space-time coded wireless systems. *IEEE J. Sel. Areas Commun.,* 21(3):281–302, 2003.

[GSV04] D. Guo, S. Shamai, and S. Verdu. Mutual information and mmse in gaussian channels. In *Proc. IEEE International Symposium on Information Theory*, page 347, Chicago, IL, USA, 2004.

[GSWM05] M. Guenacha, F. Simoens, H. Wymeersch, and M. Moeneclaey. Code-aided frame joint channel and frequency estimation for st-bicm multiuser DS-CDMA system. In *EURASIP European Signal Processing Conference, EUSIPCO*, Antalya, Turkey, September 2005.

[GVL96] G. H. Golub and C. F. van Loan. *Matrix Computations* (3rd Ed). John Hopkins University Press, Baltimore, 1996.

[GVR02] D. Guo, S. Verdu, and L. K. Rasmussen. Asymptotic normality of linear multiuser receiver outputs. *IEEE Trans. Inf. Theory*, 48(12):3080–3095, December 2002.

[GWDN05] D. Garrett, G. K. Woodward, L. Davis, and C. Nicol. A 28.8 Mb/s 4/spl times/ 4 MIMO 3G CDMA receiver for frequency selective channels. *IEEE Int. Solid-State Circuits Conf. (ISSCC)*, 40(1):320–330, January 2005.

[GWM04] M. Guenach, H. Wymeersch, and M. Moeneclaey. Joint estimation of path delay and complex gain for coded systems using the em algorithm. In *Proceedings of the International Zurich Seminar, IZS*, Zurich, Switserland, Feburary 2004.

[H70] R. A. Harshman. Foundations of the PARAFAC procedure: Model and conditions for an "explanatory" multi-mode factor analysis. *UCLA Working Papers in Phonetics*, 16: 1–84, December 1970.

[H83] A. J. Viterbi and A. M. Viterbi. Nonlinear estimation of PSK-modulated carrier phase with application to burst digital transmission. *IEEE Trans. Inf. Theory*, 29:543–551, July 1983.

[H88] M. Oerder and H. Meyr. Digital filter and square timing recovery. *IEEE Trans. Commun.*, 36:605–611, May 1988.

[HA59] H. A. Haus and R. B. Adler. *Circuit Theory of Linear Noisy Networks*. Wiley, New York, 1959.

[Hal05] P. Hallbjörner. The significance of radiation efficiencies when using S-parameters to calculate the received signal correlation from two antennas. *IEEE Antennas Wirel. Propag. Lett.*, 4:97–99, 2005.

[Han95] S. V. Hanly. An algorithm for combined cell-site selection and power control to maximize cellular spread spectrum capacity. *IEEE J. Sel. Areas Commun.*, 13(7):1332–1340, September 1995.

[HBO06] D. Hammarwall, M. Bengtsson, and B. Ottersten. On downlink beamforming with indefinite shaping constraints. *IEEE Trans. Signal Process.*, 54:3566–3580, September 2006.

[HCL+08] S. Hong, K. Chung, J. Lee, S. Jung, S. Lee, and J. Choi. Design of a diversity antenna with stubs for UWB applications. *Microw. Opt. Technol. Lett.*, 50(5):1352–1356, May 2008.

[Hea02] H. Herdin and et al. Variation of measured indoor MIMO capacity with receive direction and position at 5.2 GHz. *Electron Lett.*, 38(21):1283–1285, October 2002.

[Hea07] P. S. Hall and et al. Antennas and propagation for on-body communications. *IEEE Antennas Propag. Mag.*, 49(3):41–58, June 2007.

[HFGS05] T. Haustein, A. Forck, H. Gabler, and S. Schiffermuller. From theory to practice: MIMO real-time experiments of adaptive bit-loading with linear and non-linear transmission and detection schemes. In *Proceedings of the 61st Semiannual Vehicular Technology Conference*, volume 2, pages 1115–1119, Stockholm, Sweden, May 2005.

[HG03] L. Hanlen and A. Grant. Capacity analysis of correlated MIMO channels. In *Proceedings of the IEEE International Symposium on Information Theory*, Yokohama, Japan, July 2003.

[HG07] L. Hao and J. Guo. A MIMO-OFDM scheme over coupled multi-conductor powerline communication channel. In *Proceedings of the ISPLC*, Pisa, Italia, 2007.

[HH02] B. Hassibi and B. M. Hochwald. High-rate codes that are linear in space and time. *IEEE Trans. Inf. Theory*, 48:1804–1824, July 2002.

[HJU09] R. Hunger, M. Joham, and W. Utschick. On the MSE-duality of the broadcast channel and the multiple access channel. *IEEE Trans. Signal Process.*, 57(2):698–713, 2009.

[HMG05] A. Hutter, S. Mekrazi, and B. Getu. Alamouti-based space-frequency coding for OFDM. *Wirel. Personal Commun.*, 2005.

[HMRE03] H. Özcelik, M. Herdin, R. Prestros, and E. Bonek. How MIMO Capacity is Linked with Single Element Fading Statistics. In *Proceedings of the International Conference on Electromagnetics in Advanced Applications, ICEAA 2003*, pages 775–778, Torino, Italy, September 2003.

[HMT02] B. M. Hochwald, T. L. Marzetta, and V. Tarokh. Multi-Antenna Channel-Hardening and its implications for rate feedback and scheduling. *IEEE Trans. Inf. Theory*, 50(9), pages 1893–1909, September 2002.

[HOY02] B. Han, M. L. Overton, and T. P.-Y. Yu. Design of hermite subdivision schemes aided by spectral radius optimization. *Submitted to SIAM J. Scient. Comp.*, 2002.

[HRVM03] C. Herzet, V. Ramon, L. Vandendorpe, and M. Moeneclaey. EM algorithm-based timing synchronization in turbo receivers. *IEEE International Conference on Acoustics, Speech and Signal Processing, ICASSP*, Hong-Kong, April 2003.

[HtB03] B. M. Hochwald and S. ten Brink. Achieving near-capacity on a multiple-antenna channel. *IEEE Trans. Commun.*, 51(3):389–399, March 2003.

[HY98] C. Huang and R. Yates. Rate of convergence for minimum power assignment algorithms in cellular radio systems. *Baltzer/ACM Wirel. Netw.*, 4:223–231, 1998.

[IEE09a] IEEE, 802.16-2009 IEEE Standard for Local and Metropolitan Area Networks – Part 16: Air Interface for Broadband Wireless Access Systems, IEEE 802.16 Standards, May 2009. Available: http://www.ieee802.org/16.

[IEE09b] IEEE, 802.11n-2009 IEEE Standard for Information Technology – Part 11: Wireless LAN Medium Access Control (MAC) and Physical Layer (PHY) Specifications Amendment: Enhancements for Higher Throughput, IEEE 802.11 Standards, October 2009. Available: http://www.ieee802.org/11.

[IEEE P802.11n] IEEE P802.11n/D2.00, Draft STANDARD for 802.11: Wireless LAN Medium Access Control (MAC) and Physical Layer (PHY) specifications: Amendment: Enchancements for higher throughput, February 2007.

[IEEE Std 802.11a] IEEE Std 802.11a, 802.11: Wireless LAN Medium Access Control (MAC) and Physical Layer (PHY) specifications: High-speed physical layer in the 5 GHz band. The Institute of Electrical and Electronics Engineers, Inc., New York, 1999.

[IEEE16mSDD] IEEE 802.16 Broadband Wireless Access Working Group, IEEE 802.16m system description document, IEEE 802.16m-08/003r7delta, February 2009.

[IMW+09] R. Irmer, H.-P. Mayer, A. Weber, V. Braun, M. Schmidta, M. Ohm, N. Ahr, A. Zoch, C. Jandura, P. Marsch, and G. Fettweis. Multisite field trial for LTE and advanced concepts. *IEEE Commun. Mag.*, 92–98, February 2009.

[ITU-R08] ITU-R. Requirements related to technical system performance for IMT-Advanced radio interfaces. ITU-R Rep M2134, 2008. Available: http://www.itu.int/publ/R-REP-M.2134-2008/en.

[J01] H. Jafharkani. A quasi-orthogonal space-time block code. *IEEE Trans. Commun.*, 49(1):1–4, 2001.

[JL10] M. A. Jensen and B. K. Lau. Uncoupled matching for active and passive impedances of coupled arrays in MIMO systems. *IEEE Trans. Antennas Propag.*, in press.

[Jak71] W. C. Jakes Jr. A comparison of specific space diversity techniques for reduction of fast fading in UHF mobile radio systems. *IEEE Trans. Veh. Technol.*, VT-20(4):81–92, November 1971.

[JB04] E. A. Jorswieck and H. Boche. Performance analysis of capacity of MIMO systems under multiuser interference based on worst-case behavior. *EURASIP J. Wirel. Commun. Netw.*, 2:273–285, 2004.

[JB06] E. A. Jorwieck and H. Boche. Outage probability in multiple antenna systems. *Eur. Trans. Telecommun.*, 18:217–233, 2006.

[JLK+02] J. P. Kermoal, L. Schumacher, K. I. Pedersen, P. E. Mogensen, and F. Frederiksen. A stochastic MIMO radio channel model with experimental validation. *IEEE J. Sel. Areas Commun.*, 20(6):1211–1226, August 2002.

[JP98] J. J. Blanz and P. Jung. A flexibly configurable spatial model for mobile radio channels. *IEEE Trans. Commun.*, 46(3):367–371, March 1998.

[JQB08] M. A. Jensen, B. Quist, and N. Bikhazi. Antenna design for mobile MIMO systems. *IEICE Trans. Commun.*, E91-B(6):1702–1712, June 2008.

[JTW+09] V. Jungnickel, L. Thiele, T. Wirth, T. Haustein, S. Schiffermüller, A. Forck, S. Wahls, S. Jaeckel, S. Schubert, C. Juchems, F. Luhn, R. Zavrtak, H. Droste, G. Kadel, W. Kreher, J. Mueller, W. Stoermer, and G. Wannemacher. Coordinated multipoint trials in the downlink. In *Proceedings of the 5th IEEE Broadband Wireless Access Workshop (BWAWS)*, Honolulu, HI, November 2009.

[KBC01] M. Kobayashi, J. Boutros, and G. Caire. Successive interference cancellation with siso decoding and em channel estimation. *IEEE J. Select. Areas Commun.*, 19:1729–1743, September 2001.

[KCVW02] P. Kyritsi, D. Cox, R. Valenzuela, and P. Wolniansky. Effect of antenna polarization on the capacity of a multiple element systems in an indoor environment. *IEEE J. Select. Areas Commun.*, 20(6):1227–1239, 2002.

[Kea07] J. Koivunen and et al. Dynamic multi-link indoor MIMO measurements at 5.3 GHz. In *EuCAP*, Edinburgh, UK, November 2007.

[KG70] A. R. Kaye and D. A. George. Transmission of multiple PAM signals over multiple channel and diversity systems. *IEEE Trans. Commun.*, 18(10):520–526, October 1970.

[KHX02] M. A. Kamath, B. L. Hughes, and Y. Xinying. Gaussian approximations for the capacity of MIMO Rayleigh fading channels. *Thirty-Sixth Asilomar Conference on Signals, Systems and Computers*, November 2002.

[KJ07] P. Kyösti and T. Jämsä. Complexity comparison of MIMO channel modelling methods. In *Proceedings of the IEEE International Symposium on Wireless Communication Systems (ISWCS'07)*, Trondheim, Norway, October 2007.

[KKGK09] F. Kaltenberger, M. Kountouris, D. Gesbert, and R. Knopp. On the trade-off between feedback and capacity in measured MU-MIMO channels. *IEEE Trans. Wirel. Commun.*, 18(9):4866–4875, September 2009.

[KKP08] A. Kalis, A. Kanatas, and C. Papadias. A novel approach to MIMO transmission using a single RF front end. *IEEE J. Select. Areas Commun.*, 26(6):972–980, August 2008.

[KLF08] T. Kokkinos, E. Liakou, and A. P. Feresidis. Decoupling antenna elements of PIFA arrays on handheld devices. *Electron Lett.*, 44(25):1442–1444, December 2008.

[KMH+07] P. Kyosti, J. Meinila, L. Hentila, X. Zhao, T. Jamsa, C. Schneider, M. Narandzic, M. Milojevic, A. Hong, J. Ylitalo, V.-M. Holappa, M. Alatossava, R. Bultitude, Y. de Jong, and T. Rautiainen. WINNER II channel models.Technical Report IST-WINNER D1.1.2 ver 1.1, September 2007.

[KS07] M. Katz and S. Shamai. On the outage probability of a multiple-input single-output communication link, *IEEE Trans. Wirel. Commun.*, 6(11):4120–4128, November 2007.

[KSB09] R. Kobeissi, S. Sezginer, and F. Buda. Downlink performance analysis of full-rate STCs in 2×2 MIMO WiMAX Systems. *VTC 2009 RAS Workshop*, Barcelona, Spain, April 2009.

[KYH95] P. S. Kumar, R. Yates, and J. Holtzman. Power control based on bit error (BER) measurements. In *Proceedings of the IEEE Military Communications Conference MILCOM 95*, pages 617–620, McLean, VA, November 1995.

[LA06] B. K. Lau and J. B. Andersen. On closely coupled dipoles with load matching in a random field. *IEEE International Symposium Pers., Indoor and Mobile Radio Communications (PIMRC)*, Helsinki, Finland, September 2006.

[LA07a] B. K. Lau and J. B. Andersen. An antenna system and a method for operating an antenna system. *PCT Filed (Pub. No. WO/2008/030165)*, September 2007.

[LA07b] B. K. Lau and J. B. Andersen. Antenna system and method of providing an antenna system. *Swedish Patent Application (No. 0702307-0)*, October 2007.

[LA09] B. K. Lau and J. B. Andersen. Unleashing multiple antenna systems in compact terminal devices. *International Workshop Antenna Technology (IWAT)*, Santa Monica, CA, USA, March 2009.

[LAKM06a] B. K. Lau, J. B. Andersen, G. Kristensson, and A. F. Molisch. Antenna matching for capacity maximization in compact MIMO systems. In *Proceedings of the 3rd International Symposium Wireless Communications System (ISWCS)*, pages 253–257, Valencia, Spain, September 2006.

[LAKM06b] B. K. Lau, J. B. Andersen, G. Kristensson, and A. F. Molisch. Impact of matching network on bandwidth of compact antenna arrays. *IEEE Trans. Antennas Propag.*, 54(11):3225–3238, November 2006.

[LB10] D. Liu and A. G. Burr. Full diversity full rate tensor-based space time codes. COST 2100 TD(10)10038, 10th MCM, Athens, Greece, 3–5 February 2010.

[LBL+08] M. Li, B. Bougard, E. E. Lopez, A. Bourdoux, D. Novo, L. Van Der Perre, and F. Catthoor. Selective spanning with fast enumeration: a near maximum-likelihood MIMO detector designed for parallel programmable baseband architectures. In *Proceedings of the IEEE International Conference on Communications (ICC) 2008*, pages 737–741, May 2008.

[LBX+08] M. Li, B. Bougard, W. Xu, D. Novo, L. Van Der Perre, and F. Catthoor. Optimizing near-ML MIMO detector for SDR baseband on parallel programmable architectures. In *Proc. Design, Automation and Test in Europe (DATE)*, pages 444–449, March 2008.

[LDGW04] Z.-Q. Luo, T. N. Davidson, G. B. Giannakis, and K. Wong. Transceiver optimization for block-based multiple access through ISI channels. *IEEE Trans. Signal Process.*, 52(4):1037–1052, April 2004.

[LERBVdP09] E. Lopez-Estraviz, V. Ramon, A. Bourdoux, and L. Van der Perre. Symbol based search space constraining for complexity/performance scalable near ml detection in spatial multiplexing MIMO OFDM systems. *IEEE International Conference on Communications (ICC)*, Dresden, Germany, June 2009.

[LGF05] G. Lebrun, J. Gao, and M. Faulkner. MIMO transmission over a time-varying channel using SVD. *IEEE Trans. Wirel. Commun.*, 4:757–764, 2005.

[LGH01] Y. Li, C. Georghiades, and G. Huang. Iterative maximum-likelihood sequence estimation for space-time coded systems. *IEEE Trans. Commun.*, 49:948–951, June 2001.

[LGL01] C. Laot, A. Glavieux, and J. Labat. Turbo-equalization: adaptive equalization and channel decoding jointly optimized. *IEEE J. Sel. Areas Commun.*, 19(9):1744–1752, September 2001.

[LJ09] M. Lai and S. Jeng. Slot antennas with an extended ground for multiple-antenna systems in compact wireless devices. *IEEE Antennas Wirel. Propag. Lett.*, 8:19–22, 2009.

[LLL$^+$10] Q. Li, G. Li, W. Lee, M. Lee, D. Mazzarese, B. Clerckx, and Z. Li. MIMO techniques in WiMAX and LTE: a feature overview. *Communications Magazine, IEEE*, 48(5):86–92, May 2010. Doi: 10.1109/MCOM.2010.5458368

[LL07] S. Lim and H. Ling. Design of electrically small, pattern reconfigurable Yagi antenna. *Electron Lett.*, 43(24):1326–1327, November 2007.

[LLW$^+$07] Y. Lin, H. Lee, M. Woh, Y. Harel, S. Mahlke, T. Mudge, C. Chakrabarti, and K. Flautner. SODA: a high-performance DSP architecture for software-defined radio. *IEEE Micro*, 27(1):114–123, Feburary 2007.

[LLZR09] Q. Li, X. E. Lin, J. Zhang, and W. Roh. Advancement of MIMO technology in WiMAX: from IEEE 802.16d/e/j to 802.16m. *IEEE Commun. Mag.*, 47(6):100–107, June 2009.

[LR99] X. Li and J. A. Ritcey. Trellis-coded modulation with bit interleaving and iterative decoding. *IEEE J. Sel. Areas Commun.*, 17(4):715–724, April 1999.

[LSWL04] K. K. Leung, C. W. Sung, W. S. Wong, and T. M. Lok. Convergence theorem for a general class of power-control algorithms. *IEEE Trans. Comput.*, 52(9):1566–1574, September 2004.

[LW00] B. Lu and X. Wang. Iterative receivers for multiuser space-time coding systems. *IEEE J. Sel. Areas Commun.*, 18:2322–2335, November 2000.

[LW08] T. Lee and Y. E. Wang. Mode-based information channels in closely coupled dipole pairs. *IEEE Trans. Antennas Propag.*, 56(12):3804–3811, December 2008.

[LY06] Z.-Q. Luo and W. Yu. An introduction to convex optimization for communications and signal processing. *IEEE J. Select. Areas Commun.*, 24(8):1426–1438, August 2006.

[MÖ2] R. Müller. A random matrix model of communication via antenna arrays. *IEEE Trans. Inf. Theory*, pages 2495–2506, September 2002.

[M04] C. Marshman. Antenna array technology and MIMO systems. Final report for Ofcom contract AY4476 2004. Available: www.ofcom.org.uk.

[M2108] Guidelines for evaluation of radio interface technologies for imt-advanced. Technical Report Report ITU-R M.2135, Geneva, 2008.

[MAE01] M. Steinbauer, A. F. Molisch, and E. Bonek. The double-directional radio channel. *IEEE Mag. Antennas Propag.*, 43(4):51–63, August 2001.

[Mak02] S. M. Makarov. *Antenna and EM Modeling with MATLAB*. John Wiley and Sons, New York, 2002.

[MB09] M. Baker. LTE-Advanced physical layer. *IMT-Advanced Evaluation Workshop*, Beijing, pages 17–18, December 2009.

[MD09] P. Mookiah and K. R. Dandekar. Metamaterial-substrate antenna array for MIMO communication system. *IEEE Trans. Antennas Propag.*, 57(10):3283–3292, October 2009.

[MDG$^+$00] M. Steinbauer, D. Hampicke, G. Sommerkorn, A. Schneider, A. F. Molisch, R. Thomä, and E. Bonek. Array measurement of the double-directional mobile radio channel. In *Proceedings of the Vehicular Technology Conference, VTC 2000-Spring*, volume 3, pages 1656–1662, Tokio, Japan, May 2000.

[Mey00] C. D. Meyer. *Matrix Analysis and Applied Linear Algebra*. SIAM, 2000.

[ME04] M. Herdin and E. Bonek. A MIMO correlation matrix based metric for characterizing non-stationarity. *IST Mobile and Wireless Communications Summit 2004*, Lyon, France, June 2004.

[Men95] J. M. Mendel. *Lessons in Estimation Theory for Signal Processing, Communications, and Control*. Prentice Hall, 1995.

[MH94] U. Madhow and M. Honig. MMSE interference supression for direct-sequence spread spectrum CDMA. *IEEE Trans. Commun.*, 42:3178–3188, December 1994.

[MH99] T. L. Marzetta and B. M. Hochwald. Capacity of a mobile multiple-antenna communication link in Rayleigh flat fading. *IEEE Trans. Inf. Theory*, 45(1):139–157, January 1999.

[MJ03] M. T. Ivrlac and J. A. Nossek. Quantifying diversity and correlation of rayleigh fading MIMO channels. In *Proceedings of the IEEE International Symposium on Signal Processing and Information Technology, ISSPIT'03*, pages 158–161, Darmstadt, Germany, December 2003.

[MJ05] M. L. Morris and M. A. Jensen. Network model for MIMO systems with coupled antennas and noisy amplifiers. *IEEE Trans. Antennas Propag.*, 53(1):545–552, January 2005.

[MJW05] M. L. Morris, M. A. Jensen, and J. W. Wallace. Superdirectivity in MIMO systems. *IEEE Trans. Antennas Propag.*, 53(9):2850–2857, September 2005.

[MLM$^+$05] B. Mei, A. Lambrechts, J.-Y. Mignolet, D. Verkest, and R. Lauwereins. Architecture exploration for a reconfigurable architecture template. *Des. Test Comput., IEEE*, 22(2):90–101, March 2005.

[MLG06] H. G. Myung, J. Lim, and D. J. Goodman. Single carrier FDMA for uplink wireless transmission. *Veh. Technol. Mag., IEEE*, 1(3):30–38, September 2006. Doi:10.1109/MVT.2006.307304

[MM80] R. A. Monzingo and T. W. Miller. *Introduction to Adaptive Arrays*. Wiley, New York, 1980.

[MM98] R. J. McEliece and D. J. C. MacKay. Turbo-decoding as an instance of Pearl's 'belief propagation' algorithm. *IEEE J. Sel. Areas Commun.*, 16(2):140–152, February 1998.

[MMDdC00] P. Magniez, B. Muquet, P. Duhamel, and M. de Courville. Improved turbo-equalization with application to bit interleaved modulations. In *Proceedings of the Asilomar Conference on Signals, Systems and Computers*, Monterey (CA), USA, October 2000.

[MMF98] H. Meyr, M. Moeneclaey, and S. A. Fechtel. *Digital Communication Receivers: Synchronization, Channel Estimation and Signal Processing*. Series in Telecommunications and Signal Processing. Wiley, USA, 1998.

[Mol10] A. Molisch. UWB MIMO channel propagation modelling. In *Proceedings of the 4th European Conference on Antennas and Propagation*, Barcelona, Spain, April 2010.

[MRM08] A. C. K. Mak, C. R. Rowell, and R. D. Murch. Isolation enhancement between two closely packed antennas. *IEEE Trans. Antennas Propag.*, 56(11):3411–3419, November 2008.

[MROU10] L. Mroueh. On space time coding design and multiuser multiplexing gain over selective channels, PhD dissertation, Telecom ParisTech, January 2010.

[MS05B] J. Mebdo and P. Shramm. Channel models for HIPERLAN/2, ETSI/BRAN, 3ERI085B, 1998.

[MSS02] A. L. Moustakas, S. H. Simon, and A. M. Sengupta. MIMO Capacity Through Correlated Channels in the Presence of Correlated Interferers and Noise: A (not so) Large Analysis. Bell Laboratories, Lucent Technologies, Murray Hill, NJ 07974, May 2002. Available: *http://mars.bell-labs.com.*

[MS98] G. Montalbano and D. T. M. Slock. Matched filter bound optimization for multiuser downlink transmit beamforming. In *Proceedings of the IEEE International Conference on Universal Personal Communications (ICUPC)*, Florence, Italy, October 1998.

[MSY08] Y. Ma, W. Shieh, and Q. Yang. Bandwidth-efficient 21.4gb/s coherent optical 2×2 MIMO OFDM transmission. *OFC*, San Diego, CA, USA, 2008.

[MV01] R. Müller and S. Verdu. Design and analysis of low-complexity interference mitigation on vector channels. *IEEE J. Sel. Areas Commun.*, 1429–1441, August 2001.

[NHD+03] N. Noels, C. Herzet, A. Dejonghe, V. Lottici, H. Steendam, M. Moeneclaey, M. Luise, and L. Vandendorpe. Turbo-synchronization: an em algorithm approach. In *Proceedings of the IEEE International Conference on Communications, ICC*, Anchorage, May 2003.

[NLM98] J. Ng, K. Letaief, and R. Murch. Complex optimal sequences with constant magnitude for fast channel estimation initialization. *IEEE Trans. Commun.*, 46:305–308, March 1998.

[NSD+07] S. Nedevschi, S. Surana, B. Du, R. Patra, E. Brewer, and V. Stan. Potential of CDMA450 for rural network connectivity. *IEEE Commun. Mag.*, 45(1):128–135, January 2007.

[NTD08] A. Nilsson, E. Tell, and L. Dake. An 11 mm^2 70 mW fully-programmable baseband processor for mobile WiMAX and DVB-T/H in 0.12um CMOS. In *Proc. International Solid-State Circuits Conference (ISSCC)*, pages 266–612, Feburary 2008.

[OC07] C. Oestges and B. Clerckx. *MIMO Wireless Communications*. Academic Press, 2007.

[OC08] Y. Okano and K. Cho. Multiple antenna performance of quarter-wavelength monopole antennas for card-type terminal. *IEICE Trans. Commun.*, E91-B(9):2948–2955, September 2008.

[OCGD08] C. Oestges, B. Clerckx, M. Guillaud, and M. Debbah. Dual-polarized wireless communications: From propagation models to system performance evaluation. *IEEE Trans. Wirel. Commun.*, 7(10):4019–4031, October 2008.

[Özc04] H. Özcelik. *Indoor MIMO Channel Models*. PhD thesis, Technische Universität Wien, Vienna, Austria, 2004.

[P01] J. G. Proakis. *Digital Communications* (4th Ed). McGraw-Hill, 2001.

[PBM31] H. O. Peterson, H. H. Beverage, and J. B. Moore. Diversity telephone receiving system of RCA Communications, Inc. *Proc. IRE*, 19(4):562–584, April 1931.

[PBT05] A. S. Y. Poon, R. W. Brodersen, and D. Tse. Degrees of freedom in multiple antenna channels: a signal space approach. *IEEE Trans. Inform. Theory*, 51(2):523–536, February 2005.

[Pet92] H. J. M. Peters. *Axiomatic Bargaining Game Theory*. Kluwer Academic Publishers, Dordrecht, 1992.

[Per08] E. Perahia. IEEE 802.11n development: history, process, and technology. *IEEE Commun. Mag.*, 46(7):48–55, July 2008.

[PFA03] P. Almers, F. Tufvesson, and A. F. Molisch. Measurement of keyhole effects in a wireless multiple-input multiple-output (MIMO) channel. *IEEE Commun. Lett.*, 7(8):373–375, August 2003.

[PGNB04] A. J. Paulraj, D. A. Gore, R. U. Nabar, and H. Bolcskei. An overview of MIMO communications – a key to gigabit wireless. *Proc. IEEE*, 92(2):198–218, Feburary 2004.

[PK94] A. J. Paulraj and T. Kailath. Increasing capacity in wireless broadcast systems using distributed transmission/directional reception. Technical Report U.S. Patent, no. 5,345,599, 1994.

[PKC$^+$] A. Paier, J. Karedal, N. Czink, H. Hofstetter, C. Dumard, T. Zemen, F. Tufvesson, A. F. Molisch, and C. F. Mecklenbräuker. Car-to-car radio channel measurements at 5 GHz: Pathloss, power-delay profile, and delay-doppler spectrum. In *Proceedings of the IEEE International Symposium on Wireless Communication Systems (ISWCS 2007)*, pages 224–228, Trondheim, Norway, October.

[PL09] V. Plicanic and B. K. Lau. Impact of spacing and gain imbalance between two dipoles on HSPA throughput performance. *Electron Lett.*, 45(21):1063–1065, October 2009.

[PLDY09] V. Plicanic, B. K. Lau, A. Derneryd, and Z. Ying. Actual diversity performance of a multiband antenna with hand and head effects. *IEEE Trans. Antennas Propag.*, 57(5):1547–1556, May 2009.

[Poz05] D. M. Pozar. *Microwave Engineering* (3rd Ed). John Wiley and Sons, Hoboken, NJ, 2005.

[PPIL07] M. Payaró, A. Pascual-Iserte, and M. A. Lagunas. Robust power allocation designs for multiuser and multiantenna downlink communication systems through convex optimization. *IEEE J. Sel. Areas Commun.*, 25(7):1390–1401, September 2007.

[Pro95] J. G. Proakis. *Digital Communications* (3rd Ed). New York, 1995.

[PSLDY09] V. Plicanic Samuelsson, B. K. Lau, A. Derneryd, and Z. Ying. Channel capacity performance of multi-band dual antenna in proximity of a user. In *International Workshop Antenna Technology (IWAT)*, Santa Monica, CA, USA, March 2009.

[PTSL02] C. Pietsch, W. Teich, S. Sand, and J. Lindner. Performance evaluation of different multiuser MIMO detection techniques based on a generalized matrix transmission model. COST 273 TD(02)125, 5th MCM, Lisbon, Portugal, 19–20 September 2002.

[PV97] H. Poor and S. Verdú. Probability of error in MMSE multiuser detection. *IEEE Trans. Inf. Theory*, 43(3):858–871, May 1997.

[RC98] G. G. Raleigh and J. M. Cioffi. Spatio-temporal coding for wireless communication. *IEEE Trans. Commun.*, 46(3):357–366, 1998.

[RHV97] P. Robertson, P. Hoeher, and E. Villebrun. Optimal and suboptimal maximum a posteriori algorithms for turbo decoding. *Eur. Trans. Telecommun.*, 8(2):119–125, 1997.

[RHV99] P. Robertson, P. Hoeher, and E. Villebrun. A comparison of optimal and suboptimal map decoding algorithms operating in the log-domain. In *Proceedings of the IEEE International Conference on Communications*, volume 2, pages 1009–1013, 1999.

[RHVM04] V. Ramon, C. Herzet, L. Vandendorpe, and M. Moeneclaey. Em algorithm-based multiuser synchronization in turbo receivers. In *Proceedings of the IEEE International Conference on Acoustics, Speech and Signal Processing, ICASSP*, Montreal, April 2004.

[Ric05] A. Richter. *Estimation of Radio Channel Parameters: Models and Algorithms*. PhD thesis, Technische Universität Ilmenau, 2005.

[RLT98] F. Rashid-Farrokhi, K. J. Liu, and L. Tassiulas. Transmit beamforming and power control for cellular wireless systems. *IEEE J. Select. Areas Commun.*, 16(8):1437–1449, October 1998.

[S49] C. E. Shannon. Communication in the presence of noise. *Proc. IRE*, 37:10–21, 1949.

[SA98] M. Simon and M. S. Alouini. A unified approach to the probability of error for noncoherent and differentially coherent modulations over generalized fading channels. *IEEE Trans. Commun.*, 46(12):1625–1638, December 1998.

[Sal85] J. Salz. Digital transmission over cross-coupled linear channels. *AT&T Tech. J.*, 64(6):1147–1159, July/August 1985.

[SB02] M. Schubert and H. Boche. SIR balancing for multiuser downlink beamforming – a convergence analysis. In *Proceedings of the IEEE International Conference on Communication (ICC)*, vol. 2, pages 841–845, New York, USA, April 2002.

[SB04] M. Schubert and H. Boche. Solution of the multi-user downlink beamforming problem with individual SINR constraints. *IEEE Trans. Veh. Technol.*, 53(1):18–28, January 2004.

[SB05] M. Schubert and H. Boche. Iterative multiuser uplink and downlink beamforming under SINR constraints. *IEEE Trans. Signal Process.*, 53(7):2324–2334, July 2005.

[SB06] M. Schubert and H. Boche. *QoS-Based Resource Allocation and Transceiver Optimization*, volume 2. Foundations and Trends in Communications and Information Theory, 2005/2006.

[SBB06] C. Studer, A. Burg, and H. Bolcskei. Soft-output sphere decoding: algorithms and VLSI implementation. In *Proceedings of the 40th Asilomar Conference Signals, Systems and Computers*, pages 2071–2076, Pacific Grove, CA, USA, November 2006.

[SC92] J. W. Silverstein and P. L. Combettes. Large dimensional random matrix theory for signal detection and estimation in array processing. *Stat. Signal Array Process.*, pages 276–279, 1992.

[Sca02] A. Scaglione. Statistical analysis of the capacity of MIMO frequency selective Rayleigh fading channels with arbitrary number of inputs and outputs. In *Proceedings of the IEEE International Symposium on Information Theory*, Lausanne, Switzerland, July 2002.

[SDHM07] J. E. Suris, L. A. DaSilva, Zhu Han, and A. B. MacKenzie. Cooperative game theory for distributed spectrum sharing. In *Proceedings of the IEEE International Conference on Communication (ICC)*, Glasgow, Scotland, June 2007.

[SDL06] N. D. Sidiropoulos, T. N. Davidson, and Z.-Q. Luo. Transmit beamforming for physical layer multicasting. *IEEE Trans. Signal Process.*, June 2006.

[SH01] M. Sellathurai and S. Haykin. Joint beamformer estimation and co- antenna interference cancelation for turbo-blast. In *Proceedings of the IEEE International Conference on Acoustics, Speech and Signal Processing, ICASSP'01*, volume 4, pages 2453–2456, Salt Lake City, USA, May 2001.

[Sha48] C. E. Shannon. A Mathematical Theory of Communication. *Bell Labs Tech. J.*, 27:379–457, 623–656, July-October, 1948.

[Sha49] C. E. Shannon. Communication Theory of Secrecy Systems. *Bell Labs Tech. J.*, 28(4):656–715, May 1949.

[Sib05] A. Sibille. *EURASIP, special issue on UWB state of the art*, vol. 2005, Issue 3, chapter Time domain diversity in ultra wide band MIMO communications, pages 316–327. HINDAWI, 2005.

[Sil90] J. W. Silverstein. Weak convergence of random functions defined by the eigenvectors of sample covariance matrcies. *Ann. Probab.*, 18:1174–1194, 1990.

[SJL05] P. Stoica, Y. Jiang, and J. Li. On MIMO channel capacity: an intuitive discussion. *IEEE Signal Process. Mag.*, 22(3):83–84, May 2005.

[SM02] A. M. Sengupta and P. P. Mitra. Capacity of multivariate channels with multiplicative noise: random matrix techniques and large-N expansions for full transfer matrices. Bell Laboratories, Lucent Technologies, Murray Hill, NJ 07974, December 2002. can be downloaded at *http://mars.bell-labs.com.*

[SOR09] UL single user MIMO schemes for LTE-Advanced. Technical Report R1-090727, 3GPP, 2009.

[SR01] T. Svantesson and T. Ranheim. Mutual coupling effects on the capacity of multi-element antenna systems. In *Proceedings of the IEEE International Conference on Acoustics, Speech, and Signal Processing*, volume 4, pages 2485–2488, Salt Lake City, Utah, May 2001.

[SRS03] P. J. Smith, S. Roy, and M. Shafi. Capacity of MIMO systems with semicorrelated flat fading. *IEEE Trans. Inf. Theory*, 49(10):2781–2788, October 2003.

[SS99] S. R. Saunders. *Antennas and Propagation for Wireless Communication Systems*. John Wiley, Chichester, 1999.

[SSB07] S. Shi, M. Schubert, and H. Boche. Downlink MMSE transceiver optimization for multi-user MIMO systems: duality and sum-MSE minimization. *IEEE Trans. Signal Process.*, 55(11):5436–5446, November 2007.

[SSH04] Q. Spencer, A. Swindlehurst, and M. Haardt. Zero-forcing methods for downlink spatial multiplexing in multiuser MIMO channels. *IEEE Trans. Signal Process.*, 2004.

[SSWT00] R. K. Settaluri, G. Sundberg, A. Weisshaar, and V. K. Tripathi. Compact folded line rat-race hybrid couplers. *IEEE Microw. Guid. Wave Lett.*, 10(2):61–63, February 2000.

[STB09] S. Sesia, I. Toufik, and M. Baker. *LTE, The UMTS Long Term Evolution: From Theory to Practice*. John Wiley & Sons, 2009.

[Ste62] S. Stein. On cross coupling in multiple-beam antennas. *IEEE Trans. Antennas Propag.*, AP-10(5):548–557, September 1962.

[Stj05] A. Stjernman. Relationship between radiation pattern correlation and scattering matrix of lossless and lossy antennas. *Electron. Lett.*, 41(12):678–680, June 2005.

[Sun02] C. W. Sung. Log-convexity property of the feasible SIR region in power-controlled cellular systems. *IEEE Commun. Lett.*, 6(6):248–249, June 2002.

[SV01] S. Shamai and S. Verdu. The impact of frequency-flat fading on the spectral efficiency of CDMA. *IEEE Trans. Inf. Theory*, 1302–1326, May 2001.

[SW03] T. Svantesson and J. Wallace. On signal strength and multipath richness in multi-input multi-output systems. *Proc. Int. Conf. Commun.*, 4:2683–2687, 2003.

[SW85] J. Salz and A. Wyner. On data transmission over cross coupled linear channels. Technical Report AT&T Bell Labs Tech. Memo, 1985.

[SW94] N. Seshadri and J. H. Winters. Two signaling schemes for improving the error performance of frequency-division-duplex (FDD) transmission systems using transmitter antenna diversity. *Int. J. Wireless Inf. Netw.*, 1:508–511, 1994.

[SWB06] S. Stanczak, M. Wiczanowski, and H. Boche. *Theory and Algorithms for Resource Allocation in Wireless Networks*. Lecture Notes in Computer Science (LNCS). Springer-Verlag, 2006.

[SWB08] S. Stanczak, M. Wiczanowski, and H. Boche. *Fundamentals of Resource Allocation in Wireless Networks: Theory and Algorithms*, volume 3 of *Foundations in Signal Processing, Communications and Networking*. Springer, second expanded edition, 2008.

[SYK07] R. Shariat-Yazdi and T. Kwasniewski. Reconfigurable K-best MIMO detector architecture and FPGA implementation. In *Proceedings of the International Symposium on Intelligent Signal Processing and Communication Systems (ISPACS)*, pages 349–352, December 2007.

[TAB07] A. H. Sayed, A. Tarighat, R. Hsu, and B. Jalali. Fundamentals and challenges of optical multiple-input multiple-output multi-mode fiber links. *IEEE Commun. Mag.*, May 2007.

[TCW02] T. Zwick, C. Fischer, and W. Wiesbeck. A stochastic multipath channel model including path directions for indoor environments. *IEEE J. Select. Areas Commun.*, 20(6):1178–1192, August 2002.

[TDHS07] H. Taoka, K. Dai, K. Higuchi, and M. Sawahashi. Field experiments on ultimate frequency efficiency exceeding 30 bit/second/Hz using MLD signal detection in MIMO-OFDM broadband packet radio access. In *Proceedings of the IEEE Vehicular Technology Conference on Spring*, pages 2129–2134, Dublin, Ireland, April 2007.

[Tea05] R. Thomä and et al. Multidimensional high resolution channel sounding. *Smart Antennas in Europe, State-of-the-Art, EURASIP – Book Series*, 2005.

[Tel95] I. E. Telatar. Capacity of multi-antenna Gaussian channels. Technical report, AT & T Bell Labs, 1995.

[Tel99] İ. Emre Telatar. Capacity of multi-antenna Gaussian channels. *European Trans. Telecommun.*, 10(6):585–595, November/December 1999.

[TH07] H. Taoka and K. Higuchi. Field experiment on 5-Gbit/s ultra-high-speed packet transmission using MIMO multiplexing in broadband packet radio access. *NTT DoCoMo Tech. J.*, 9(2):25–31, September 2007.

[TH99] D. N. C. Tse and S. Hanly. Linear multi-user receiver: effective interference, effective bandwidth and user capacity. *IEEE Trans. Inf. Theory*, 641–657, March 1999.

[TJC99] V. Tarokh, H. Jafarkhani, and A. R. Calderbank. Space-time block codes from orthogonal designs. *IEEE Trans. Inf. Theory*, 45:1456–1467, 1999.

[TKS02] M. Tüchler, R. Koetter, and A. Singer. Turbo equalization: principles and new results. *IEEE Trans. Commun.*, 50:754–767, May 2002.

[TL08] R. Tian and B. K. Lau. Uncoupled antenna matching for performance optimization in compact MIMO systems using unbalanced load impedance. In *Proceedings of the IEEE Vehicular Technology Conference on Spring*, pages 299–303, Singapore, May 2008.

[Ton00a] A. M. Tonello. Space-time bit-interleaved coded modulation over frequency selective fading channels with iterative decoding. In *Proceedings of the IEEE Global Communications Conference, GLOBECOM'00*, San Francisco, USA, November 2000.

[Ton00b] A. M. Tonello. Space-time bit-interleaved coded modulation with an iterative decoding strategy. In *Proceedings of the IEEE Vehicular Technology Conference, VTC'00 Fall*, Boston, USA, September 2000.

[TPC07] C. W. Tan, D. P. Palomar, and M. Chiang. Exploiting hidden convexity for flexible and robust resource allocation in cellular networks. In *Proceedings of the IEEE Infocom*, pages 964–972, May 2007.

[TPL+09] R. Tian, V. Plicanic, B. K. Lau, J. Långbacka, and Z. Ying. MIMO performance of diversity-rich compact six-port dielectric resonator antenna arrays in measured indoor environments at 2.65 GHz. In *Proceedings of the 2nd COST2100 Workshop – Multiple Antenna Systems on Small Terminals (Small and Smart)*, Valencia, Spain, May 2009. Available: http://www.cost2100.org/.

[TPLY10] R. Tian, V. Plicanic, B. K. Lau, and Z. Ying. A compact six-port dielectric resonator antenna array: MIMO channel measurements and performance analysis. *IEEE Trans. Antennas Propag.*, 58(4):1369–1379, April 2010.

[TR308] Requirements for technical advances to E-UTRA (LTE-Advanced). Technical Report TR 36.913, 3GPP, 2008.

[TR309] Feasibility study for further enhancements to E-UTRA (LTE-Advanced). Technical Report TR 36.912, 3GPP, 2009.

[TR310] Further advances for E-UTRA: physical layer aspects. Technical Report TR 36.814, 3GPP, 2010.

[TSC98] V. Tarokh, N. Seshadri, and A. R. Calderbank. Space-time codes for high data rate wireless communication: performance criteria and code construction. *IEEE Trans. Inf. Theory*, 44:744–765, March 1998.

[TSK02] M. Tüchler, A. Singer, and R. Koetter. Minimum mean squared error equalization using a priori information. *IEEE Trans. Signal Process.*, 50(3):673–683, March 2002.

[TV07] D. Tse and P. Viswanath. *Fundamentals of Wireless Communication.* Cambridge University Press, Cambridge, UK, 2005.

[TVZ04] D. N. C. Tse, P. Viswanath, and L. Zheng. Diversity-multiplexing tradeoff in multiple-access channels. *IEEE Trans. Inf. Theory*, 50(9):1859–1874, September 2004.

[TZ00] D. N. C. Tse and O. Zeitouni. Linear multiuser receivers in random environments. *IEEE Trans. Inf. Theory*, 46(1):171–188, January 2000.

[UER09] UE-RS patterns for ranks 5 to 8 of LTE-A. Technical Report R1-094212, 3GPP, 2009.

[UMT08] Towards global mobile broadband — standardizing the future of mobile communications with LTE. Technical report, UMTS Forum, 2008.

[UY98] S. Ulukus and R. Yates. Adaptive power control and MMSE interference suppression. *ACM Wirel. Netw.*, 4(6):489–496, 1998.

[VA87] R. G. Vaughan and J. B. Andersen. Antenna diversity in mobile communications. *IEEE Trans. Veh. Technol.*, 36(4):149–172, November 1987.

[Vau88] R. Vaughan. On optimum combining at the mobile. *IEEE Trans. Veh. Technol.*, 37(4), November 1988.

[VBA03] R. Vaughan and J. B. Andersen. *Channels, Propagation and Antennas for Mobile Communications.* The IEE, London, 2003.

[VBW98] L. Vandenberghe, S. Boyd, and S.-P. Wu. Determinant maximization with linear matrix inequality constraints. *SIAM J. Matrix Anal. Appl.*, 19(2):499–533, 1998.

[vBHM+05] K. van Berkel, F. Heinle, P. Meuwissen, K. Moerman, and M. Weiss. Vector processing as an enabler for software-defined radio in handheld devices. *J. Appl. Signal Process. (EURASIP)*, (16):2613–2625, September 2005.

[VDCM08] R. Verdone, D. Dardari, A. Conti, and G. Mazzini. *Wireless Sensor and Actuator Networks Technologies, Analysis and Design.* Academic Press, 2008.

[Ver98] S. Verdú. *Multiuser Detection.* Cambridge University Press, 1998.

[VG97] M. K. Varanasi and T. Guess. Achieving vertices of the capacity region of the Gaussian correlated-waveform multiple-access channel with decision feedback receivers. In *Proceedings of the IEEE International Symposium on Information Theory (ISIT)*, page 270, Ulm, Germany, June 1997.

[VJG03] S. Vishwanath, N. Jindal, and A. Goldsmith. Duality, achievable rates, and sum-rate capacity of Gaussian MIMO broadcast channels. *IEEE Trans. Inf. Theory*, 49(10):2658–2668, October 2003.

[VOKK02] P. Vainikainen, J. Ollikainen, O. Kivekas, and K. Kelander. Resonator-based analysis of the combination of mobile handset antenna and chassis. *IEEE Trans. Antennas Propag.*, 50(10):1433–1444, October 2002.

[VOKV06] J. Villanen, K. Ollikainen, O. Kivekäs, and P. Vainikainen. Coupling element based mobile terminal antenna structures. *IEEE Trans. Antennas Propag.*, 54(7):2142–2153, July 2006.

[VS99] S. Verdu and S. Shamai. Spectral efficiency of CDMA with random spreading. *IEEE Trans. Inf. Theory*, 622–640, March 1999.

[VSW+08] C. Volmer, M. Sengul, J. Weber, R. Stephan, and M. A. Hein. Broadband decoupling and matching of a superdirective two-port antenna array. *IEEE Antennas Wirel. Propag. Lett.*, 7:613–616, 2008.

[VTA01] P. Viswanath, D. N. C. Tse, and V. Anantharam. Asymptotically optimal water-filling in vector multiple-access channels. *IEEE Trans. Inf. Theory*, 47(1):241–267, January 2001.

[VT03] P. Viswanath and D. Tse. Sum capacity of the vector Gaussian broadcast channel and uplink-downlink duality. *IEEE Trans. Inf. Theory*, 49(8):1912–1921, August 2003.

[VWS+08] C. Volmer, J. Weber, R. Stephan, K. Blau, and M. A. Hein. An eigen-analysis of compact antenna arrays and its application to port decoupling. *IEEE Trans. Antennas Propag.*, 56(2):360–370, February 2008.

[VWS+09] C. Volmer, J. Weber, R. Stephan, and M. A. Hein. Mutual coupling in multi-antenna systems: figures-of-merit and practical verification. In *Proceedings of the European Conference on Antennas and Propagation (EUCAP)*, Berlin, Germany, March 2009.

[W00] A. Reial and S. G. Wilson. Concatenated space-time coding. In *Proceedings of the Conference of Information Science and Systems*, 1:710–715 Princeton, NJ, March 2000.

[W04] X. Wautelet, A. Dejonghe, and L. Vandendorpe. MMSE-based fractional turbo receiver for space-time BICM over frequency selective MIMO fading channels. *IEEE Trans. Signal Process.*, 52:1804–1809, June 2004.

[W93] A. Wittneben. Base station modulation diversity for digital SIMULCAST. *IEEE Veh. Technol. Conf.*, May 1991.

[Way10] Way forward on OCC mapping for DL DMRS. Technical Report R1-105058, 3GPP, 2010.

[WC07a] T. J. Willink and G. W. K. Colman. Diversity-multiplexing trade-off for mobile urban environments. *Electron. Lett.*, 43:321–232, February 2007.

[WC07b] T. J. Willink and G. W. K. Colman. Measured Div-Mux Trade-off for urban environments. COST 2100 TD(07)255, 2nd MCM, Lisbon, February 26–28, 2007.

[WE04] Z. Wireless, V. Erceg and et al. IEEE P802.11 Wireless LANs TGn channel models. IEEE 802.11-03/940r4, 2004.

[Wea06] W. Weichselberger and et al. A stochastic MIMO channel model with joint correlation of both link ends. *IEEE Trans. Wirel. Commun.*, 5(1), January 2006.

[Wei03] W. Weichselberger. *Spatial Structure of Multiple Antenna Radio Channels*. PhD thesis, Institut für Nachrichtentechnik und Hochfrequenztechnik, Vienna University of Technology, December 2003.

[WEL09] D. Wu, J. Eilert, and D. Liu. Implementation of a high-speed MIMO soft-output symbol detector for software defined radio. *J. Signal Process. Syst.*, Springer, New York, April 2009.

[WES06] A. Wiesel, Y. C. Eldar, and S. Shamai (Shitz). Linear precoding via conic optimization for fixed MIMO receivers. *IEEE Trans. Signal Process.*, 54(1):161–176, 2006.

[WG04] R. Wang and G. B. Giannakis. Approaching MIMO channel capacity with reduced-complexity soft sphere decoding. *IEEE Wirel. Commun. Networking Conf. (WCNC)*, 3:1620–1625, March 2004.

[WGH07] M. Weis, G. Del Galdo, and M. Haardt. A correlation tensor-based model for time variant frequency selective MIMO channels. *WSA Workshop on Smart Antennas*, February 2007.

[WH07] L. Wood and W. Hodgkiss. MIMO channel models and performance metrics. In *Proceedings of the IEEE Global Telecommunications Conference (Globecom)*, pages 3740–3744, Washington DC, USA, November 2007.

[Wie48] N. Wiener. Cybernetics, or control and communication in the animal and the machine. *Herman et Cie/The Technology Press*, 1948.

[Wie49] N. Wiener. *Extrapolation, Interpolation and Smoothing of Stationary Time Series*. J. Wiley and Sons, Inc, New York, 1949.

[Wil10] T. J. Willink. Observation-based time-varying MIMO channel model. *IEEE Trans. Veh. Technol.*, 59(1):3–15, January 2010.

[WIN+D17] M. Bengtsson, E. Björnson, L. Brunel, P. Komulainen, Y. Liu, A. Osseiran, L. Rasmussen, F. Roemer, M. Schellmann, S. Sezginer, B. Shankar, B. Song, L. Thiele, A. Tölli, G. Vivier, and M. Xiao. CELTIC/CP5-026 WINNER+, D1.7 Intermediate report on advanced antenna schemes, November 2009. Available: http://projects.celtic-initiative.org/winner+/deliverables_winnerplus.html.

[Win87] J. H. Winters. On the capacity of radio communication systems with diversity in a Rayleigh fading environment. *IEEE J. Select. Areas Commun.*, SAC-5(5), June 1987.

[WJ04a] J. W. Wallace and M. A. Jensen. Mutual coupling in MIMO wireless systems: a rigorous network theory analysis. *IEEE Trans. Wirel. Commun.*, 3(4):1317–1325, July 2004.

[WJ04b] J. W. Wallace and M. A. Jensen. Termination-dependent diversity performance of coupled antennas: network theory analysis. *IEEE Trans. Antennas Propag.*, 52(1): 98–105, January 2004.

[WMA+08] S. Wyne, A. F. Molisch, P. Almers, G. Eriksson, J. Karedal, and F. Tufvesson. Outdoor-to-indoor office MIMO measurements and analysis at 5.2 GHz. *IEEE Trans. Veh. Technol.*, 57(3):1374–1386, May 2008.

[WP99] X. Wang and H. V. Poor. Iterative (turbo) soft interference cancellation and decoding for coded CDMA. *IEEE Trans. Commun.*, 47(7):1046–1061, July 1999.

[WSS04] H. Weingarten, Y. Steinberg, and S. Shamai (Shitz). The capacity region of the Gaussian MIMO broadcast channel. In *Proceedings of the 38th Annual Conference on Information Sciences and Systems (CISS)*, 2004.

[Wu83] C. F. J. Wu. On the convergence properties of the EM algorithm. *Ann. Stat.*, 11(1):95–103, 1983.

[WVB+06] J. Weber, C. Volmer, K. Blau, R. Stephan, and M. A. Hein. Miniaturized antenna arrays using decoupling networks with realistic elements. *IEEE Trans. Microwave Theory Tech.*, 54(6):2733–2740, June 2006.

[WXG03] Y. Xin, Z. Wang, and G. B. Giannakis. Space-time diversity systems based on linear constellation precoding. *IEEE Trans. Wirel. Commun.*, 2:294–309, March 2003.

[Xea04] H. Xu and et al. A generalized space-time multiple-input multiple-output (MIMO) channel model. *IEEE Trans. Wirel. Commun.*, 3(3):966–975, 2004.

[Yat95] R. D. Yates. A framework for uplink power control in cellular radio systems. *IEEE J. Select. Areas Commun.*, 13(7):1341–1348, September 1995.

[YC95] R. D. Yates and H. Ching-Yao. Integrated power control and base station assignment. *IEEE Trans. Veh. Technol.*, 44(3):638–644, August 1995.

[YC04] W. Yu and J. M. Cioffi. Sum capacity of Gaussian vector broadcast channels. *IEEE Trans. Inf. Theory*, 50(9):1875–1892, 2004.

[YD02] Z. Ying and A. Dahlstrom. Multi frequency-band antenna. *US Patent 6,456,250*, September 2002.

[YL07] W. Yu and T. Lan. Transmitter optimization for the multi-antenna downlink with per-antenna power constraints. *IEEE Trans. Signal Process.*, 55(6):2646–2660, June 2007.

[YSI+06] A. Yamamoto, T. Sakata, H. Iwai, K. Ogawa, J. Takada, K. Sakaguchi, and K. Araki. BER measurement on a handset adaptive antenna array in a Rayleigh-fading channel by a variable-XPR spatial fading emulator. In *Proceedings of the IEEE Antennas and Propagation Society International Symposium*, pages 4569–4572, Albuquerque, NM, USA, July 2006.

[YW03] H. Yao and G. Wornell. Achieving the full MIMO diversity-multiplexing frontier with rotation-based space-time codes. In *Proceedings of the Allerton Conference Communication, Control, and Computing*, pages 400–409, Illinois, 2003.

[Zan92] J. Zander. Performance of optimum transmitter power control in cellular radio systems. *IEEE Trans. Veh. Technol.*, 41(1):57–62, February 1992.

[ZCT01] J. Zhang, E. Chong, and D. Tse. Output MAI distributions of linear MMSE multiuser receivers in CDMA systems. *IEEE Trans. Inf. Theory*, 1128–1144, March 2001.

[ZE10] J. Zhu and G. V. Eleftheriades. A simple approach for reducing mutual coupling in two closely spaced metamaterial-inspired monopole antennas. *IEEE Antennas Wirel. Propag. Lett.*, 9:379–382, 2010.

[ZF94] J. Zander and M. Frodigh. Comment on performance of optimum transmitter power control in cellular radio systems. *IEEE Trans. Veh. Technol.*, 43(3):636, August 1994.

[Zeh92] E. Zehavi. 8-PSK trellis codes for a rayleigh channel. *IEEE Trans. Commun.*, 40(5):873–884, May 1992.

[ZG08] A. G. Zajič and G. L. Stüber. Three-dimensional modelling, simulation and capacity analysis of space-time correlated mobile-to-mobile channels. *IEEE Trans. Veh. Technol.*, 57(4):2042–2054, July 2008.

[Zho97] L. Zhou. The Nash bargaining theory with non-convex problems. *Econometrica, Econometric Soc.*, 3(65):681–686, May 1997.

[ZJP99] H. Zamiri-Jafarian and S. Pasupathy. EM-based recursive estimation of channel parameters. *IEEE Trans. Commun.*, 47:1297–1302, September 1999.

[ZT03] L. Zheng and D. Tse. Diversity and multiplexing: a fundamental tradeoff in multiple antenna channels. *IEEE Trans. Inf. Theory*, 49(5):1073–1096, May 2003.

[ZWYH08] H. Zhang, Z. Wang, J. Yu, and J. Huang. A compact MIMO antenna for wireless communication. *IEEE Antennas Propag. Mag.*, 50(6):104–107, December 2008.

Index

Printed in the United States
By Bookmasters